AF352114

METHODS IN MOLECULAR BIOLOGY

Series Editor
John M. Walker
School of Life Sciences
University of Hertfordshire
Hatfield, Hertfordshire, AL10 9AB, UK

For further volumes:
http://www.springer.com/series/7651

Neuronal Cell Death

Methods and Protocols

Edited by

Laura Lossi and Adalberto Merighi

Department of Veterinary Sciences, University of Torino, Grugliasco, Torino, Italy

 Humana Press

Editors
Laura Lossi
Department of Veterinary Sciences
University of Torino
Grugliasco, Torino, Italy

Adalberto Merighi
Department of Veterinary Sciences
University of Torino
Grugliasco, Torino, Italy

ISSN 1064-3745   ISSN 1940-6029 (electronic)
ISBN 978-1-4939-2151-5   ISBN 978-1-4939-2152-2 (eBook)
DOI 10.1007/978-1-4939-2152-2
Springer New York Heidelberg Dordrecht London

Library of Congress Control Number: 2014955570

Humana Press is a brand of Springer
Springer is part of Springer Science+Business Media (www.springer.com)

Preface

The ongoing research on neuronal cell death is rapidly expanding after the recognition that not only do neurons die as a consequence of external insults, but also following activation of stereotyped genetic programs/intracellular pathways leading to death. However, a long time passed before the term "cell death" took over in the scientific literature: in the nineteenth century, when Rudolf Virchow published his famous book entitled *Lectures on Cellular Pathology in its Grounds on Physiological and Pathological Histology* first reporting on the occurrence of cell death in damaged tissues, the dominating idea was that only injured tissues and cells underwent some sort of degeneration when damaged. Thus, the concept that cell death took place also during the course of normal development or during growth and aging was still very far from general acceptance, and required a considerable amount of experimental work to convince the academic community. A significant step forward along this direction was done in the 1960s of the last century, when electron microscopists started describing the different modes of cell death. Since then the field has dramatically widened, particularly with the recognition of several forms of neuronal degeneration, such as necrosis, apoptosis, autophagy, pyroptosis, oncosis, etc. and of the close relationship of many of these with cell proliferation and aging. Today, the ultimate frontiers in neuronal cell death research lie in the development of novel approaches to monitor the phenomenon by the use of in vivo and/or ex vivo preparations, such as organotypic cultures, that are more closely related to the intact mammalian brain than primary cultures, and to better exploit the use of non-mammalian model organisms. In parallel, there is a need for understanding the type and role of cell death in neurodegenerative diseases, to develop pharmacologically active compounds that are capable to exert their biological role(s) in vivo, and to construct genetic vectors to be employed in gene therapy.

With such a wide array of exciting and rapidly expanding fields of research, this book, from its initial conception, had obviously to be limited in the choice of subjects, but we believe it represents a valuable and readily reproducible collection of established and emerging techniques for neuronal cell death research. Such a collection is preceded by a general introductory chapter (Chapter 1) that recalls the history of cell death and, to put things into perspective, discusses the main morphological features of the most diffuse types of cell death in neurons, in parallel with relevant cellular pathways and current assays for a proper recognition. The methods presented include immunocytochemical localization at light and electronic levels, biochemical characterization, and functional analysis in vivo or ex vivo by novel types of microscopy, as well as protocols for development and production of genetic probes. Although this book is primarily devoted to approaches for analysis of the mammalian brain, a few non-mammalian species are also taken into consideration to demonstrate specific methodologies that are of great value to boost cell death research by taking advantage from the use of less complex models.

As a general indication to the readers, the book is divided into four parts.

Part I (Chapters 2–12) is focused on a series of techniques for the molecular, structural, functional, and genomic characterization of dying neurons. They cover a broad range of protocols, such as epifluorescence and digital holography to monitor the cell volume

(Chapter 2); a series of techniques to study DNA synthesis/alterations and the morphological signs of nuclear sufferance (Chapters 3 and 4); a number of approaches to monitor parameters of primary importance in cell viability (Chapters 5–8), such as oxygen and calcium concentration, mitochondrial function, and activation of caspase-3 in single alive cells; and a series of molecular approaches for RNA silencing, genomic analysis, and high-throughput cell death assays (Chapters 9–12).

Part II (Chapters 13–18) groups together a number of protocols that are of primary interest in neuropathology (Chapters 13 and 14) and in experimental neuropathology (Chapters 15–18) by describing the current ameliorations to well-established diagnostic techniques such as the Golgi method for study of neuronal and glial death in autopsy material (Chapter 13), the use of optimized protocols and image analysis algorithms for reliable analysis of cell death in human and animal samples (Chapter 14), some specific experimental approaches such as oxygen-glucose deprivation (Chapter 15), single axon lesioning by laser microbeam targeting (Chapter 16), in vivo imaging of retinal apoptosis (Chapter 17), and use of neurotoxins to model neuronal death in Parkinson's disease (Chapter 18).

Part III (Chapters 19–22) is devoted to a series of gene engineering techniques to obtain and manipulate neuronal stem cells and progenitors (Chapter 19), to prepare HSV-1 vectors for the gene therapy (Chapters 20 and 21), and to CNS transplantation of bone marrow stem cells (Chapter 22).

Part IV (Chapters 23–26) describes some very interesting protocols for study of cell death in non-mammalian models, such as the analysis of caspase-3 activation in lamprey (Chapter 23), the generation of zebrafish models by genome editing (Chapter 24), and the assessment of cell death (Chapter 25) and phagocytosis (Chapter 26) in *Drosophila*.

All scientists who have excellently contributed to this book have a direct experience in one or more fields of neuronal cell death research. We are very much indebted to all of them for their time, the high standards of their contributions, and for successful effort in emphasizing the description of the more common pitfalls in the techniques that they have described, and of the hints to reduce the possibility of failure for beginners.

The collection of protocols that forms this book is surely not exhaustive of the wide range of approaches that today can be employed in top-level cell death research. Yet it is intended for a large audience of scientists, including histologists, biochemists, cellular and molecular biologists, and electrophysiologists that are currently active in the field or are willing to enter such an exciting and still expanding area of neurobiology.

As the two of us have been the first to benefit from such an excellent assemblage of information, we are confident that readers too will find this book very useful for their future work.

Grugliasco, Torino, Italy   *Laura Lossi, DVM, PhD*
Adalberto Merighi, DVM, PhD

Contents

Contributors

SILVIA ALASIA • *Department of Veterinary Sciences University of Turin, Turin, Grugliasco, Italy*

JOSE RAMÓN ALONSO • *Laboratory of Neuronal Plasticity and Neurorepair, Institute for Neuroscience of Castilla y León (INCyL), Universidad de Salamanca, Salamanca, Spain; IBSAL – Institute of Biomedical Research of Salamanca, Universidad de Salamanca, Salamanca, Spain; Instituto de Alta Investigación, Universidad de Tarapacá, Arica, Chile*

VELLAREDDY ANANTHARAM • *Parkinson's Disorder Research Laboratory, Department of Biomedical Sciences, IA Center for Advanced Neurotoxicology, IA State University, Ames, IA, USA*

MACARENA S. ARRÁZOLA • *Centro de Envejecimiento y Regeneración (CARE), Departamento de Biología Celular y Molecular, Facultad de Ciencias Biológicas, Pontificia Universidad Católica de Chile, Santiago, Chile*

STAVROS J. BALOYANNIS • *Thessaloniki and Institute for Research on Alzheimer's Disease and Ageing, Aristotelian University, Iraklion, Langada, Greece*

FERNANDO C. BALTANÁS • *Institute for Molecular and Cell Biology of the Cancer, CSIC – Universidad de Salamanca, Salamanca, Spain*

ANTÓN BARREIRO-IGLESIAS • *Centre for Neuroregeneration, School of Biomedical Sciences, University of Edinburgh, Edinburgh, UK*

MARIA TERESA BERCIANO • *Department of Anatomy and Cell Biology, Centro de Investigación Biomédica en Red sobre Enfermedades Neurodegenerativas (CIBERNED), Universidad de Cantabria-IDIVAL, Santander, Spain*

MICHAEL W. BERNS • *Beckman Laser Institute, University of California, Irvine, CA, USA; Department of Bioengineering, Institute for Engineering in Medicine, University of California, San Diego, La Jolla, CA, USA*

MARÍA CALVO • *Instituto de Biología y Genética Molecular (IBGM), Consejo Superior de Investigaciones Científicas y Universidad de Valladolid, Valladolid, Spain*

NADIA CANU • *Department of System Medicine, University of "Tor Vergata" Rome, Rome, Italy; National Council Research, Institute of Cell Biology and Neurobiology, Rome, Italy*

CLAUDIA CASTAGNA • *Department of Veterinary Sciences, University of Turin, Turin, Italy*

SEBASTIANO CAVALLARO • *Functional Genomics Center, Institute of Neurological Sciences, Italian National Research Council, Catania, Italy*

SOOK HYUN CHUNG • *Macular Research Group, Clinical Ophthalmology and Eye Health, The University of Sydney, Sydney, NSW, Australia*

HELENA CIMAROSTI • *Reading School of Pharmacy, University of Reading, Reading, UK*

CAROLINA COCITO • *Department of Veterinary Sciences, University of Turin, Turin, Italy*

M. FRANCESCA CORDEIRO • *Glaucoma and Retinal Degeneration Research Group, Visual Neurosciences, UCL Institute of Ophthalmology, London, UK; Imperial College Healthcare Trust, The Western Eye Hospital, London, UK*

THARINE DAL-CIM • *Departamento de Bioquímica, Centro de Ciências Biológicas, Universidade Federal de Santa Catarina, Florianópolis, SC, Brazil*

Mohammad H. Dehabadi • *Glaucoma and Retinal Degeneration Research Group, Visual Neurosciences, UCL Institute of Ophthalmology, London, UK*

David Díaz • *Laboratory of Neuronal Plasticity and Neurorepair, Institute for Neuroscience of Castilla y León (INCyL), Universidad de Salamanca, Salamanca, Spain; IBSAL – Institute of Biomedical Research of Salamanca, Salamanca, Spain*

Ruslan I. Dmitriev • *School of Biochemistry and Cell Biology, University College Cork, Cork, Ireland*

Pierre Dourlen • *Unité de Santé Publique et d'Epidémiologie moléculaire des maladies liées au vieillissement, Institut Pasteur de Lille, INSERM UMR744, Lille, France*

Alberto L. Epstein • *Centre International de Recherche en Infectiologie (CIRI), INSERM U1111 – CNRS UMR5308, Ecole Normale Supérieure Lyon, Lyon, France*

José M. Frade • *Department of Molecular, Cellular and Developmental Neurobiology, Cajal Institute, IC-CSIC, Madrid, Spain*

Cornel Fraefel • *Institute of Virology, University of Zurich, Zurich, Switzerland*

Mark Gillies • *Macular Research Group, Clinical Ophthalmology and Eye Health, The University of Sydney, Sydney, NSW, Australia*

Li Guo • *Glaucoma and Retinal Degeneration Research Group, Visual Neurosciences, UCL Institute of Ophthalmology, London, UK*

Gabriel G. Haddad • *Division of Respiratory Medicine, Departments of Pediatrics and Neuroscience, University of California, San Diego, La Jolla, USA; The Rady Children's Hospital-San Diego, San Diego, USA*

Dilshan S. Harischandra • *Parkinson's Disorder Research Laboratory, Department of Biomedical Sciences, IA Center for Advanced Neurotoxicology, Iowa State University, Ames, IA, USA*

Alexander Hruscha • *German Center for Neurodegenerative Diseases (DZNE), Munich, Germany*

Nicholas Hyun • *Department of Bioengineering and Institute of Engineering in Medicine, University of California, San Diego, La Jolla, CA, USA*

Nibaldo C. Inestrosa • *Centro de Envejecimiento y Regeneración (CARE), Departamento de Biología Celular y Molecular, Facultad de Ciencias Biológicas, Pontificia Universidad Católica de Chile, Santiago, Chile; Center for Healthy Brain Ageing, School of Psychiatry, Faculty of Medicine, University of New South Wales, Sydney, NSW, Australia*

Shilpa Iyer • *Department of Chemical and Life Science Engineering, Virginia Commonwealth University, Richmond, VA, USA*

Huajun Jin • *Parkinson's Disorder Research Laboratory, Department of Biomedical Sciences, IA Center for Advanced Neurotoxicology, Iowa State University, Ames, IA, USA*

Anumantha Kanthasamy • *Parkinson's Disorder Research Laboratory, Department of Biomedical Sciences, IA Center for Advanced Neurotoxicology, Iowa State University, Ames, IA, USA*

Arthi Kanthasamy • *Parkinson's Disorder Research Laboratory, Department of Biomedical Sciences, IA Center for Advanced Neurotoxicology, Iowa State University, Ames, IA, USA*

Tashmia Khan • *Sanford-Burnham Medical Research Institute, La Jolla, CA, USA*

Maryla Krajewska • *Sanford-Burnham Medical Research Institute, La Jolla, CA, USA; Cellestan-Immunoquant, Inc., Oceanside, CA, USA*

Stan Krajewski • *Sanford-Burnham Medical Research Institute, La Jolla, CA, USA; Cellestan-Immunoquant, Inc., Oceanside, CA, USA*

ESTEE KURANT • *Department of Anatomy and Cell Biology, The Rappaport Family Institute for Research in the Medical Sciences, Faculty of Medicine, Technion – Israel Institute of Technology, Haifa, Israel*

MIGUEL LAFARGA • *Department of Anatomy and Cell Biology, Centro de Investigación Biomédica en Red sobre Enfermedades Neurodegenerativas (CIBERNED), Universidad de Cantabria-IDIVAL, Santander, Spain*

FLONIA LEVY-ADAM • *Department of Anatomy and Cell Biology, The Rappaport Family Institute for Research in the Medical Sciences, Faculty of Medicine, Technion – Israel Institute of Technology, Haifa, Israel*

JONATHAN LIU • *Sanford-Burnham Medical Research Institute, La Jolla, CA, USA*

NOELIA LÓPEZ-SÁNCHEZ • *Department of Molecular, Cellular and Developmental Neurobiology, Cajal Institute, IC-CSIC, Madrid, Spain*

LAURA LOSSI • *Department of Veterinary Sciences, University of Torino, Grugliasco, Torino, Italy; Istituto Nazionale di Neuroscienze (INN), Turin, Italy*

PEGGY MARCONI • *Section of Applied Microbiology and Pathology, Department of Life Sciences and Biotechnology, University of Ferrara, Ferrara, Italy*

PIERRE MARQUET • *Département de Psychiatrie, Centre de Neurosciences Psychiatriques, Centre Hospitalier Universitaire Vaudois (CHUV), Lausanne, Switzerland; Ecole Polytechnique Fédérale Lausanne (EPFL), Brain Mind Institute (BMI), Lausanne, Switzerland*

ADALBERTO MERIGHI • *Department of Veterinary Sciences, University of Torino, Grugliasco, Torino, Italy; Istituto Nazionale di Neuroscienze (INN), Turin, Italy*

EDUARDO M. NORMANDO • *Glaucoma and Retinal Degeneration Research Group, Visual Neurosciences, UCL Institute of Ophthalmology, London, UK; Imperial College Healthcare Trust, The Western Eye Hospital, London, UK*

LUCÍA NÚÑEZ • *Instituto de Biología y Genética Molecular (IBGM), Consejo Superior de Investigaciones Científicas y Universidad de Valladolid, Valladolid, Spain; Departamento de Bioquímica y Biología Molecular y Fisiología, Universidad de Valladolid, Valladolid, Spain*

MATTHEW E. PAMENTER • *Department of Zoology, University of British Columbia, Vancouver, Canada*

DMITRI B. PAPKOVSKY • *School of Biochemistry and Cell Biology, University College Cork, Cork, Ireland*

NICOLAS PAVILLON • *Biophotonics Laboratory, Immunology Frontier Research Center (IFReC), Osaka University, Suita, Osaka, Japan; Microvision and Microdiagnostics Group (MVD), STI, Ecole Polytechnique Fédérale de Lausanne (EPFL), Lausanne, Switzerland*

GAIA POLLORSI • *Glaucoma and Retinal Degeneration Research Group, Visual Neurosciences, UCL Institute of Ophthalmology, London, UK*

AJAY RANA • *Department of Molecular Pharmacology and Therapeutics, Loyola University Chicago, Maywood, IL, USA; Hines Veterans Affairs Medical Center, Hines, IL, USA*

RAJ R. RAO • *Department of Chemical and Life Science Engineering, Virginia Commonwealth University, Richmond, VA, USA*

BETTINA SCHMID • *German Center for Neurodegenerative Diseases (DZNE), Munich, Germany; Munich Cluster for Systems Neurology (SyNergy), Munich, Germany*

WEIYONG SHEN • *Macular Research Group, Clinical Ophthalmology and Eye Health, The University of Sydney, Sydney, NSW, Australia*

Linda Z. Shi • *Department of Bioengineering and Institute of Engineering in Medicine, University of California, San Diego, La Jolla, CA, USA*

Michael I. Shifman • *Shriners Hospitals Pediatric Research Center (Center for Neural Repair and Rehabilitation), Temple University School of Medicine, Philadelphia, PA, USA*

Boris Shklyar • *Department of Anatomy and Cell Biology, The Rappaport Family Institute for Research in the Medical Sciences, Faculty of Medicine, Technion – Israel Institute of Technology, Haifa, Israel*

Chia-Hung Sze • *Sanford-Burnham Medical Research Institute, La Jolla, CA, USA*

Carla I. Tasca • *Departamento de Bioquímica, Centro de Ciências Biológicas, Universidade Federal de Santa Catarina, Florianópolis, SC, Brazil*

Lisa A. Turner • *Glaucoma and Retinal Degeneration Research Group, Visual Neurosciences, UCL Institute of Ophthalmology, London, UK*

Jorge Valero • *Center for Neuroscience and Cell Biology of Coimbra, University of Coimbra, Coimbra, Portugal*

Carlos Villalobos • *Instituto de Biología y Genética Molecular (IBGM), Consejo Superior de Investigaciones Científicas y Universidad de Valladolid, Valladolid, Spain*

Jeffrey Wang • *Sanford-Burnham Medical Research Institute, La Jolla, CA, USA*

Eduardo Weruaga • *Laboratory of Neuronal Plasticity and Neurorepair, Institute for Neuroscience of Castilla y León (INCyL), Universidad de Salamanca, Salamanca, Spain; IBSAL – Institute of Biomedical Research of Salamanca, Salamanca, Spain*

Chapter 1

Neuronal Cell Death: An Overview of Its Different Forms in Central and Peripheral Neurons

Laura Lossi, Claudia Castagna, and Adalberto Merighi

Abstract

The discovery of neuronal cell death dates back to the nineteenth century. Nowadays, after a very long period of conceptual difficulties, the notion that cell death is a phenomenon occurring during the entire life course of the nervous system, from neurogenesis to adulthood and senescence, is fully established. The dichotomy between apoptosis, as the prototype of programmed cell death (PCD), and necrosis, as the prototype of death caused by an external insult, must be carefully reconsidered, as different types of PCD: apoptosis, autophagy, pyroptosis, and oncosis have all been demonstrated in neurons (and glia). These modes of PCD may be triggered by different stimuli, but share some intracellular pathways such that different types of cell death may affect the same population of neurons according to several intrinsic and extrinsic factors. Therefore, a mixed morphology is often observed also depending on degrees of differentiation, activity, and injury. The main histological and ultrastructural features of the different types of cell death in neurons are described and related to the cellular pathways that are specifically activated in any of these types of PCD.

Key words Apoptosis, Autophagy, Pyroptosis, Oncosis, Necrosis

1 Brief History of Neuronal Cell Death

1.1 Discovery of Cell Death

The notion that large numbers of cells die during the course of normal development not only in the nervous system but also in many other organs and tissues of the body is nowadays fully accepted. However, recognition of the importance of this phenomenon required more than a century to be fully appreciated in evolutionary and functional terms.

The first description of cell death dates to 1842 [1] when Vogt described the *evolution*[1] of the anuran notochord and the

---

[1] Several years before Darwin published his famous book *On the Origin of Species by Means of Natural Selection* (1859), very interestingly Vogt used the German term *Evolution*, but in the meaning of *development*, which is more appropriately indicated with the German word *Entwicklung*.

Laura Lossi and Adalberto Merighi (eds.), *Neuronal Cell Death: Methods and Protocols*, Methods in Molecular Biology, vol. 1254, DOI 10.1007/978-1-4939-2152-2_1, © Springer Science+Business Media New York 2015

elimination of its cells during the course of development to be eventually replaced by new elements from the surrounding cartilage.

Nonetheless, a long time passed before the term cell death took over in the scientific literature; although the idea that tissues and cells were undergoing some sort of degeneration was at that time widely diffused and accepted among pathologists, who were among the first to realize that some form of cell death was likely to occur in damaged tissues, as first reported by Virchow in 1858 [2]. However, the concept of cell death taking place also during the course of normal development (or during growth) and aging was still very far from general acceptance, and required a considerable amount of experimental work to convince the academic community. Research on the morphological changes associated with metamorphosis in insects and amphibians was the initial trigger to acceptance of the phenomenon of cell death also as a nonpathological event [3]. In parallel, observations of specific organs and tissues started unraveling forms of cell death in higher vertebrates. These studies included the replacement of cartilage by bone during the course of so-called endochondral ossification whereby cartilage cells (chondrocytes) undergo a process of cell death to be replaced by the osteoblasts that eventually give rise to the mature bone [4], the degeneration of the granulosa in mature ovary follicles, and the death of mammary gland cells during lactation in several species [5]. Flemming's studies on dying cells of the mammary gland date back to 1885, but they were particularly important being the first observations made from tissues fixed such that morphological details were better preserved, leading to coinage of the new term "chromatolysis" to indicate the loss of staining in cells during the course of their death.

1.2 Cell Death in the Nervous System

In his book titled Developmental Neurobiology [6], Marcus Jacobson wrote: *The tyranny of theory over the evidence is nowhere more glaringly evident than in the history of the delayed discovery of neuronal death during normal development. The tyranny in this case was imposed by the theory that both ontogeny and phylogeny are progressive, from lower and less organized to higher and more organized nervous systems. Evidence of neuronal death during normal development was reported but was ignored because it was in conflict with the idea of progressive development. Reports of neuronal death were buried in the literature, to be unearthed much later as curious historical relics. Such reports come back to haunt us as they haunted previous generations who could not accept evidence that conflicted with their cherished theories.*

Thus, the first descriptions of cell death in neurons date back to the 1880s when Beard described the loss of a specific population of sensory neurons in fish and skate [7, 8]. These cells that are today referred to as Rohon-Beard neurons are specialized mechanoreceptors occurring during embryonic development in the dorsal spinal cord in fish and amphibians. A demonstration of neuron cell death in birds was given a few years later by Collin who, in 1906, reported the occurrence of dorsal root ganglion (DRG) and motor neuron cell death during embryonic development of chicks [9]. Then, in 1926, Ernst was the first to recognize that an overproduction of neurons was followed by death of a substantial fraction in the retina, the trigeminal, facial, and DRGs, and in the anterior horn of the spinal cord [10]. He was also the first to propose a general theory of neuron death during normal development of the nervous system, to obtain supporting evidence, and to describe the existence of three main types of cell death during normal development: the first occurring during the regression of vestigial organs; the second during morphological modifications of organ anlages; the third during tissue remodeling. These were subsequently named phylogenetic, morphogenetic, and histogenetic cell death by Glücksmann [11].

The long delay in accepting the evidence for developmental neuronal death has been regarded as an historical enigma, as discussed in [6]. Deterministic theories of development made it difficult to conceive of the demolition of temporary structures as a part of the normal development, as it was hard to imagine, at that time, that construction and destruction could be proceeding simultaneously.

Years later, the idea of tissue organization by means of selective deletion of temporary structures started to gain attention in the 1970s. Before that time, the dominant idea, boosted by the neurotrophic theory that followed the original work of Levi-Montalcini and Cohen [12], was that matching between the number of central neurons and their peripheral targets is achieved by programs of cell proliferation, migration, and differentiation in which orderly progress was always prevalent.

A strong refractoriness to the idea of an intervention of cell death during development was, in fact, still present in textbooks of the 1930s dealing with the interpretation of one of the most widely used experimental paradigms to investigate the development of the nervous system, i.e. the removal or grafting of limbs to alter peripheral sensory and motor fields. These experiments were shown to result in hypoplasia or hyperplasia, respectively, and those results were consistently interpreted as consequences of altered proliferation of the nerve cells [6].

Glücksmann's 1951 review of cell death during normal development was indeed the first publication to prime discussion of the importance of neuronal cell death in experimental limb amputation [11]. A few years later, Hamburger was able to show that the number of motoneurons in the chick embryo decreased after limb amputation as a result of increased cell death, and not as the consequence of failure of mitosis or motoneuron differentiation [13]. Then, Hughes showed for the first time that a large overproduction of motoneurons occurred during normal development and that motoneuron death was a critical factor for the regulation of their final number [14, 15], and Prestige demonstrated the same phenomenon in spinal DRG neurons [16].

After these studies, the concept that programmed cell death (PCD) already expanded in abstract was occurring in the nervous system at different stages of development and growth independently of external insults started rapidly catching on. Thus PCD was recognized as a highly phylogenetically conserved general mechanism by which eukaryotic cells die following a stereotyped series of molecular and cellular events in nervous tissues.

2 Morphological Types of Cell Death in Neurons

Today, the concept that developmental cell death occurs during the course of central and peripheral nervous system maturation is fully established. In parallel, evidence is accumulating to indicate a major role of PCD during physiological aging and in neurodegenerative diseases. These concepts are nowadays so deep-rooted that to strengthen the significance of cell death during the course of the *normal* life cycle of central and peripheral neurons, the term naturally occurring neuronal death (NOND) has come into use to indicate PCD in the nervous system [17].

It is now widely accepted that about half of the neurons produced during neurogenesis die before completion of central (CNS) and peripheral nervous system (PNS) maturation, and that nearly all classes of neurons are produced in excess during development. These oversized populations of neurons are then significantly reduced during NOND, by a relatively limited number of widely investigated modalities that are not exclusive to neurons but affect all eukaryotic cells.

The importance of morphological observation for correct identification of the mode of cell death in neurons is linked to the historical roots of the distinction between apoptosis—the first morphological form of cell death to be recognized (see below)—and necrosis, a term that was originally employed by Virchow to indicate the death of a cell in consequence of a passive pathological event [18]. After PCD came to be recognized as a widespread phenomenon in development, an ultrastructural study on several embryonic tissues proposed that there were three main types of cell death during normal development, on the basis of the role of lysosomes in cell disruption [19]. In the first type, cell death occurs without any detectable activation of endogenous lysosomes, but cells are eventually destroyed by phagocytosis and secondary lysosome activation by tissue macrophages. This process has also been referred to as heterophagocytosis. In autophagocytosis cells are, instead, eliminated after activation of their own lysosomal enzymes, whereas in the third type, there is no obvious lysosome intervention. The first two types are by far the more common, and have been described by various authors starting from the 1960s. The ultrastructural features of type 1 cell death in Schweichel and Merker classification [19] correspond to the current definition of apoptosis, while at least a variant of type 3 shares several features with necrosis [20].

2.1 Apoptosis

Although often considered the prototype form of PCD—and often erroneously indicated *as* PCD—apoptosis is only one of the several types of NOND in neurons. Such a misinterpretation is derived from the initial recognition of apoptosis as the most widely diffused mode of cell death.

Apoptosis was originally defined as a distinct mode of cell death on the basis of a series of characteristic ultrastructural features according a sequence of events starting from nuclear and cytoplasmic condensation, and leading to cell fragmentation and phagocytosis [21]. Initially Kerr and coauthors [21] used the term "shrinkage necrosis" to describe this form of cell death. Subsequently they coined the word "apoptosis" (from the Greek = falling of the leaves), which indicates the dropping of leaves from trees or petals from flowers, to emphasize the occurrence in normal tissue turnover.

Apoptosis involves a series of stereotyped, morphologically well-defined phases that can be only in part appreciated by light microscopy but are most clearly evident at the transmission electron microscope (TEM) level (Figs. 1 and 2a–c).

Changes in the nucleus represent the first unequivocal sign of apoptosis: at onset, chromatin condensation and segregation into sharply delineated masses that abut the nuclear envelope are typically observed. These masses are very electrondense and often show a characteristic crescent-like appearance (Fig. 2a). High-magnification TEM images reveal that they are made up of closely packed, fine granular material. This initial condensation then leads to true nuclear pyknosis. In parallel, cytoplasmic condensation also occurs (Fig. 2b), and the cell—at least in some cases—assumes a stellate appearance, as its membrane becomes convoluted with the onset of protuberances of various sizes. As cytoplasmic density increases, some vacuoles can be seen, but the cell organelles remain undamaged; although they become closely packed, likely as a consequence of significant cytosolic loss (Fig. 2c). Ribosomes can be detached from the rough endoplasmic reticulum and from polysomes, eventually to disappear. As the process continues, the cell and its nucleus assume a more irregular shape and nuclear budding occurs to produce discrete fragments, still surrounded by an intact nuclear envelope (Fig. 2c). The subsequent steps in nuclear and chromatin condensation that can be observed ultrastructurally are paralleled by a well-characterized series of molecular events (*see* Subheading 3.1). Eventually the cell fragments into

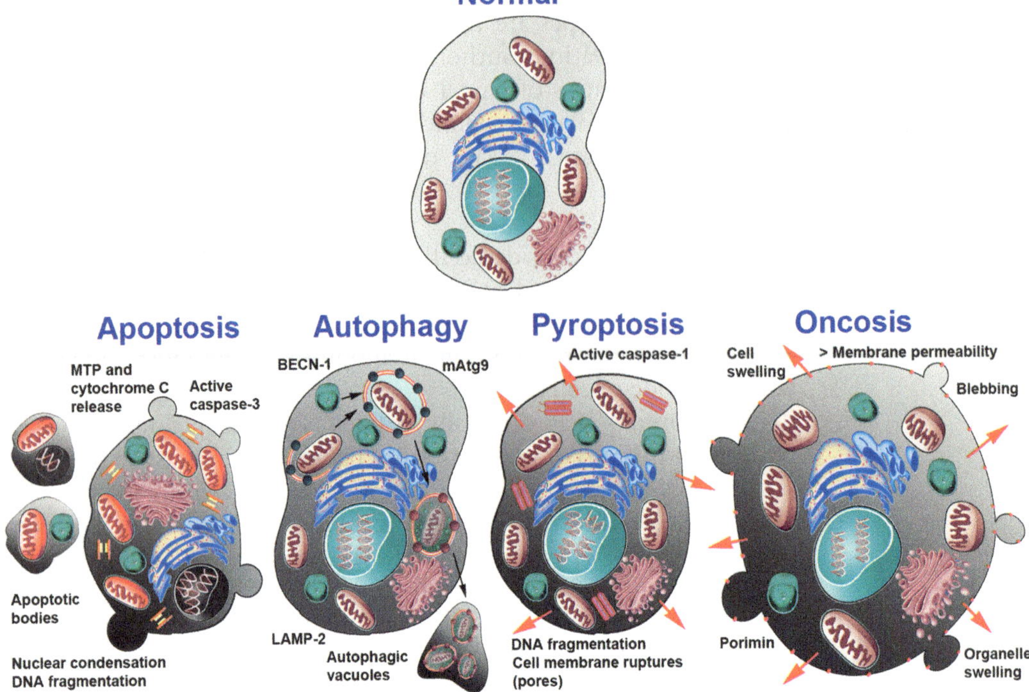

Fig. 1 Schematic representation of the main different types of cell death occurring in neurons. For simplicity only the main structural features and molecules involved are indicated. See text for further explanations

membrane-bounded apoptotic bodies (Fig. 2c). Outside CNS, apoptotic bodies are rapidly cleared by macrophages or neighboring cells, and degraded within heterophagosomes. In the brain this clearing is carried out by the resident microglia [22].

As this series of events appears to be triggered in the absence of external insults and the main elements of the apoptotic machinery are constitutively expressed or generated by the cell itself, apoptosis is also commonly referred to as a form of "cell suicide."

Apoptosis is accompanied by a specific pattern of DNA damage that leads to the formation of low molecular weight DNA oligomers. These can be visualized after DNA extraction and electrophoresis in tissue extracts, or directly in tissue sections by the use of the TUNEL technique [23].

Fig. 2 (continued) during the course of its migration is undergoing cell death of the autophagic type. Note that the cell displays an unaltered nucleus in the presence of numerous vacuoles and mitochondrial disintegration (*inset*). (**e**) Autophagic cell death in a cerebellar interneuron. The high electronic density of mitochondria (*inset*) is indicative of early stages in mitophagia. *Abbreviations*: *ab* apoptotic body, *ch* chromatin, *ly* lysosomes, *m* mitochondria, *nu* nucleus. Bars = 1 μm

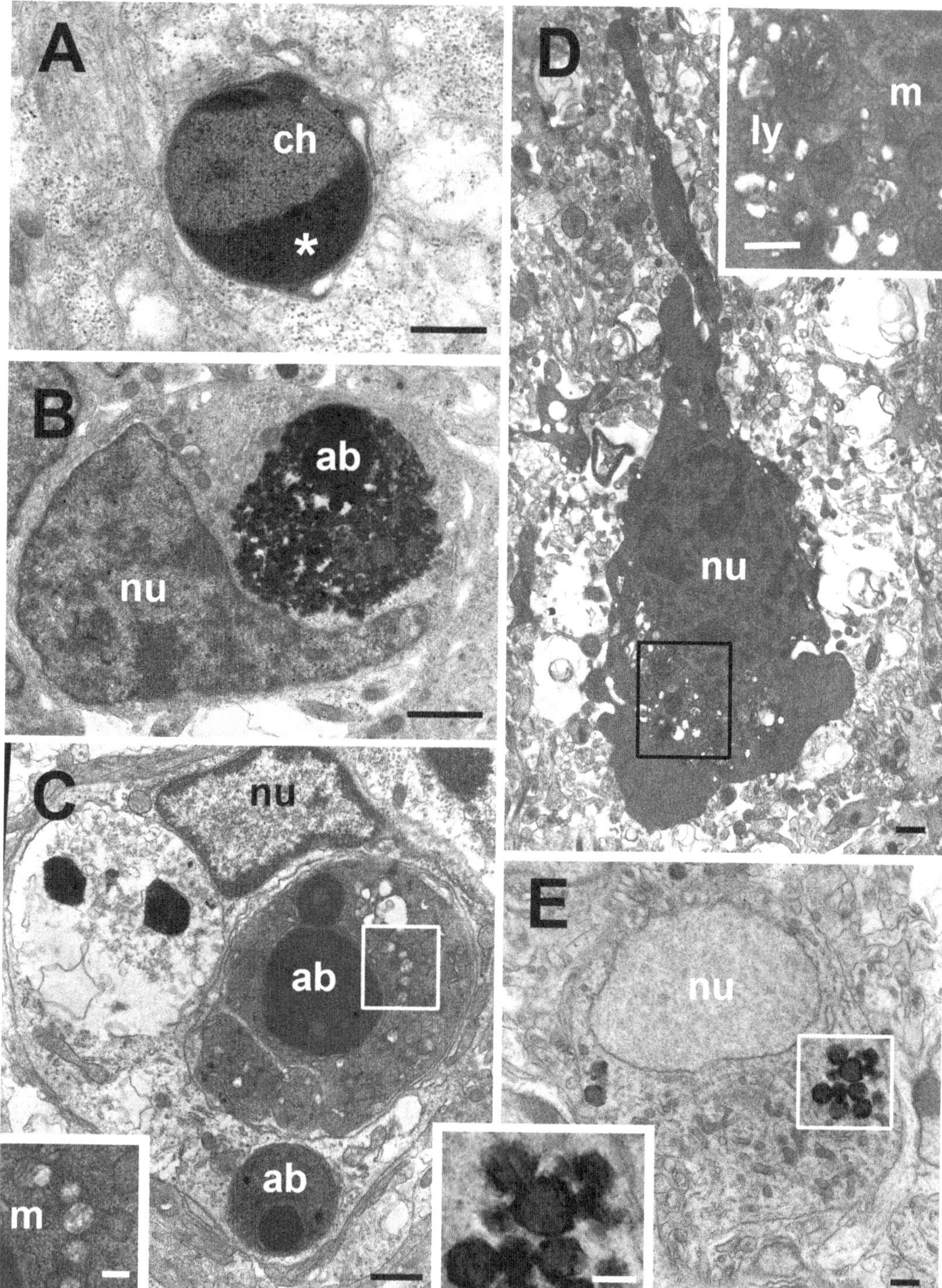

Fig. 2 Exemplificative ultrastructural features of apoptotic (**a–c**) and autophagic (**d, e**) cell death in cerebellar neurons. (**a**) Nuclear condensation in an early apoptotic granule cells in the external layer of the postnatal day 5 (P5) cerebellar cortex. The chromatin is highly condensed and shows a typical crescent-like appearance at nuclear periphery (*asterisk*). (**b**) A late apoptotic granule cell displays high nuclear and cytoplasm condensation and has been engulfed by a microglial cell. (**c**) A microglial cell with engulfed apoptotic bodies at different stages of intracellular digestion. Note the preservation of mitochondrial ultrastructure (*inset*). (**d**) A granule cell

2.2 Autophagy

The term autophagy (from the Greek = self eating), coined by de Duve in 1963 [24], refers to a type of cell death in which the cytoplasm is actively destroyed by lysosomal enzymes well before nuclear changes become visible (Fig. 1). Three different types of autophagy have been described so far: macro-, micro-, and chaperone-mediated autophagy [25]. Whereas the two last types are directly related to degradation of intracellular cytosolic components (microautophagy) or to soluble proteins in lysosomes (chaperone-mediated authophagy), macroautophagy affects large parts of the cell with the characteristic appearance of autophagic vacuoles of lysosomal origin in the cytoplasm, eventually leading to death (Fig. 2d, e).

Macroautophagy is also known as a type II PCD or autophagic cell death [26]. Excessive activity of the autophagy machinery may lead to self-destruction and cell death, or, conversely, also an insufficient autophagic activity (or an imbalanced autophagic flux) may contribute to cell death [27].

In cells undergoing autophagy, many of the characteristic changes of apoptosis eventually become evident, but they are notably delayed; substantial cellular degradation is clearly observed before the typical nuclear alterations of apoptosis occur (Fig. 2d, e). When about three quarters of the cytoplasm has been destroyed, it begins to condense, and chromatin coalesces to subsequently display margination at the nuclear periphery. Notably, although autophagy is per se a form of cell death or accompanies cell death it has also been proposed to have a role in cell survival [25].

2.3 Pyroptosis

Pyroptosis (from the Greek = falling of fire) is a pro-inflammatory form of cell death that causes an infected macrophage to kill itself, and at the same time to release interleukin-1β (IL-1β) [28]. Processing of IL-1β is carried out by caspase-1 that is specifically activated during pyroptosis (Fig. 1). In the mammalian brain, caspase-1 activation also has a direct role in noninfectious cell death processes [29]. The most extensive evidence for both the inflammatory and direct cell death activities of caspase-1 exists in neuronal injury. The caspase family can be further subdivided into two subfamilies, consisting of apoptotic and inflammatory caspases. Caspase-1 belongs to the inflammatory group, and, in mammals, it cannot be considered a typical regulator of apoptosis, as caspase-1 knockout mice develop normally and respond to apoptotic stimuli [30, 31]. However, activation and/or experimental manipulation of the caspase-1 pathway causes cell death in neurons, at least under certain circumstances. For example, caspase-1 activates mitochondrial cell death pathways in cortical neuron cultures following hypoxia/ischemia [32] and is involved in cell death in SOD1 models of ALS [33]. Conversely, transgenic mice overexpressing a dominant-negative caspase-1 under the control of a neuron-specific promoter exhibit reduced ischemic brain injury compared with wild-type controls [34], and caspase-1 inhibition protects DRG neurons against trophic factor-induced cell death independently of IL-1β processing [35].

2.4 Oncosis

Oncosis (from the Greek = swelling) is a form of "cell murder" that eventually leads to necrosis. There are only a few examples of this type of cell death in neurons, although the term was already proposed at the beginning of the last century by von Recklinghausen [36]. In cells other than neurons, oncosis (Fig. 1) is characterized by cellular and organelle swelling, blebbing, and increased membrane permeability caused by the failure of the ionic pumps of the plasma membrane [37], and a rearrangement of cytoskeletal proteins [38], while the DNA damage is nonspecific [39]. It has been postulated that oncotic cell death involves a progressive three-stage membrane injury process [40, 41]: in stage 1, cells become committed to oncosis because of a selective membrane leakage of ions and water as a consequence to ATP depletion. This leads to swelling without a generalized increase in cell membrane permeability. During stage 2, the cell membrane becomes leaky to propidium iodide (PI) and Trypan blue, indicating the development of a non-selective increase in membrane permeability. Stage 3 coincides with the final physical disruption of the cell membrane. It is generally accepted that early necrotic cells lose their plasma membrane integrity, thus permitting the entry of Trypan blue and PI into the cell [42]. Therefore, the exclusion of Trypan blue and PI indicates that oncosis cannot be considered a form of early necrosis.

Oncosis has recently been indicated as the possible cell death pathway that affects astrocytes after focal brain ischemia [43, 44].

3 Neuronal Cell Death Pathways and Their Recognition in Death Assays

The different morphological types of cell death that have been so far described in neurons are associated with a series of cellular biochemical changes that display a certain degree of specificity. Therefore, it follows that it is theoretically possible to distinguish between each different type by the localization of specific components of the respective intracellular machineries. However, it must be remembered that PCD is a highly dynamic process and very often a cross-talk between different types of PCD is likely to occur, under either normal or pathological conditions, so that a combination of different approaches is always necessary to obtain an unequivocal proof of a given type of cell death taking place in specific neuronal populations and/or experimental contexts.

A detailed description of the intracellular pathways specifically associated with any of the possible modalities of PCD in neurons is beyond the purpose of this introductory chapter. Below, the most relevant features of each of the types of cell death described in the

Table 1
Main features of the different types of cell death in the nervous system

Types of cell death		Apoptosis	Autophagy	Pyroptosis	Oncosis
Characteristics	Cell lysis	No	No	Yes	No
	Cell swelling	No	No	Yes/no	Yes
	Pore formation	No	No	Yes	No
	Membrane blebbing	Yes	No	No	Yes
	DNA fragmentation	Yes	No	Yes but *not* in oligonucleosomes	No
	Autophagic vacuoles	No/late	Yes	No	No
Mechanisms pathways	Caspase-1	No	No	Yes	No
	Caspase-3	Yes	No	No	No
	Cytochome c release	Yes	No	No	No
	MTP	Yes	No	No	No
	ICAD	Yes	No	No	No
	ATP depletion	No	No	No	Yes
	Atgs/BECN-1 LAMP-2	No	Yes	No	No
	Bcl-2	Protective	Inhibition of BECN-1	Unknown	Unknown
NOND		Yes	Yes	Injured neurons	Injured astrocytes

Atgs autophagy-related proteins, *Bcl-2* B-cell lymphoma/leukemia-2, *BECN-1* Beclin-1, *ICAD* inhibitor of caspase-activated DNase, *LAMP-2* lysosome-associated membrane protein-2, *MTP* mitochondrial permeability transition

previous section will be briefly considered to set the ground for better understanding the arrays of technical approaches available to the study of PCD in neurons (Table 1).

3.1 Apoptosis

The current view indicates that there are two main apoptotic pathways in neurons: the extrinsic or death receptor pathway, and the intrinsic or mitochondrial pathway. However, it is now clear that these two pathways are intimately linked and that molecules in the death receptor pathway can influence the mitochondrial pathway and vice versa. Both pathways then converge on the same execution (or terminal pathway) that is initiated by the cleavage of caspase-3 and eventually results in DNA fragmentation, degradation of cytoskeletal and nuclear proteins, cross-linking of proteins, formation of apoptotic bodies, expression of ligands for phagocytic cell receptors and, finally, uptake by phagocytic cells [17].

Caspases are a family of proteins with proteolytic activity and, in general, are capable of cleaving proteins at aspartic acid residues, although different caspases have different specificities involving recognition of neighboring amino acids [45, 46]. After initial caspase activation it appears that neurons are irreversibly committed to death. Caspases are widely expressed in an inactive proenzyme form and once activated they can, in turn, activate other procaspases, allowing the initiation of a protease cascade. Such a cascade amplifies the apoptotic signal and thus leads to rapid cell death. Activation of caspases is therefore a hallmark of apoptosis, and the localization of activated caspases by immunocytochemistry/Western blotting with specific antibodies raised against the cleaved forms of individual caspases is one of the most widely employed approaches in the detection of apoptotic cell death in neurons [23].

As a specific additional feature, apoptotic cells display extensive protein cross-linking. This is achieved through the expression and activation of tissue transglutaminase [47]. DNA breakdown by Ca^{2+}- and Mg^{2+}-dependent endonucleases also occurs, resulting in the cleavage of DNA into fragments of 180–200 base pairs [48]. A characteristic "DNA ladder" can thus be visualized by agarose gel electrophoresis and ethidium bromide staining. DNA fragmentation can also be visualized by the use of a series of in situ end labeling (ISEL) techniques among which the most popular one is the so called terminal deoxynucleotidyl transferase (TdT)-mediated nick end labeling (TUNEL) of the DNA fragments generated by the apoptotic process [23].

Expression of specific cell surface markers is another biochemical feature of apoptosis. It results in the early recognition of apoptotic cells by neighboring phagocytic cells, thus permitting quick phagocytosis with minimal tissue damage. Phagocyte recognition is achieved by exposure of phosphatidylserine on the outer layer of the plasma membrane of cells committed to death [49]. Annexin V is a recombinant phosphatidylserine-binding protein that interacts strongly and specifically with phosphatidylserine residues and can be used for the detection of apoptosis [50]. Although externalization of phosphatidylserine is a well-known phagocyte recognition ligand on the surface of apoptotic cells, other proteins are also exposed on the cell surface during apoptotic cell clearance. These include annexin A1 [51] and calreticulin [52]. In the CNS, annexin A1 is restricted to microglia and has recently been implicated in controlling the noninflammatory phagocytosis of apoptotic neurons and in promoting the resolution of inflammatory microglial activation. Microglial-derived annexin A1 tags apoptotic neurons, serving as both signal and bridge between the dying cell and the phagocytosing microglia [53].

Other biochemical modifications occurring in neurons in the course of apoptosis are related to the type of pro-apoptotic stimulus that, in turn, specifically activates the extrinsic or intrinsic pathways. These two major pathways can be differentiated by the relative timing of caspase activation and mitochondrial release of cytochrome c. In the first, which is exemplified by activation of death receptors, an effector caspase is activated prior to mitochondrial alterations. In the other, cytochrome c is instead released from the mitochondrial intermembrane space prior to caspase activation [17].

While the death receptor extrinsic pathway is triggered by a relatively small number of structurally related ligands, mitochondrial apoptosis along the intrinsic pathway can be triggered in neurons by a variety of structurally unrelated agents [54]. This implies that mitochondrial apoptosis may be induced by more than one single stimulus. All of these stimuli produce changes in the inner mitochondrial membrane, resulting in the opening of the mitochondrial permeability transition (MPT) pore, the loss of mitochondrial transmembrane potential, and the cytosolic release of two main sets of pro-apoptotic proteins that are normally sequestered inside the mitochondria [55]. The first set of proteins is released immediately after the mitochondrial pathway is triggered. It consists of cytochrome c, Smac/DIABLO, and HtrA2/Omi [56–58]. Once released from mitochondria, cytochrome c binds and activates Apaf-1 as well as procaspase-9, forming a macromolecular complex called an "apoptosome" that activates caspase-9 [46]. Smac/DIABLO and HtrA2/Omi are reported to promote apoptosis by inhibiting a family of inhibitors of apoptosis proteins (IAPs) [59, 60].

The second set of pro-apoptotic proteins that are released at a later phase, when the cell is irreversibly committed to die, consists of apoptosis-inducing factor (AIF), endonuclease G and DNA fragmentation factor 40 (DFF40) also referred to as caspase-activated DNase (CAD). AIF translocates to the nucleus and causes DNA fragmentation and condensation of peripheral chromatin [61]. This early form of nuclear condensation is referred to as "stage I" condensation [62]. Endonuclease G also translocates to the nucleus and cleaves nuclear chromatin to produce smaller oligonucleosomal DNA fragments [63]. Both AIF and endonuclease G operate independently of caspase activation. DFF40/CAD is then also released from the mitochondria and reaches the nucleus where, after cleavage by caspase-3, it leads to oligonucleosomal DNA fragmentation and further chromatin condensation [64]. This later and more pronounced chromatin condensation is referred to as "stage II" condensation [65].

The control and regulation of the apoptotic mitochondrial events occurs through members of the B-cell lymphoma/leukemia-2 (Bcl-2) family of proteins [66, 67]. The family consists of a total 25 members that regulate mitochondrial membrane permeability and can be either pro-apoptotic or anti-apoptotic. These proteins are of particular relevance since they can determine whether a cell becomes committed to apoptosis or can interrupt the death process. It is thought that the main mechanism of action of the Bcl-2 family of proteins is the regulation of cytochrome c release from the mitochondria via alteration of mitochondrial membrane permeability. Recent studies in cerebellar neurons indicate that the regulation of cellular levels of the BCL-2 protein occurs at the posttranslational level and involves the activation of the autophagic machinery [68, 69].

3.2 Autophagy

The core machinery of autophagy is represented in simple organism models by a family of autophagy-related (Atg) proteins that are differently involved in the distinct steps that can be recognized in the autophagic process: induction, cargo recognition and selection, vesicle formation, autophagosome–vacuole fusion, and breakdown of the cargo followed by release of the degradation products back into the cytosol (Fig. 1).

In mammals the functions of Atgs 1 and 17 are carried out by the Unc-51-like kinase 1 (ULK1) and 2 (ULK2), and by the focal adhesion kinase family-interacting protein of 200 kDa (FIP200) which forms a complex with ULKs and Atg13, and localizes to the phagophore [70, 71].

Cargos are recognized through interactions with specific receptor proteins. The formation of the sequestering vesicles is likely the most complex step of autophagy. Various organelles, including the mitochondria, the Golgi complex, and the endoplasmic reticulum, have been proposed as sources of the autophagosomal membrane. Numerous Atg proteins participate in autophagosome formation and are recruited to the phagophore. Formation of the initial phagophore membrane requires the class III phosphatidylinositol 3-kinase (PtdIns3K) complex, which is composed of p150, barkor (mAtg14) and beclin 1 [72, 73]. The function of beclin 1 in autophagy is negatively regulated by Bcl-2, an anti-apoptotic protein that binds and sequesters beclin 1 [69]. To date, Atg9 is the only identified integral membrane protein required for autophagosome formation. Mammalian Atg9 (mAtg9) moves from the *trans*-Golgi network (TGN) to the late endosomes, which are also labeled with microtubule-associated protein 1 light chain 3 (LC3), when autophagy is induced. The redistribution of mAtg9 from the TGN to late endosomes is dependent on ULK1 and human Atg13 [74, 75].

Autophagosome–lysosome fusion requires the lysosome-associated membrane protein 2 (LAMP-2) [76] and the small GTPase Rab7 [77]. After fusion, degradation of the inner vesicle is dependent on a series of lysosomal/vacuolar acid hydrolases, including cathepsin B, D, and L [78]. The resulting small molecules from the degradation, particularly amino acids, are transported back to the cytosol for protein synthesis and maintenance of cellular functions.

Noncanonical PtdIns3K/beclin 1-independent induction pathways have been described for injury-induced autophagy and mitophagy [79].

3.3 *Pyroptosis*

Pyroptosis is distinguished from other forms of cell death not only by its unique morphological characteristics but also on biochemical bases [80]. DNA cleavage during pyroptosis results from caspase-1-stimulated nuclease activity [81], but does not involve the degradation of the inhibitor of CAD (ICAD) that, in apoptosis, operates as a brake to the activity of the enzyme [64]. Fragmented DNA in pyroptotic cells is diffusely present within the nucleus (Fig. 1), which contrasts with the typical condensed nuclear morphology observed during apoptosis [82]. In addition, the cleaved DNA does not display the oligonucleosomal pattern characteristic of apoptosis, but pyroptotic cells (as those undergoing apoptosis) are stained with the TUNEL technique [81].

Differently from apoptosis, in pyroptosis poly-adenyl-ribose polymerase (PARP) activity and DNA fragmentation do not intervene in cell lysis that is, instead, a consequence of the formation of plasma membrane pores following actin polymerization and caspase-1 activation [81]. Finally, during pyroptosis, there is not an activation of apoptotic caspases, and mitochondrial integrity remains intact; although mitochondria lose their membrane potential [82, 83].

3.4 *Oncosis*

The pro-oncosis receptor inducing membrane injury protein (porimin), also referred to as keratinocyte-associated transmembrane protein-3 (KCT-3), is encoded by the gene TMEM123 and is believed to be specifically associated with oncosis [84]. Biochemically, this type of cell death appears to be primarily linked to the reduction of mitochondrial respiration and ATP synthesis [37], as intracellular ATP levels determine the cell-death fate by apoptosis or oncosis, and depletion of intracellular ATP can irreversibly induce a cell death of the oncotic type [85]. Oxygen-glucose deprivation (OGD) in astrocytes induces oncosis in parallel with an upregulation of the anti-apoptotic gene *bcl-2* [86] which inhibits MPT and the release of apoptogenic proteins from mitochondria [87].

4 Concluding Remarks and Future Perspectives

Neuronal cell death is today a very important and continuously expanding area of investigation in the neurosciences. NOND, whereby cells are discretely removed without interfering with the further development of remaining cells is now a universally recognized form of PCD that occurs during normal development of central and peripheral neurons in vertebrates (and invertebrates). In several brain areas and in different species, NOND occurs in two subsequent temporal waves. In the first wave a large number of neuronal progenitors are eliminated during the peak of neurogenesis. The second wave is, instead, the death of fully differentiated neurons while migrating toward their target locations or connecting to target cells. Very likely, these two waves respectively regulate the neuronal precursor pool size and ensure the proper wiring of developing neuronal networks, although additional functions can be envisaged for the first wave of PCD [88]. Interestingly caspase-3, the main effector caspase in vertebrates, appears to intervene only in late PCD but not in the death of the neuronal precursor pool [89]. This example clearly highlights the need of further work to get better insights into the regulatory mechanisms of cell death in different populations of neurons and in different areas of CNS and PNS. Such a need is reinforced by increasingly emerging "crosstalk" between the molecular pathways that are typically activated in apoptosis, autophagy, pyroptosis and oncosis. Also, there are reasons to consider PCD occurring after a pathological insult as different from NOND [90]. The main reason for this assumption is that under pathological conditions several cell death pathways (such as the intrinsic pathway of apoptosis MPT that follows an excessive elevation of intracellular calcium, ionic imbalances, and intracellular edema due to excessive Na and Cl influx into the cell, and oxidative stress) can be activated simultaneously.

Therefore, although histology and molecular biology have up to now been able to provide considerable information on the different types of cell death that occur in the nervous system under normal and pathological conditions (Table 1), the future NOND and "pathological apoptosis" research lies in the development of genetically encoded in vitro and in vivo probes allowing for real-time visualization and analysis of cell death in alive individual neurons in complex multicellular organisms.

Acknowledgements

We wish to thank Dr. Matthew Bradman for his critical reading of the text and useful suggestions.

References

1. Vogt C (1842) Untersuchungen uber die Entwicklungsgeschichte der Geburtshelferkröte (Alytes obstetricans). Jent und Gassmann, Solothurn

2. Virchow R (1858) Vorlesungen über Cellularpathologie in ihrer Begründung auf physiologischer und pathologischer Gewebelehre. Verlag August Hirschwald, Berlin

3. Clarke PG, Clarke S (2012) Nineteenth century research on cell death. Exp Oncol 34: 139–145

4. Stieda L (1872) Die Bildung des Knochengewebes: Festschrift des Naturforschervereins zu Riga, zur Feier des Fünfzigjährigen Bestehens der Gesellschaft Practischer Aerzte zu Riga. Engelmann, Leipzig

5. Flemming W (1885) Über die Bildung von Richtungsfiguren in Säugethiereiern beim Untergang Graaf'scher Follikel. Arch Anat Entwickl:221–244

6. Rao MS, Jacobson M (2005) Developmental neurobiology, IVth edn. Kluwer Academic/

Plenum Publishers, New York, Boston, Dordrecht, London, Moscow

7. Beard J (1889) On the early development of Lepidosteus osseus—preliminary notice. Proc Roy Soc Lond 46:108–118

8. Beard J (1892) The transient ganglion cells and their nerves in Raja batis. Anat Anz 7:191–206

9. Collin R (1906) Histolyse de certains neuroblastes au cours du développement du tube nerveux chez le poulet. C R Soc Biol 60:1080–1081

10. Ernst M (1926) Über Untergang von Zellen während der normalen Entwicklung bei Wirbeltieren. Z Anat Entwicklungsgesch 79:228–262

11. Glücksmann PD (1951) Cell deaths in normal vertebrate ontogeny. Biol Rev 26:59–86

12. Levi-Montalcini R, Cohen S (1960) Effects of the extract of the mouse submaxillary salivary glands on the sympathetic system of mammals. Ann N Y Acad Sci 85:324–341

13. Hamburger V (1958) Regression versus peripheral control of differentiation in motor hypoplasia. Am J Anat 102:365–409

14. Hughes A (1961) Cell degeneration in the larval ventral horn of Xenopus laevis (Daudin). J Embryol Exp Morphol 9:269–284

15. Hughes AF, Lewis PR (1961) Effect of limb ablation on neurones in Xenopus larvae. Nature 189:333–334

16. Prestige MC (1965) Cell turnover in the spinal ganglia of Xenopus laevis tadpoles. J Embryol Exp Morphol 13:63–72

17. Lossi L, Merighi A (2003) In vivo cellular and molecular mechanisms of neuronal apoptosis in the mammalian CNS. Prog Neurobiol 69:287–312

18. Virchow R (1858) Cellularpathologie in ihre Begrundung auf Physiologische und Pathologische Gewebelehre. A. Hirschwal, Berlin

19. Schweichel JU, Merker HJ (1973) The morphology of various types of cell death in prenatal tissues. Teratology 7:253–266

20. Clarke PGH (1990) Developmental cell death: morphological diversity and multiple mechanisms. Anat Embryol 181:195–213

21. Kerr JF, Wyllie AH, Currie AR (1972) Apoptosis: a basic biological phenomenon with wide-ranging implications in tissue kinetics. Br J Cancer 26:239–257

22. Lossi L, Mioletti S, Merighi A (2002) Synapse-independent and synapse-dependent apoptosis of cerebellar granule cells in postnatal rabbits

occur at two subsequent but partly overlapping developmental stages. Neuroscience 112:509–523

23. Lossi L, Mioletti S, Aimar P, Bruno R, Merighi A (2002) In vivo analysis of cell proliferation and apoptosis in the CNS. In: Merighi A, Carmignoto G (eds) Cellular and molecular methods in neuroscience research. Springer Verlag, New York, pp 235–258

24. de Duve C (1963) The lysosome. Sci Am 208:64–72

25. Rosello A, Warnes G, Meier UC (2012) Cell death pathways and autophagy in the central nervous system and its involvement in neurodegeneration, immunity and central nervous system infection: to die or not to die—that is the question. Clin Exp Immunol 168:52–57

26. Maiuri MC, Zalckvar E, Kimchi A et al (2007) Self-eating and self-killing: crosstalk between autophagy and apoptosis. Nat Rev Mol Cell Biol 8:741–752

27. Galluzzi L, Maiuri MC, Vitale I et al (2007) Cell death modalities: classification and pathophysiological implications. Cell Death Differ 14:1237–1243

28. Bergsbaken T, Fink SL, Cookson BT (2009) Pyroptosis: host cell death and inflammation. Nat Rev Microbiol 7:99–109

29. Denes A, Lopez-Castejon G, Brough D (2012) Caspase-1: is IL-1 just the tip of the ICEberg? Cell Death Dis 3:e338

30. Li P, Allen H, Banerjee SK et al (1995) Mice deficient in IL-1 beta converting enzyme are defective in production of mature IL-1 beta and resistant to endotoxin shock. Cell 80:401–411

31. Kuida K, Lippke JA, Ku G et al (1995) Altered cytokine export and apoptosis in mice deficient in interleukine-1-beta converting enzyme. Science 267:2000–2003

32. Zhang WH, Wang X, Narayanan M et al (2003) Fundamental role of the Rip2/caspase-1 pathway in hypoxia and ischemia-induced neuronal cell death. Proc Natl Acad Sci U S A 100:16012–16017

33. Friedlander RM, Brown RH, Gagliardini V et al (1997) Inhibition of ICE slows ALS in mice. Nature 388:31

34. Friedlander RM, Gagliardini V, Hara H et al (1997) Expression of a dominant negative mutant of interleukin-1 beta converting enzyme in transgenic mice prevents neuronal cell death induced by trophic factor withdrawal and ischemic brain injury. J Exp Med 185:933–940

35. Gagliardini V, Fernandez PA, Lee RK et al (1994) Prevention of vertebrate neuronal death by the crmA gene. Science 263:826–828

36. Weerasinghe P, Buja LM (2012) Oncosis: an important non-apoptotic mode of cell death. Exp Mol Pathol 93:302–308

37. Majno G, Joris I (1995) Apoptosis, oncosis, and necrosis. An overview of cell death. Am J Pathol 146:3–15

38. Trump BF, Berezesky IK, Chang SH et al (1997) The pathways of cell death: oncosis, apoptosis, and necrosis. Toxicol Pathol 25:82–88

39. Jugdutt BI, Idikio HA (2005) Apoptosis and oncosis in acute coronary syndromes: assessment and implications. Mol Cell Biochem 270:177–200

40. Buja LM, Eigenbrodt ML, Eigenbrodt EH (1993) Apoptosis and necrosis. Basic types and mechanisms of cell death. Arch Pathol Lab Med 117:1208–1214

41. Buja LM (2005) Myocardial ischemia and reperfusion injury. Cardiovasc Pathol 14:170–175

42. O'Brien MC, Healy SF Jr, Raney SR et al (1997) Discrimination of late apoptotic/necrotic cells (type III) by flow cytometry in solid tumors. Cytometry 28:81–89

43. Cao X, Zhang Y, Zou L et al (2010) Persistent oxygen-glucose deprivation induces astrocytic death through two different pathways and calpain-mediated proteolysis of cytoskeletal proteins during astrocytic oncosis. Neurosci Lett 479:118–122

44. Chu X, Fu X, Zou L et al (2007) Oncosis, the possible cell death pathway in astrocytes after focal cerebral ischemia. Brain Res 1149:157–164

45. Blatt NB, Glick GD (2001) Signaling pathways and effector mechanisms pre-programmed cell death. Bioorg Med Chem 9:1371–1384

46. Adams JM, Cory S (2002) Apoptosomes: engines for caspase activation. Curr Opin Cell Biol 14:715–720

47. Fesus L, Nemes Z, Piredda L et al (1987) Induction and activation of tissue transglutaminase during programmed cell death. FEBS Lett 224:104–108

48. Bortner CD, Oldenburg NB, Cidlowski JA (1995) The role of DNA fragmentation in apoptosis. Trends Cell Biol 5:21–26

49. Fadok VA, Voelker D, Campbell PA et al (1992) Exposure of phosphatidylserine on the surface of apoptotic lympmhocytes triggers specific recognition and removal by macrophages. J Immunol 148:2207–2216

50. van Engeland M, Nieland LJ, Ramaekers FC et al (1998) Annexin V-affinity assay: a review on an apoptosis detection system based on phosphatidylserine exposure. Cytometry 31:1–9

51. Arur S, Uche UE, Rezaul K et al (2003) Annexin I is an endogenous ligand that mediates apoptotic cell engulfment. Dev Cell 4:587–598

52. Martins I, Kepp O, Galluzzi L et al (2010) Surface-exposed calreticulin in the interaction between dying cells and phagocytes. Ann N Y Acad Sci 1209:77–82

53. McArthur S, Cristante E, Paterno M et al (2010) Annexin A1: a central player in the anti-inflammatory and neuroprotective role of microglia. J Immunol 185:6317–6328

54. Sastry PS, Rao K (2000) Apoptosis in the nervous system. J Neurochem 74:1–20

55. Saelens X, Festjens N, Vande WL et al (2004) Toxic proteins released from mitochondria in cell death. Oncogene 23:2861–2874

56. Cai J, Yang J, Jones DP (1998) Mitochondrial control of apoptosis: the role of cytochrome c. Biochim Biophys Acta 1366:139–149

57. Du C, Fang M, Li Y et al (2000) Smac, a mitochondrial protein that promotes cytochrome c-dependent caspase activation by eliminating IAP inhibition. Cell 102:33–42

58. van Loo G, van Gurp M, Depuydt B et al (2002) The serine protease Omi/HtrA2 is released from mitochondria during apoptosis. Omi interacts with caspase-inhibitor XIAP and induces enhanced caspase activity. Cell Death Differ 9:20–26

59. van Loo G, Saelens X, van Gurp M et al (2002) The role of mitochondrial factors in apoptosis: a Russian roulette with more than one bullet. Cell Death Differ 9:1031–1042

60. van Loo G, Saelens X, Matthijssens F et al (2002) Caspases are not localized in mitochondria during life or death. Cell Death Differ 9:1207–1211

61. Joza N, Susin SA, Daugas E et al (2001) Essential role of the mitochondrial apoptosis-inducing factor in programmed cell death. Nature 410:549–554

62. Susin SA, Lorenzo HK, Zamzami N et al (1999) Molecular characterization of mitochondrial apoptosis-inducing factor. Nature 397:441–446

63. Li LY, Luo X, Wang X (2001) Endonuclease G is an apoptotic DNase when released from mitochondria. Nature 412:95–99

64. Enari M, Sakahira H, Yokoyama H et al (1998) A caspase-activated DNase that degrades DNA during apoptosis, and its inhibitor ICAD. Nature 391:43–50

65. Susin SA, Daugas E, Ravagnan L et al (2000) Two distinct pathways leading to nuclear apoptosis. J Exp Med 192:571–580

66. Adams JM, Cory S (1998) The Bcl-2 protein family: arbiters of cell survival. Science 281:1322–1326

67. Adams JM, Cory S (2001) Life-or-death decisions by the Bcl-2 protein family. Trends Biochem Sci 26:61–66

68. Lossi L, Gambino G, Ferrini F et al (2009) Posttranslational regulation of BCL2 levels in cerebellar granule cells: a mechanism of neuronal survival. Dev Neurobiol 69:855–870

69. Lossi L, Gambino G, Salio C et al (2010) Autophagy regulates the post-translational cleavage of BCL-2 and promotes neuronal survival. ScientificWorldJournal 10:924–929

70. Hara T, Takamura A, Kishi C et al (2008) FIP200, a ULK-interacting protein, is required for autophagosome formation in mammalian cells. J Cell Biol 181:497–510

71. Jung CH, Jun CB, Ro SH et al (2009) ULK-Atg13-FIP200 complexes mediate mTOR signaling to the autophagy machinery. Mol Biol Cell 20:1992–2003

72. Itakura E, Kishi C, Inoue K et al (2008) Beclin 1 forms two distinct phosphatidylinositol 3-kinase complexes with mammalian Atg14 and UVRAG. Mol Biol Cell 19:5360–5372

73. Sun Q, Fan W, Chen K et al (2008) Identification of Barkor as a mammalian autophagy-specific factor for Beclin 1 and class III phosphatidylinositol 3-kinase. Proc Natl Acad Sci U S A 105:19211–19216

74. Chan EY, Longatti A, McKnight NC et al (2009) Kinase-inactivated ULK proteins inhibit autophagy via their conserved C-terminal domains using an Atg13-independent mechanism. Mol Cell Biol 29:157–171

75. Young AR, Chan EY, Hu XW et al (2006) Starvation and ULK1-dependent cycling of mammalian Atg9 between the TGN and endosomes. J Cell Sci 119:3888–3900

76. Tanaka Y, Guhde G, Suter A et al (2000) Accumulation of autophagic vacuoles and cardiomyopathy in LAMP-2-deficient mice. Nature 406:902–906

77. Jager S, Bucci C, Tanida I et al (2004) Role for Rab7 in maturation of late autophagic vacuoles. J Cell Sci 117:4837–4848

78. Tanida I, Minematsu-Ikeguchi N, Ueno T et al (2005) Lysosomal turnover, but not a cellular level, of endogenous LC3 is a marker for autophagy. Autophagy 1:84–91

79. Zhu JH, Horbinski C, Guo F et al (2007) Regulation of autophagy by extracellular signal-regulated protein kinases during 1-methyl-4-phenylpyridinium-induced cell death. Am J Pathol 170:75–86

80. Fink SL, Cookson BT (2005) Apoptosis, pyroptosis, and necrosis: mechanistic description of dead and dying eukaryotic cells. Infect Immun 73:1907–1916

81. Fink SL, Cookson BT (2006) Caspase-1-dependent pore formation during pyroptosis leads to osmotic lysis of infected host macrophages. Cell Microbiol 8:1812–1825

82. Brennan MA, Cookson BT (2000) Salmonella induces macrophage death by caspase-1-dependent necrosis. Mol Microbiol 38:31–40

83. Jesenberger V, Procyk KJ, Yuan J et al (2000) Salmonella-induced caspase-2 activation in macrophages: a novel mechanism in pathogen-mediated apoptosis. J Exp Med 192:1035–1046

84. Ma F, Zhang C, Prasad KV et al (2001) Molecular cloning of Porimin, a novel cell surface receptor mediating oncotic cell death. Proc Natl Acad Sci U S A 98:9778–9783

85. Eguchi Y, Shimizu S, Tsujimoto Y (1997) Intracellular ATP levels determine cell death fate by apoptosis or necrosis. Cancer Res 57:1835–1840

86. Huang Q, Zhang R, Zou L et al (2013) Cell death pathways in astrocytes with a modified model of oxygen-glucose deprivation. PLoS One 8:e61345

87. Kroemer G (1997) The proto-oncogene Bcl-2 and its role in regulating apoptosis. Nat Med 3:614–620

88. Lossi L, Cantile C, Tamagno I et al (2005) Apoptosis in the mammalian CNS: lessons from animal models. Vet J 170:52–66

89. Lossi L, Tamagno I, Merighi A (2004) Molecular morphology of neuronal apoptosis: activation of caspase 3 during postnatal development of mouse cerebellar cortex. J Mol Histol 35:621–629

90. Blomgren K, Leist M, Groc L (2007) Pathological apoptosis in the developing brain. Apoptosis 12:993–1010

Part I

Molecular, Structural, Functional, and Genomic Changes in Dying Neurons

Chapter 2

Cell Volume Regulation Monitored with Combined Epifluorescence and Digital Holographic Microscopy

Nicolas Pavillon and Pierre Marquet

Abstract

Quantitative phase imaging emerged recently as a valuable tool for cell observation, by enabling label-free imaging through the intrinsic phase-contrast provided by transparent living cells, thus greatly simplifying observation protocols. The quantitative phase signal, unlike the one provided by the widely used phase-contrast microscope, can be related to relevant biological indicators including dry mass, cell volume regulation or transmembrane water movements. Here, we present quantitative phase imaging coupled with live fluorescence, making it possible to follow the phase signal in time to monitor the cell volume regulation, an early indicator of cell viability, along with specific information such as intracellular Ca^{2+} imaging with Fura-2 ratiometric fluorescence.

Key words Microscopy, Quantitative phase imaging, Digital holography, Fluorescence, Ca^{2+} imaging, Cell biology, Cell volume regulation

1 Introduction

Phase imaging has been historically one of the first tools available to visualize transparent living cells, since the invention of the phase-contrast microscope in the forties of the last century [1]. Nowadays, a microscope employing phase-contrast or differential interference contrast (DIC) microscopy [2] is a fundamental tool of any wet laboratory, for example to assess cell cultures condition.

Outside routine observation, these tools have then been mostly supplanted by more specific imaging approaches such as staining, which permits to differentiate tissues in histological slides, or functional imaging through fluorescence. There is however a renewed interest for phase imaging in the recent years, thanks to the rapid development of a new type of imaging called quantitative phase microscopy (QPM).

Unlike most microscopy methods which rely on intensity contrast, such as the absorption of a specimen or the emission of a specific fluorophore, phase imaging is based on the measurement of

Laura Lossi and Adalberto Merighi (eds.), *Neuronal Cell Death: Methods and Protocols*, Methods in Molecular Biology, vol. 1254, DOI 10.1007/978-1-4939-2152-2_2, © Springer Science+Business Media New York 2015

the variation of a light wave front passing through a specimen, making it sensitive to its refractive index, a physical property intimately linked with the density of intracellular components, but primarily determined by the intracellular protein concentration. One of its main characteristics is that it does not require labeling, making experimental protocols far simpler than for most recent imaging approaches. Another feature worth mentioning is its recording speed capability (10–100 frames per second with standard cameras). On the other hand, the signal it delivers is typically less specific than fluorescence, which can be used for instance with specific chemical bonding dyes, so that its interpretation is more challenging.

It has been demonstrated in the past years that QPM could be used, for example, to monitor cell morphology and life cycle [3, 4], to measure the cell dry mass [5, 6], or to assess cell viability [7, 8]. This chapter discusses a particular application of phase imaging, where it is used in conjunction with wide-field fluorescence [9], in order to obtain, in parallel, the phase information along with specific information from a chosen dye. The fluorophore presented here consists of Fura-2, a ratiometric dye which makes it possible to monitor the concentration of free intracellular Ca^{2+} [10, 11].

1.1 Principle of Quantitative Phase Measurement

While a complete description of the technical aspects involved in the measurement of QPM is beyond the scope of this chapter, it is worthwhile mentioning some fundamental elements. For more details, interested readers can refer to existing reviews mentioning, in particular, the numerous methods allowing to obtain a quantitative phase signal (*see* refs. [12–14]).

In the application presented here, phase imaging is coupled with fluorescence, by employing a spectral separation scheme; this implies that the light used for QPM must not overlap with the excitation and emission bands of the employed fluorophore. For this reason, we focus on QPM with laser light, and more particularly on a technique called off-axis digital holographic microscopy (DHM) [15, 16].

Off-axis DHM is a specific implementation which uses coherent light to generate holograms, from which quantitative phase images can be extracted through numerical reconstruction. Its main feature is a fast imaging capability by enabling phase retrieval with a single hologram. Interested readers can refer to existing literature for a detailed overview of the theoretical steps involved in hologram reconstruction [17, 18]. In this context, the laser light employed for DHM is well defined in the optical spectrum and can be separated from the fluorescence emission, yielding two independent signals [9], as employed in this chapter.

1.2 Phase Signal Interpretation

One key point when employing phase measurement on biological samples such as cells is to interpret the measurement correctly. The phase is essentially proportional to the refractive index of the

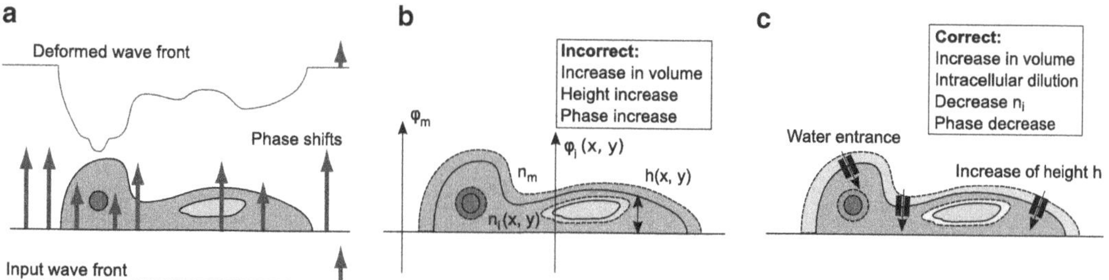

Fig. 1 (**a**) Schematic representation of the generation of the phase signal, where the input wave front is deformed by the optical density of the different organelles. (**b**) Naive interpretation of the effect of a cell volume increase on the phase signal, through the increase in height $h(x,y)$. (**c**) Correct interpretation, when taking into account intracellular dilution during volume increase

intracellular content of a cell and to its thickness. The generation of a phase wave front is depicted in Fig. 1a, where the light of an input planar wave front passing through a cell is deformed as it is retarded by the cellular content which has a larger optical density (refractive index) than the extracellular medium. Organelles can be identified by an additional phase shift, being either positive or negative depending on their refractive index.

Mathematically, the phase signal can be expressed as

$$\Delta\varphi(x,y) = \varphi_i(x, y) - \varphi_{\text{ref}}$$
$$= \frac{2\pi}{\lambda}\left(n_i(x, y) - n_m\right)h(x,y), \tag{1}$$

where $\varphi_i(x,y)$ is the phase shift at a location in the cell, φ_{ref} is the phase shift outside the cell (background signal), λ is the DHM laser wavelength, $n_i(x, y)$ is the average refractive index value along the optical axis, n_m is the refractive index of the perfusion medium, and $h(x, y)$ is the cell height at a given location, as represented in Fig. 1b.

The phase signal can be counter-intuitive at first sight, since a cell volume increase, as depicted in Fig.1b, leading to an increase of the height $h(x, y)$, does not correspond to a phase increase, as Eq.(1) could indicate. This is due to the fact that through cell volume homeostasis, a volume increase is concomitant with water entry through various membrane pathways, as depicted in Fig. 1c, which implies an intracellular content dilution, and consequent decrease of the refractive index $n_i(x, y)$ and phase. In the case of cellular bodies, the changes in refractive index $n_i(x,y)$ are indeed more significant than the height $h(x,y)$ on the phase signal. This is due to the fact that a change in height, which purely modifies the respective contributions of n_i and n_m, has a low influence as the intracellular n_i is classically close to the one of the medium n_m.

The phase signal should thus be understood generally as an indication of intracellular dilution for proper interpretation.

In this fashion, the phase signal can be used as an indicator of cell volume regulation (CVR), which is related to the transmembrane water movements occurring for instance during ionic homeostasis [7, 19]. In this context, the CVR can be employed as an early indicator of cell viability [20], since a prolonged dysregulation is known to trigger several cellular death pathways [21].

1.3 Ratiometric Fluorescence

Live fluorescence can be employed to monitor specific biological parameters. In this chapter, we employ intracellular fluorescent probes [22] to measure the free Ca^{2+} concentration. Ionic concentration dyes are based on the activation of the fluorophore upon bonding with the investigated ion, so that the intensity of the measured fluorescence signal depends on the amount of ions attached to fluorescent molecules, in function of the dissociation constant of the bonding chemical reaction [11].

Several dyes denoted as ratiometric have the property of changing their absorption spectrum, so that their maximum of absorption shifts in wavelength upon bonding with their affinity ion [10]. In this case, the signal can be calculated as a ratio between two excitation wavelengths as

$$R = \frac{F_{\lambda_1}}{F_{\lambda_2}},\qquad(2)$$

where F_{λ_1} is the fluorescent signal at the first excitation wavelength, and F_{λ_2} at the second. A ratiometric measurement presents several advantages, such as making the final measurement independent of the intracellular concentration of fluorophore molecules, and more robust towards photobleaching.

Fura-2 is a ratiometric fluorophore sensitive to $[Ca^{2+}]$ at low concentrations (typically 100 nM). It is excited in the ultraviolet with typical wavelengths $\lambda_1 = 340$ nm and $\lambda_2 = 380$ nm, and emits light in the green region.

2 Materials

On a general point of view, QPM is characterized by rather simple protocols, thanks to its label-free imaging capability. This implies that nearly any sample can be imaged directly out of the cell incubator. However, several steps described in this chapter can greatly increase the image and signal quality.

Furthermore, it should be reminded that the phase signal can be indiscriminately coupled with any fluorophore, as long as the DHM wavelength laser does not overlap with the excitation and emission spectra of the dye. The materials and methods provided here for fluorescence, namely intracellular $[Ca^{2+}]$ dynamics monitored with a ratiometric probe, could thus be interchanged with other standard protocols for other fluorophores, in order to observe another cell property along with the phase signal (*see* **Note 1**). Several examples employing other fluorophores such as Fluo-4 or propidium iodide can be found in the literature [23].

Similarly, the excitation solution provided here is meant for observation of the response of neuronal cells under glutamate perfusion, but the protocol can be modified in order to observe other types of responses.

While the protocols below assume the type of equipment detailed in the following, they should still be valid or easily adaptable for different configurations.

Unless specified otherwise, all operations are supposed to be performed at room temperature. For the preparation of solutions, employ ultrapure sterilized water. For each employed chemical, follow the handling instructions provided by the manufacturer.

2.1 Equipment

1. DHM T1000® microscope (Lyncée Tec, Lausanne, Switzerland).

2. Epi-fluorescence module for DHM T1000® [9] (Lyncée Tec).

3. Koala® software (Lyncée Tec).

4. MetaFluor® software (Molecular Devices, Sunnyvale, CA).

2.2 Cell Culture Handling

1. Poly-L-ornithine (PLO) pre-coated microscopy glass coverslips (Sigma Chemicals, St. Louis, MO). *See* **Notes 2** and **3**.

2. Six- or twenty-four-well plates.

3. Perfusion chamber. *See* **Note 4**.

4. Perfusion system.

2.3 Solutions

1. L-glutamate stock solution: dilute a vial of L-glutamate (Tocris Bioscience, Bristol, UK) to 25 mM with ultrapure water. Store in small aliquots at –20 °C.

2. Perfusion solution: 140 mM NaCl, 3 mM KCl, 3 mM $CaCl_2$, 2 mM $MgCl_2$, 5 mM glucose, 10 mM HEPES. Dilute in ultrapure water (75 % of the final volume). Then, adjust the pH to 7.3 by gradually adding 1 M NaOH. Adjust finally the volume to the desired quantity (*see* **Notes 5** and **6**). If not used at once, store at 4 °C.

3. Excitatory solution: right before the experiment, dilute the L-glutamate stock solution to 30 μM in a small volume (typically 10 mL) of the perfusion solution.

2.4 Loading of Calcium Indicator

1. Calcium indicator stock solution: dilute one vial of Fura-2 acetoxymethyl ester (AM) salt (Molecular Probes®, Life Technologies™, Carlsbad, CA) at 2 mM with dimethyl sulfoxide (DMSO). Store at –20 °C and protect from light.

2. Loading solution: dilute the calcium indicator stock solution at 4 μM in 1 mL of perfusion solution.

3. Place cells in a 35 mm Petri dish with the loading solution and incubate at 37 °C for 30 min.

4. Wash 2–3 times with the perfusion solution to remove the loading solution and incubate at 37 °C for 10 min to ensure full de-esterification of the dye before observation.

3 Methods

The method described below consists of two main steps, first being the measurement itself, where holograms and fluorescence images are acquired in a given set of conditions. The second step consists in the data treatment, in order to extract the signals from the measurements. In case of fluorescence and DHM measurements, the offline processing of the data is also of key importance, as careless treatment can lead to incorrect results, making it a step of similar importance to the measurement itself to derive reliable results.

3.1 Measurements

1. Mount cells in the perfusion chamber (*see* **Note 7**) and place them under the microscope (*see* **Note 8**).

2. Connect the chamber to the perfusion system. The perfusion chamber and system must ensure rapid delivery and exchange of the perfusion solutions. Start flux with the perfusion solution. *See* **Note 9**.

3. Select a magnification of 10–20× (*see* **Notes 10** and **11**), and choose a zone of interest (*see* **Note 12**) to be measured. The field of view should contain enough cells for imaging, but avoid too dense regions, as cells may be difficult to identify afterwards.

4. Open the fluorescence lamp shutter, and focus the fluorescence image. It is important to ensure a proper focus for a good fluorescence signal (*see* **Note 10**), but this should be performed rapidly, to avoid bleaching of the fluorophore (*see* **Note 13**).

5. Close the fluorescence lamp shutter, and set it to open only during image acquisition.

6. Set the fluorescence acquisition to successive excitation at $\lambda_1 = 340$ nmand $\lambda_2 = 380$ nm (*see* **Note 14**).

7. Start acquisition of both holograms (with the Koala® software) and fluorescence images (with the MetaFluor® software) in order to record a baseline of at least 5–10 min, with an acquisition rate of 0.5 Hz in fluorescence, and typically 1–5 Hz with DHM (*see* **Note 15**).

8. After baseline measurement, increase the acquisition rate of fluorescence to 2 Hz to ensure being able to follow the rapid $[Ca^{2+}]$ rise, and perfuse the excitatory solution for 30 s.

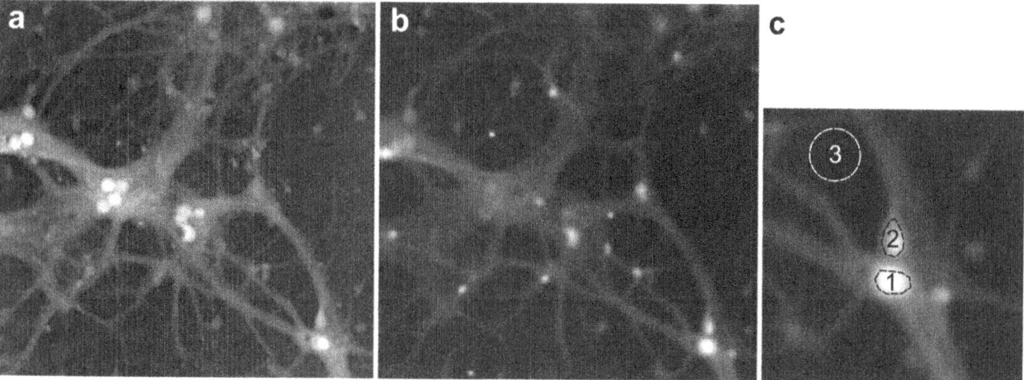

Fig. 2 Typical images of hippocampal neuron cells in culture measured with (**a**) label-free quantitative unwrapped phase, representing a dynamic range of [0, 9.6] radians and (**b**) Fura-2 fluorescence, excited at 380 nm. The field of view represents a region of 360×350 μm. (**c**) Illustration of the selection of 2 cell soma (1, 2) and a background region (3) for computation of the average fluorescence intensity

9. After calcium reached its maximum, reduce the frame rate of fluorescence back to 0.5 Hz.

10. Continue acquisition to follow the rest of the response, during typically 30–40 min.

11. If required, perform a second excitation, otherwise terminate the experiment.

An example of typical measurements taken with the protocol given above is shown in Fig. 2, with the quantitative phase image reconstructed from the hologram (*see* Fig. 2a) and the Fura-2 fluorescence image at 380 nm (*see* Fig. 2b).

One can appreciate the difference of contrast between the two images, where all neurons can be easily identified in phase, while their visibility in fluorescence depends on the intracellular Ca^{2+} concentration, but also on the fluorescent dye loading efficiency for each cell.

3.2 Treatment of Data

1. With fluorescence images use the MetaFluor® software. Compute the mean fluorescence intensity value of the measured cells by selecting the soma. Ensure that the selection region is well placed within the cell in order to avoid signal disruption from pixels outside the cell (*see* Fig. 2c). Also select a background region outside any cell (*see* **Note 16**).

2. Extract the temporal fluorescence signal on the chosen regions by running through all measured images.

3. Subtract the background signal from the different cell signals, in order to suppress the background fluorescence and potential temporal drifts. For a ratiometric measurements, this implies to make the subtraction on both measured wavelengths.

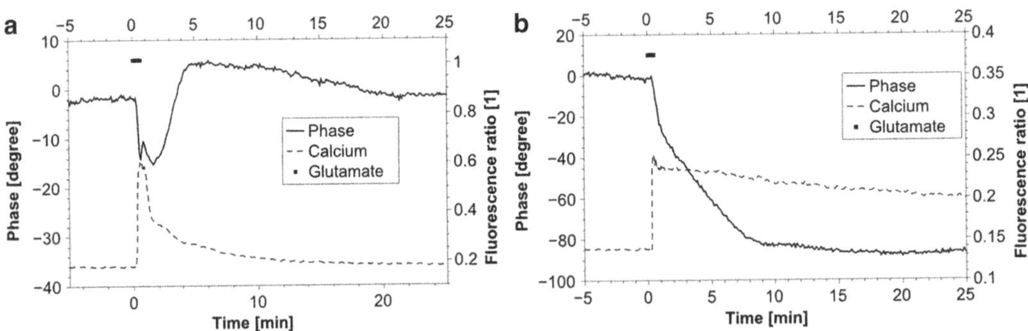

Fig. 3 Typical glutamate-mediated responses obtained by monitoring quantitative phase and Fura-2 fluorescence on neuron cells soma. (**a**) Physiological response, characterized by transient signals which are rapidly regulated. (**b**) Pathological response, associated with a dysregulation in both phase (indicative of an absence of CVR) and $[Ca^{2+}]$

4. Compute the ratiometric signal by dividing the two signals according to Eq. (2).

5. From the acquired holograms, reconstruct the phase signals with the Koala® software (*see* **Note 17**).

6. Perform a similar analysis on phase images by repeating points 1–3, i.e., by selecting similar regions on the cells and a background signal, and subtracting the background.

7. The two signals can be compared by matching the two temporal axes, as shown for example in Fig. 3.

An example of typical measurements as obtained from the protocol given above is provided in Fig. 3, where the temporal signals of phase and ratiometric Fura-2 are presented for both a physiological response (*see* Fig. 3a) and a pathological one (*see* Fig. 3b). Glutamate, the main excitatory neurotransmitter in the brain, induces an early response characterized by an intracellular increase of Na^+ and Ca^{2+}, as well as a neuronal swelling resulting from a transmembrane influx of water associated with the ions entrance for osmotic reasons. The physiological case is characterized by a transient phase decrease, indicative of a cell swelling, which is regulated in the minutes following the stimulation. This decrease is concomitant with the intracellular $[Ca^{2+}]$ increase, which rapidly returns to basal levels. On the other hand, the pathological response is characterized by an irreversible decrease in phase and rise in $[Ca^{2+}]$ levels, which are not regulated. This absence of CVR was shown to be an early marker of the subsequenty neuronal death [7], linked here with the known toxicity of intracellular $[Ca^{2+}]$ dysregulation during glutamate stimulation [24].

4 Notes

1. As the principle for combined imaging presented in this chapter only relies on the absence of overlap between the laser light of DHM and the excitation/emission spectra of the fluorophore, nearly any dye could be employed with a similar protocol. Changing the dye thus essentially corresponds to change the filter cube, as in any standard fluorescence microscope. As an example, the laser wavelength used in this chapter is 680 nm, so that any dye emitting in the blue, green, yellow, or orange could be used. Additionnally, the DHM laser light is not fixed, so that another laser could also be employed to extend the usable range of fluorophores.

2. As phase is sensitive to the surface quality of the different interfaces, glass substrates are usually preferred to ensure the best signal stability. Plastic can indeed have scratches which can appear in the image, and add noise to the signal. For this reason, in case Petri dishes are used for observation, it is usually preferable to use glass-bottom dishes, or to culture cells on microscopy slides, which can be stored in Petri dishes or well plates during cell culture.

3. The protocol described in this chapter is exemplified with measurements on astrocytes-neurons cocultures. PLO is known to provide a better adherence for astrocytes, but the coating should be adapted depending on the cell type, such as poly-L-lysine or collagen as a replacement for PLO.

4. It is assumed here that a closed perfusion chamber is used, which implies that both interfaces in the sample are made of glass. This ensures the best results for phase measurements, and a better control of the flux during perfusion. The closed perfusion chamber can however be replaced with an open chamber such as a Petri dish. In this case, the air–liquid interface may vary during the experiment and thus induce additional noise, which may have to be compensated with additional procedures [18, 25].

5. It has to be noted that it is also possible to perform DHM measurements in other solutions than isotonic ones, such as growth medium. However, fluorescence requires to avoid media with dyes such as phenol red, which can cause autofluorescence. Furthermore, isotonic solutions have a lower refractive index, so that a slight improvement in contrast may be observed also in phase.

6. As the phase is sensitive to volume changes and transmembrane water movements, it is important to ensure that the osmolarity of the perfusion solution is close to the one of the growth medium, in order to avoid hyper- or hypotonic shocks which could disrupt the baseline of the measurement.

7. Phase imaging is sensitive to surface quality, so that cleaning the bottom of the dish or of the coverslip with alcohol before the experiment can tremendously improve the image and signal quality by removing potential dust or dirt resulting from cell growth or previous manipulation.

8. It is good practice to turn on the different devices of the microscope a while before the experiment (at least 30 min), to let the light sources stabilize to ensure constant illumination, and let the whole setup come at functioning temperature to avoid mechanical drifts.

9. It is very important to avoid having any bubble in the perfusion system, as even small bubbles in tubes can gather while passing in the flux, and eventually wash cells when passing in the perfusion chamber.

10. A magnification which is too small makes it difficult to identify the cell soma during the processing of the experiment, and a too high one significantly reduces the field of view, so that only a few cells can be measured during the experiment. A magnification between 10× and 20× is a good compromise in case of neurons, but may have to be adjusted depending on the cell size and distribution.

11. High magnifications have the issue of inducing a short depth of field, so that small mechanical drifts can lead to strongly defocused images, which severely degrade the fluorescence signal. In this case, mechanical refocus during the experiment may be necessary to ensure a reliable fluorescence signal.

12. It is usually very useful to ensure that a part of the measured field of view is free of cells, in order to provide a region on which a background signal can be recorded, for suppressing the autofluorescence and stray light contribution in fluorescence, and for retaining a reliable phase reference to ensure measurement stability.

13. The mechanical focus should be adjusted according to the fluorescence image, as DHM possesses a feature of digital focusing [26], so that cells can be brought into focus during off-line DHM reconstruction, making it possible to avoid perturbating the experiment with manual refocusing.

14. The excitation wavelengths for Fura-2 are classically chosen at $\lambda_1 = 340$ nm and $\lambda_2 = 380$ nm, corresponding to the maximum absorption wavelengths of respectively the bonded and free form of the molecule. However, an excitation as low as 340 nm can require specific optical components to ensure a sufficient transmittance, which are not always available. For this reason, it is also possible to use $\lambda_1 = 360$ nm, which corresponds to the isosbectic point of Fura-2—the absorption point which stays constant upon bonding with Ca^{2+}—to ensure a better transmittance of the excitation light.

15. The temporal resolution of the phase measurement can be set at will, as there is no concern about phototoxicity, bleaching or too long exposure times. The usual practical limit can be determined in order to adequately sample the studied phenomenon without employing too much space on hard drives. In contrast, acquisition of fluorescence images must be limited in order to avoid too much bleaching. For this reason, a shutter system is also necessary to expose cells only during image acquisition.

16. It can happen that cells move during the measurement. It may therefore be necessary to adjust the regions throughout the treatment of the measurements, or track the cell soma over all images in order to retrieve a proper signal.

17. Phase shifts induced by cells can be rather large, and go beyond the accessible dynamic range of the phase signal, which is bound between $[0, 2\pi]$ radians. It may thus be necessary to employ unwrapping procedures to obtain a continuous signal. There are several methods for unwrapping phase signals, but typical examples of use and implementations can be found in refs. [27, 28].

Acknowledgements

This research was supported by the Swiss National Science Foundation grant #CR32I3-132993 and by the Commission for Technology and Innovation (CTI/KTI) project #9389.1.

References

1. Zernike F (1955) How I discovered phase contrast. Science 121:345–349

2. Allen R, David G, Nomarski G (1969) The Zeiss-Nomarski differential interference equipment for transmitted-light microscopy. Z Wiss Mikrosk 69:193–221

3. Rappaz B, Marquet P, Cuche E et al (2005) Measurement of the integral refractive index and dynamic cell morphometry of living cells with digital holographic microscopy. Opt Express 13:9361–9373

4. Girshovitz P, Shaked NT (2012) Generalized cell morphological parameters based on interferometric phase microscopy and their application to cell life cycle characterization. Biomed Opt Express 3:1757–1773

5. Rappaz B, Cano E, Colomb T et al (2009) Noninvasive characterization of the fission yeast cell cycle by monitoring dry mass with digital holographic microscopy. J Biomed Opt 14:034049

6. Mir M, Wang Z, Shen Z et al (2011) Optical measurement of cycle-dependent cell growth. Proc Natl Acad Sci U S A 108:13124–13129

7. Pavillon N, Kühn J, Moratal C et al (2012) Early cell death detection with digital holographic nicroscopy. PLoS ONE 7:e30912

8. Khmaladze A, Matz RL, Epstein T et al (2012) Cell volume changes during apoptosis monitored in real time using digital holographic microscopy. J Struct Biol 178:270–278

9. Pavillon N, Benke A, Boss D et al (2010) Cell morphology and intracellular ionic homeostasis explored with a multimodal approach combining epifluorescence and digital holographic microscopy. J Biophotonics 3:432–436

10. Grynkiewicz G, Poenie M, Tsien R (1985) A new generation of Ca^{2+} indicators with greatly improved fluorescence properties. J Biol Chem 260:3440–3450

11. Tsien RY (1989) Fluorescent probes of cell signaling. Annu Rev Neurosci 12:227–253

12. Kreis T (2005) Handbook of holographic interferometry. Wiley-VCH, Weinheim, Germany

13. Popescu G (2011) Quantitative phase imaging of cells and tissues. McGraw-Hill, New York, NY

14. Magistretti P, Marquet P, Depeursinge C (2013) Neural cell dynamics explored with digital holographic microscopy. Annu Rev Biomed Eng 15:407–431

15. Cuche E, Marquet P, Depeursinge C (1999) Simultaneous amplitude–contrast and quantitative phase–contrast microscopy by numerical reconstruction of Fresnel off–axis holograms. Appl Opt 38:6994–7001

16. Marquet P, Rappaz B, Magistretti P et al (2005) Digital holographic microscopy: A noninvasive contrast imaging technique allowing quantitative visualization of living cells with subwavelength axial accuracy. Opt Lett 30:468–470

17. Montfort F, Charrière F, Colomb T et al (2006) Purely numerical compensation for microscope objective phase curvature in digital holographic microscopy: Influence of digital phase mask position. J Opt Soc Am A 23: 2944–2953

18. Colomb T, Kühn J, Charrière F et al (2006) Total aberrations compensation in digital holographic microscopy with a reference conjugated hologram. Opt Express 14: 4300–4306

19. Jourdain P, Pavillon N, Moratal C et al (2011) Determination of transmembrane water fluxes in neurons elicited by glutamate ionotropic receptors and by the co-transporters KCC2 and NKCC1: a digital holographic microscopy study. J Neurosci 31:11846–11854

20. Hoffmann E, Lambert I, Pedersen S (2009) Physiology of cell volume regulation in vertebrates. Physiol Rev 89:193–277

21. Chen M, Sepramaniam S, Armugam A et al (2008) Water and ion channels: crucial in the initiation and progression of apoptosis in central nervous system? Curr Neuropharmacol 6:102–116

22. Tsien RY (1981) A non-disruptive technique for loading calcium buffers and indicators into cells. Nature 290:527–528

23. Pavillon N (2011) Cellular dynamics and three-dimensional refractive index distribution studied with quantitative phase imaging. Ph.D. thesis no 5100, Ecole Polytechnique Fédérale de Lausanne

24. Stout A, Raphael H, Kanterewicz B et al (1998) Glutamate-induced neuron death requires mitochondrial calcium uptake. Nat Neurosci 1:366–373

25. Colomb T, Montfort F, Kühn J et al (2006) Numerical parametric lens for shifting, magnification, and complete aberration compensation in digital holographic microscopy. J Opt Soc Am A 23:3177–3190

26. Langehanenberg P, Kemper B, Dirksen D et al (2008) Autofocusing in digital holographic phase contrast microscopy on pure phase objects for live cell imaging. Appl Opt 47: D176–D182

27. Takajo H, Takahashi T (1988) Noniterative method for obtaining the exact solution for the normal equation in least-squares phase estimation from the phase difference. J Opt Soc Am A 5:1818–1827

28. Su X, Chen W (2004) Reliability-guided phase unwrapping algorithm: a review. Optic Laser Eng 42:245–261

Chapter 3

Flow Cytometric Analysis of DNA Synthesis and Apoptosis in Central Nervous System Using Fresh Cell Nuclei

Noelia López-Sánchez and José M. Frade

Abstract

The use of flow cytometry in vertebrate nervous tissues is hampered by the morphological complexity and high level of interconnectivity intrinsic to their cellular constituents. Here we describe a simplified procedure for the identification and quantitative analysis of neural cells by flow cytometry based on the isolation and immunolabeling of fresh cell nuclei. We have applied this procedure for the quantitative analysis of apoptosis and DNA synthesis in the embryonic brain.

Key words Caspase-3, TUNEL, EdU, Cell cycle, DNA content, Nuclei labeling, Click reaction, DNA dye

1 Introduction

Flow cytometry is widely used for the quantitative analysis of different parameters, including apoptosis and cell cycle, in single cells from various origins. The morphological complexity and high level of interconnectivity intrinsic to the cellular constituents of the central nervous system (CNS) has limited the use of this technique for single cell analysis in neural cells. Some methods, based on the isolation and fixation of cell nuclei, and their subsequent use for flow cytometric analysis, have recently been applied to specific areas of the CNS as an attempt to solve these problems [1]. Nevertheless, these flow cytometric methods based on the analysis of fixed cell nuclei have a number of drawbacks, including extensive handling, low recovery, and poor fluorescence signals, which lead to ample coefficients of variation and high background levels. Here we describe a simplified and improved method for the flow cytometric analysis in neural tissues by using freshly isolated, non-fixed cell nuclei. This procedure has many advantages, including minor sample handling, thus resulting in high recovery, and optimal fluorescence signals that lead to narrow coefficients of variation and strong reproducibility [2]. Cell nuclei samples

Laura Lossi and Adalberto Merighi (eds.), *Neuronal Cell Death: Methods and Protocols*, Methods in Molecular Biology, vol. 1254, DOI 10.1007/978-1-4939-2152-2_3, © Springer Science+Business Media New York 2015

obtained with this method can be immunostained with neural specific markers, provided they localize to the nucleus. We have used this method for a comprehensive study of apoptosis using nuclear markers such as active caspase 3. Moreover, other nuclear labeling procedures such as terminal deoxynucleotidyl transferase (TdT) dUTP nick end labeling (TUNEL) staining and labeling of cells undergoing S phase with 5-ethynyl-2′-deoxyuridine (EdU) are also compatible with flow cytometric analysis using freshly isolated cell nuclei. This procedure can be applied to the analysis of cell cycle progression and apoptosis known to occur during the development of the nervous system [3], as well as to study cell cycle reentry and apoptosis in neurons affected by neurodegenerative processes [4].

2 Materials

2.1 Reagents

1. Fresh or frozen tissue samples (E16-17 murine cortex).

2. Phosphate-buffered saline (PBS).

3. Homogenization buffer: PBS containing 0.1 % Triton® X-100 (Sigma Chemicals, St. Louis, MO) and one tablet (per 10 mL) of protease inhibitor cocktail (cOmplete, Mini, EDTA-free; Roche Applied Science, Penzberg, Germany).

4. Immunostaining buffer: PBS containing 10 % bovine serum (Life Technologies™, Gaithersburg, MD) and 500 ng/µL of bovine serum albumin (BSA, stock solution 30 mg/mL in PBS) (Sigma Chemicals).

5. Rabbit cleaved caspase 3 (Asp175) antibody (Cell Signaling Technology, Inc., Danvers, MA).

6. Alexa Fluor® 488 donkey anti-rabbit IgG (H + L) antibody (Life Technologies™).

7. Propidium iodide (PI, stock 1 mg/mL) (Sigma Chemicals).

8. Ribonuclease A (RNAse A, stock 1 mg/mL): inactivated by boiling as indicated by manufacturer (Sigma Chemicals).

9. 4′,6-Diamidino-2-phenylindole dihydrochloride (DAPI, stock 100 µg/mL) (Sigma Chemicals).

10. DNase I (Roche Applied Science).

11. In situ Cell death detection kit, POD (Roche Applied Science).

12. DRAQ5 (Stock 5 mM) (Abcam, Cambridge, UK).

13. EdU (Life Technologies™).

14. 1 M Tris pH 8.5 (Roche Applied Science).

15. 100 mM CuSO₄ (Sigma Chemicals).

16. 2 mM Alexa Fluor® 488 azide (Invitrogen™).

17. 1 M ascorbic acid (Sigma Chemicals): Stock at 4 °C.

18. Click reaction buffer (for 500 μL): 1 M Tris pH 8.5 (50 μL), 100 mM $CuSO_4$ (10 μL), 2 mM Alexa Fluor® 488 azide (2.5 μL), 1 M ascorbic acid (50 μL) in distilled water (387.5 μL) (*see* **Note 1**).

2.2 Equipment and Software

1. 1 mL Dounce tissue grinder (Wheaton®, Rochdale, UK).

2. Syringe (1 mL) and needle (25–30 G).

3. Surgical material.

4. Fluorescence microscope.

5. Refrigerated microfuge.

6. Autoclaved 40 μm nylon filters.

7. Thermoblock.

8. FACSAria cytometer (BD Biosciences, Franklin Lakes, NJ) equipped with an argon (488 nm) and helium-neon (633 nm) laser and emission filters FITC, PE-Texad Red and APC-Cy7.

9. FACSDiva software (BD Biosciences) and/or comparable flow cytometric analysis software.

3   Methods

3.1 Active Caspase 3 Immunolabeling and Detection

1. Transfer one fresh (or frozen) telencephalic vesicle from an E16-17 mouse embryo into a Dounce homogenizer containing 1 mL of homogenization buffer and homogenize it sequentially using first a "loose" and then a "tight" pestle (*see* **Note 2**). If possible, process in a similar manner a tissue where increased levels of apoptosis are known to occur (positive control).

2. Split the extract into aliquots of 400 μL (i.e., nuclei samples) for subsequent labeling with specific markers and use the rest of the volume for the negative control (*see* **Note 3**). Then, block nonspecific immunostaining in both samples and controls by adding bovine serum and BSA at the final concentrations indicated for the "immunostaining buffer."

3. Add anti-cleaved caspase 3 (1:200) and secondary (1:500) antibodies to the samples and positive control, and only secondary antibody to the negative control. Mix by gently inversion.

4. Incubate samples and controls in the dark (without shaking) for 2 h at room temperature or overnight at 4 °C.

5. Carefully resuspend sedimented nuclei by gently inversion. Remove a small aliquot, add DAPI (100 ng/mL) or PI (25 μg/mL) and check the immunostaining signal under the microscope (Fig. 1a).

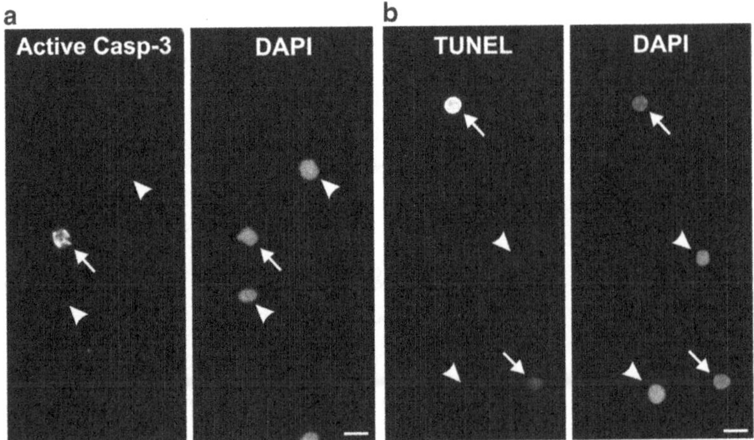

Fig. 1 Representative images of labeled nuclei with active caspase 3 (**a**) or with TUNEL staining (**b**). In both cases the *arrows* indicate positive nuclei and *arrowheads* negative ones. Bar 10 μm

6. Filter samples through an autoclaved nylon filter.

7. Add RNAse A (25 μg/mL) and PI (25–50 μg/mL) and incubate on ice for 20 min in the dark.

8. Analyze the samples and controls by flow cytometry.

9. First identify the nuclei population in the forward scatter (FSC) vs. side scatter (SSC) plot and define the gate *P1* (Fig. 2a).

10. Simultaneously eliminate the debris (enucleated particles without PI staining) by creating a *P2* gate in the DNA content-PI-A vs. DNA content-PI-H plot (Fig. 2b). This allows us to better adjust the P1 gate (Fig. 2c)

11. Create a *P3* gate in DNA content-PI-A vs. DNA content-PI-H plot to select the singlet population and exclude doublets (Fig. 2d) (*see* **Note 4**). This population consists of singlet nuclei population for subsequent analysis.

12. In the negative control, determine the background threshold of the fluorescence emission either in the FITC-A/SSC-A biplot or in the individual signal histogram (Control in Fig. 3a, b, respectively).

13. Analyze the active caspase 3 positive nuclei in immunolabeled samples (Apoptosis induction in Fig. 3a, b).

3.2 TUNEL Staining

1. To obtain the DNase I-treated sample (positive control), centrifuge one 400 μL aliquot from **step 1** of Subheading 3.1 for 4 min at 400 × *g* (*see* **Note 5**).

2. Discard supernatant and carefully add 100 μL of DNase I reaction solution as indicated by the manufacturer. Without disturbing the pellet, incubate the nuclei for 5 min at room

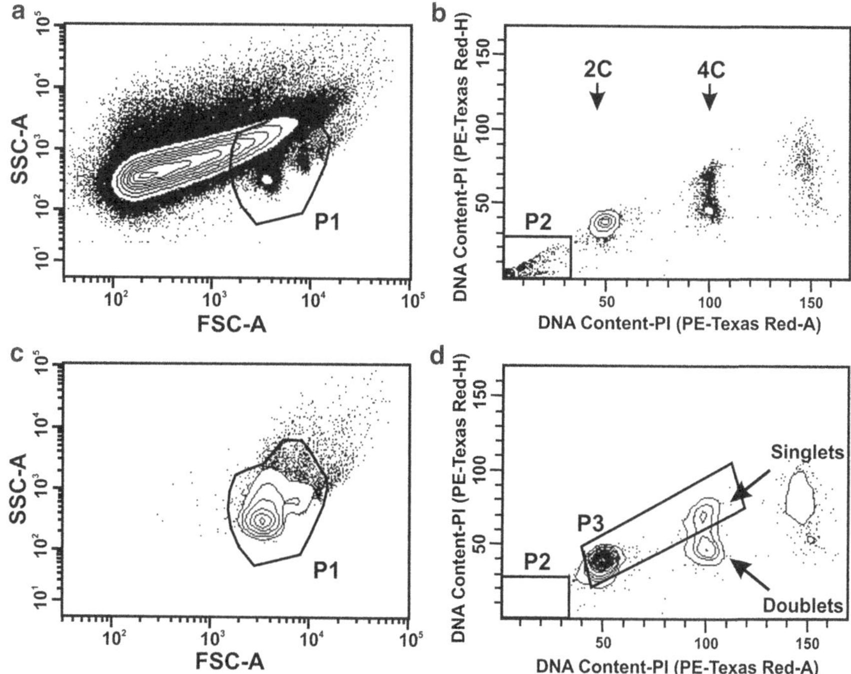

Fig. 2 Contour plots and gating procedure used to identify the singlet events of the embryonic tissue extracts described in the text

temperature, and then gently resuspend them by inversion (*see* **Note 6**). Incubate the nuclei for an additional 5 min.

3. Dilute the DNase I reaction solution by adding 1 mL of PBS.

4. Centrifuge all samples (including the negative and positive controls) at $400 \times g$ for 4 min at 4 °C.

5. Discard supernatant and add 50 μL of label solution with (positive and TUNEL samples) or without (negative control sample) TdT enzyme as indicated by the manufacturer. Wait at least 10 min to resuspend the pellet.

6. Incubate the reactions for 90 min at 37 °C in the dark. Mix by gently inversion twice during the incubation period.

7. Dilute TUNEL reactions by adding 1.4 mL of PBS and centrifuge the samples at $400 \times g$ for 4 min at 4 °C.

8. Discard supernatant and carefully add 500–600 μL of PBS. Keep on ice 15 min in the dark and then resuspend by gently inversion.

9. Remove a small aliquot, add DAPI (100 ng/mL) or PI (25 μg/ mL) to check the fluorescent signal under the microscope (Fig. 1b).

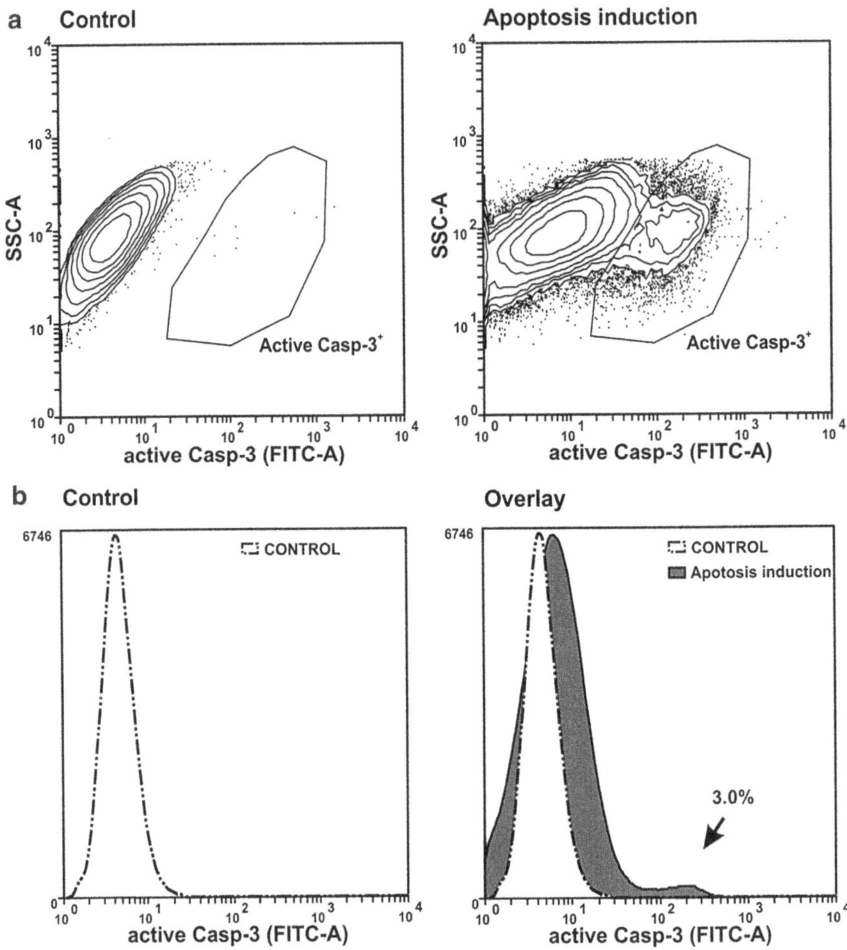

Fig. 3 Representative *contour plots* (**a**) and *histograms* (**b**) of control and apoptosis-induced samples immuno-labeled with anti-cleaved caspase 3 antibody (Active casp-3)

10. Filter the nuclei sample through an autoclaved nylon filter and add both RNAse A (25 μg/mL) and DRAQ5 (5 μM) (*see* **Note 7**). Incubate on ice for 20 min in the dark.

11. Analyze the samples by flow cytometry. First, identify the singlet nuclei population as indicated in **steps 9–11** of Subheading 3.1.

12. As in **step 12** of Subheading 3.1, determine the background threshold of the fluorescein emission in the negative control sample using individual signal histogram of FITC filter or alternatively using FITC-A/SSC-A biplot. Finally analyze the TUNEL positive nuclei in labeled samples (Fig. 4).

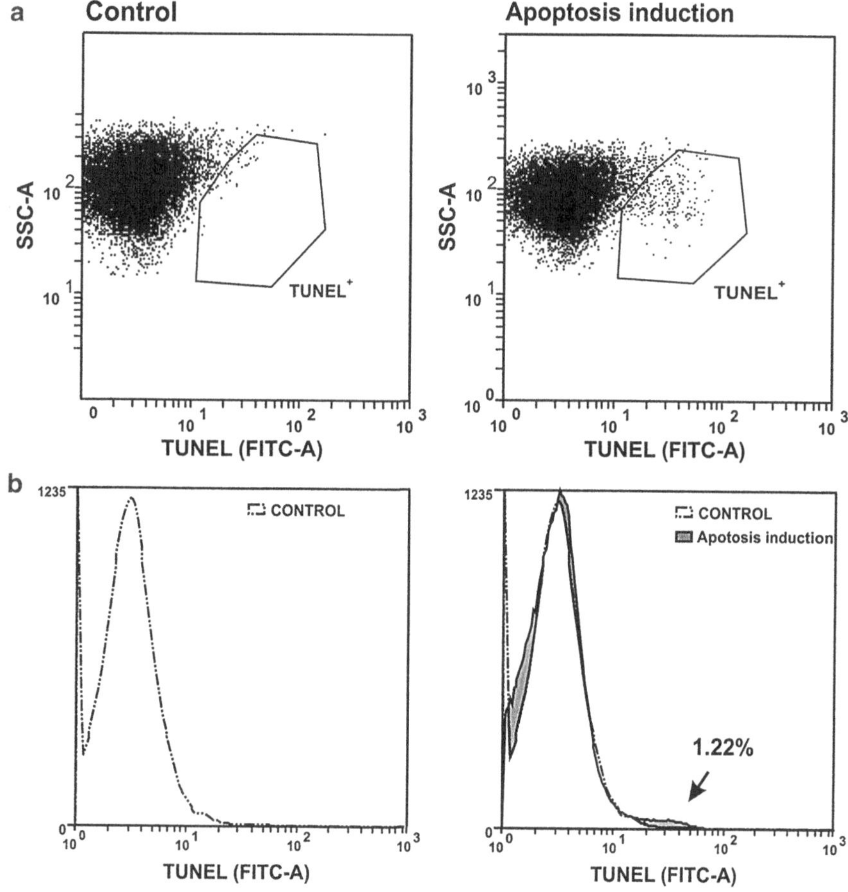

Fig. 4 Representative *dot plots* (**a**) and *histograms* (**b**) of control and apoptosis-induced samples labeled with TUNEL

3.3 EdU Incorporation Assay and Detection

1. Inject (intraperitoneally) a saline solution containing 80 µg of EdU to a pregnant female mouse at embryonic day 16–17. Sacrifice by cervical dislocation 6 h later. Then, remove the embryos from the uterus.

2. Dissect the telencephalic vesicles on PBS and proceed immediately with the labeling or, alternatively, store at −80 °C.

3. Process telencephalic vesicles as described in **step 1** of Subheading 3.1 and remove an aliquot for the negative control (*see* **Note 3**).

4. Centrifuge all samples (include negative control) at $400 \times g$ for 4 min at 4 °C.

5. Discard the supernatant and add 200 µL of Click reaction buffer to the samples and 200 µL of Click reaction buffer lacking Alexa Fluor® 488 azide to the negative control. Do not resuspend

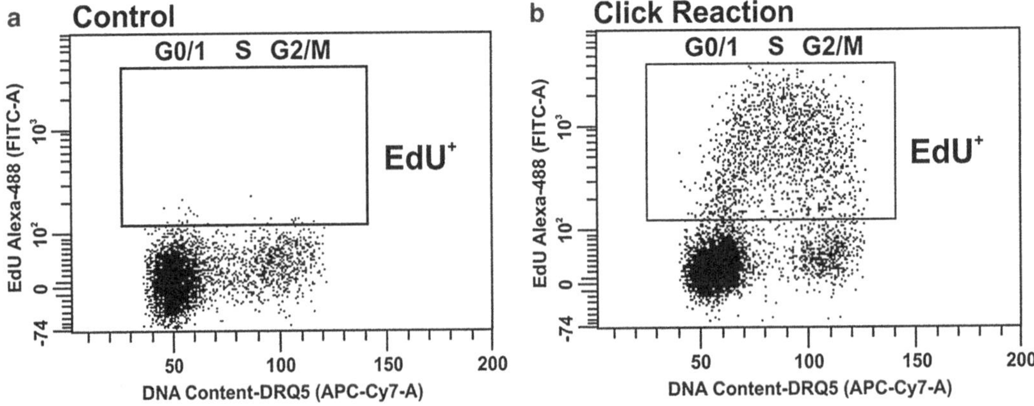

Fig. 5 *Dot plots* of control and Click reaction-labeled nuclei for the detection and quantification of incorporated EdU in nuclei of mouse embryonic brains. The nuclei of different cell cycle phases (G0/G1, S, and G2/M) are indicated based on its respective amount of DNA

pellet at this point, wait at least 10 min to resuspend the pellet by gently inversion (*see* **Note 6**). Incubate in the dark for 30 min at room temperature.

6. Add 1 mL of PBS and centrifuge at $400 \times g$ for 4 min at 4 °C.

7. Discard supernatant and carefully add 500–600 µL of PBS. Keep on ice 15 min in the dark and then resuspend by gently inversion.

8. Remove a small aliquot, add DAPI (100 ng/mL) or PI (25 µg/mL) to check the fluorescent signal under the microscope.

9. Filter through an autoclaved nylon filter, add both RNAse A (25 µg/mL) and DRAQ5 (5 µM) and incubate on ice for 20 min in the dark.

10. Using DRAQ5 signal (APC-Cy7 filter) analyze the singlet population of nuclei as indicated in **steps 9–11** of Subheading 3.1.

11. Using the negative control, determine the threshold for positive nuclei in the DNA content-DRQ5-A vs. EdU Alexa Fluor® 488-A plot and quantify the EdU positive nuclei in the samples (Fig. 5).

4 Notes

1. Click reaction buffer must be prepared immediately before use. The ascorbic acid stock must be maintained at 4 °C and discarded when it becomes oxidized (the solution turns from clear to yellow/orange).

2. Volumes should be scaled according to the density of the nuclei in the tissue (or the density of nuclei that is needed for a

particular analysis). The complexity of some tissues (for instance, the adult brain), results in samples with high levels of debris that requires additional washing steps in large PBS volumes. For the adult cerebral cortex, homogenize one hemicortex in 3–4 mL of homogenization buffer, and centrifuge it at $200 \times g$ for 1.5 min (4 °C). Then, discard the pellet and add PBS-0.1 % Triton® X-100 to the supernatant to a final volume of 12 mL. Centrifuge at $400 \times g$ for 4 min (4 °C). Discard the supernatant and allow the pellet (i.e., isolated nuclei) to slowly dissolve for 30 min in 0.5–1 mL of ice-cold PBS. For further details *see* refs. 2, 5.

3. Usually we process several nuclei samples or replicates in each experiment and we prepare the negative control (nuclei sample treated only with secondary antibody) by pooling small aliquots from these samples (400 µL in total).

4. The method used for doublet discrimination is based on the pulse-processing method [6]. The signal recorded from a 4C nucleus has a double DNA-content-H value compared with a 2C nucleus since the former contains double amount of DNA. *See* refs. [2, 5] for further explanation of the pulse-processing method. Other doublets discrimination methods such as the association between high SSC particles and doublets is only appropriate to samples with very homogeneous size and complexity which usually is not the case of nervous system samples due to the large differences between neuronal and glial nuclei morphology and DNA/chromatin structure.

5. The apoptosis detection kit used in this protocol is based on the incorporation of fluorescein-labeled nucleotides, catalyzed by TdT, into the free 3′-OH DNA ends of apoptotic cells. The manufacturer's protocol recommends the preparation of both negative and positive samples for the reaction. The negative control consists in a sample treated with buffer lacking TdT. The positive control must be treated with DNase I to induce breaks in DNA mimicking those found in apoptotic cells, and then processed as the rest of the samples.

6. The correct resuspension of pelleted nuclei is critical in order to preserve both quality and high recovery rates. Never try to resuspend the pelleted nuclei immediately after centrifugation, this could result in compaction of pellet and/or nuclei break.

7. PI can bleach the emission signal of EdU/Azide-Alexa Fluor® 488 or fluorescein-dependent TUNEL signal, due to their close proximity to each other (they all are associated with DNA). In these cases, other stoichiometric DNA dyes and/or fluorophores with compatible excitation spectra (such as ultraviolet or far red) should be used. We obtain very good results with Alexa Fluor® 488 or fluorescein in combination with the DNA dye DRAQ5.

Acknowledgements

This study was supported by grants from the "Ministerio de Ciencia e Innovación" (SAF2012-38316) and "Fundación Areces" (CIVP16A1815).

References

1. Westra JW, Rivera RR, Bushman DM et al (2010) Neuronal DNA content variation (DCV) with regional and individual differences in the human brain. J Comp Neurol 518:3981–4000

2. López-Sánchez N, Frade JM (2013) Cell cycle analysis in the vertebrate brain using immunolabeled fresh cell nuclei. Bio-Protocol 3:e973. http://www.bio-protocol.org/e973

3. De la Rosa EJ, de Pablo F (2000) Cell death in early neural development: beyond the neurotrophic theory. Trends Neurosci 23:454–458

4. Wang W, Bu B, Xie M et al (2009) Neural cell cycle dysregulation and central nervous system diseases. Prog Neurobiol 89:1–17

5. López-Sánchez N, Frade JM (2013) Genetic evidence for p75[NTR]-dependent tetraploidy in cortical projection neurons from adult mice. J Neurosci 33:7488–7500

6. Nunez R (2001) DNA measurement and cell cycle analysis by flow cytometry. Curr Issues Mol Biol 3:67–70

Chapter 4

Nuclear Signs of Pre-neurodegeneration

Fernando C. Baltanás, Jorge Valero, Jose Ramón Alonso,
Maria Teresa Berciano, and Miguel Lafarga

Abstract

Nuclear architecture is highly concerted including the organization of chromosome territories and distinct nuclear bodies, such as nucleoli, Cajal bodies, nuclear speckles of splicing factors, and promyelocytic leukemia nuclear bodies, among others. The organization of such nuclear compartments is very dynamic and may represent a sensitive indicator of the functional status of the cell. Here, we describe methodologies that allow isolating discrete cell populations from the brain and the fine observation of nuclear signs that could be insightful predictors of an early neuronal injury in a wide range of neurodegenerative disorders. The tools here described may be of use for the early detection of pre-degenerative processes in neurodegenerative diseases and for validating novel rescue strategies.

Key words Cell dissociation, Confocal microscopy, Immunofluorescence, Nucleus, Pre-neurodegeneration, Squash

1 Introduction

Cell nucleus is organized in structural and functional compartments that are mainly involved in DNA transcription and RNA processing. They include chromosome territories, which are occupied by interphasic chromosomes with their euchromatin and transcriptionally silencing heterochromatin domains, and the interchromatin region. The latter harbors the nucleolus, the site of transcription and processing of pre-rRNAs, nuclear speckles, where pre-mRNA splicing factors are assembled and stored, and Cajal bodies, which are nuclear organelles involved in the biogenesis of small nuclear and nucleolar ribonucleoproteins (snRNPs and snoRNPs) required for pre-mRNA and pre-rRNA processing, respectively. The organization of such nuclear compartments is very dynamic and represents a sensitive indicator of the functional status of the cell.

In the case of mammalian neurons, reorganization of nuclear compartments has been observed in response to changes in gene

Laura Lossi and Adalberto Merighi (eds.), *Neuronal Cell Death: Methods and Protocols*, Methods in Molecular Biology,
vol. 1254, DOI 10.1007/978-1-4939-2152-2_4, © Springer Science+Business Media New York 2015

expression associated with physiological or experimental conditions, as well as with certain human neurodegenerative disorders. In addition, there is growing evidence of a broad participation of nuclear foci containing components of the DNA damage/repair signaling pathway in the pathophysiology of human neurodegenerative diseases. Interestingly, these pre-degenerative changes appear in neurons with a well-preserved general cytology and could be sensitive indicators of an early and potentially reversible stage of neuronal injury in a broad range of neurodegenerative disorders. In this context, methodological approaches allowing the fine observation of nuclear compartments appear as useful tools in the early identification of degenerative disorders.

Here, the "squash" technique is described in detail. This procedure allows excellent preservation of the global morphology and cytological organization of cell perikarya, and represents the best option for studying neuronal nuclear compartments in whole nuclei using confocal laser microscopy [1]. In addition, we suggest a list of potential nuclear proteins that can be used to identify nuclear hallmarks of pre-degenerative stages in mammalian neurons. In this case, we show different nuclear signs of pre-neurodegeneration in squash preparation samples of Purkinje cells from the cerebellar cortex of the PCD (Purkinje Cell Degeneration) mutant mice.

2 Materials

2.1 Tissue Fixation

We recommend preparing fresh all solutions.

1. Anesthetic: 3:4 mixture of ketamine hydrochloride (Imalgene, Merial, Lyon, France) and xylazine (Rompum, Bayer, Kiel, Germany).

2. Saline solution (1 L): Dissolve 9 g NaCl in 1 L of water.

3. Heparinized saline (1 %).

4. 0.2 M phosphate buffer (PB): Prepare 0.2 M solutions of both Na_2HPO_4 and NaH_2PO4, mix and adjust pH to 7.4 with concentrated HCl (25 °C).

5. Fixative solution (1 L): Add 500 mL of distilled water in an Erlenmeyer and warm up to 80 °C. Weigh 40 g paraformaldehyde, transfer to the Erlenmeyer, and mix. Add PB up to 1 L and mix, cool the solution in ice and filter.

2.2 Cellular Dissociation "Squash Technique"

1. Vibrating blade microtome.

2. Light microscope (low magnification objectives 10× or 20×).

3. Peristaltic pump or syringe. *See* **Note 1**.

4. Histological needle.

5. Lab forceps.

6. Siliconized slides (SuperFrostPlus, Menzel-Gläser, Germany).

7. Coverslips (18 × 18 cm).

8. Dry ice.

9. Ethanol 96 % (v/v).

10. Phosphate buffered saline (PBS): Dilute 8 g of NaCl, 0.2 g of KCl, 1.44 g of Na_2HPO_4, and 0.21 g of KH_2PO_4 in distilled water (1 L). Adjust pH to 7.4 with concentrated HCl (25 °C).

2.3 Immuno-fluorescence Cell Staining

1. PBS pH to 7.4 (25 °C).

2. 0.1 M glycine solution: Dilute 3.75 g glycine in 500 mL of PBS.

3. PBS-Triton™ solution (0.5 %): Add 5 mL of Triton™ X-100–995 mL of PBS and mix.

4. PBS-Tween™ solution (0.05 %): Add 500 μL of Tween™ 20–999.5 mL of PBS and mix.

5. Primary antibody solutions: Dilute primary antibodies in PBS. *See* **Note 2**.

6. Secondary antibody solutions: Dilute secondary antibodies in PBS. Please follow the recommendations of manufacturers. *See* **Note 3**.

7. Propidium iodide (PI) stock solution: dilute 1 mg of PI per mL of PBS and store as frozen stock aliquots for several months.

8. Antifade medium.

2.4 Imaging Nuclear Signs of Pre-neurodegeneration

1. Confocal microscope with high numerical aperture objectives.

---

3 Methods

Procedures are carried out at room temperature unless otherwise stated.

3.1 Tissue Fixation

1. Anaesthetize the animal with the anesthetic mixture. The precise amount of anesthetic should be determined (1 μL/g body weight of anesthetic mixture is recommended).

2. Quickly expose the heart.

3. Inject heparinized saline (100 μL/10 g body weight) in the left ventricle to reduce coagula.

4. Perfuse through aorta with the saline solution for 1 min, and then the fixative solution (1 mL/g body weight). *See* **Note 1**.

5. Dissect the brain area of interest and postfix it in the same fixative solution for 20 min using gentle shaking.

6. Rinse three times with PB.

3.2 Cellular Dissociation "Squash Technique"

1. Cut 200–400 μm-thick slices of the brain area of interest containing the desired cell type/s using a vibratome, and maintain the blocks in PBS.

2. Put the slices over a slide with a drop of PBS (prevent the tissue from becoming dry).

3. Cut the slices using blades into small fragments under light microscope and transfer each tissue fragment to a drop of PBS on a siliconized slide. Add a coverslip at the top of the slide.

4. "Squash" the tissue by mechanical percussion with a histological needle to dissociate neuronal perikarya. Push carefully (avoid breaking the coverslip) 3–4 times with the needle over the sample. A cellular dispersion from the fragment should be observed. *See* **Note 4**.

5. Visualize the "squash" in the light microscope. Discard samples that could show affectation of cell morphology as consequence of mechanical percussion (Fig. 1).

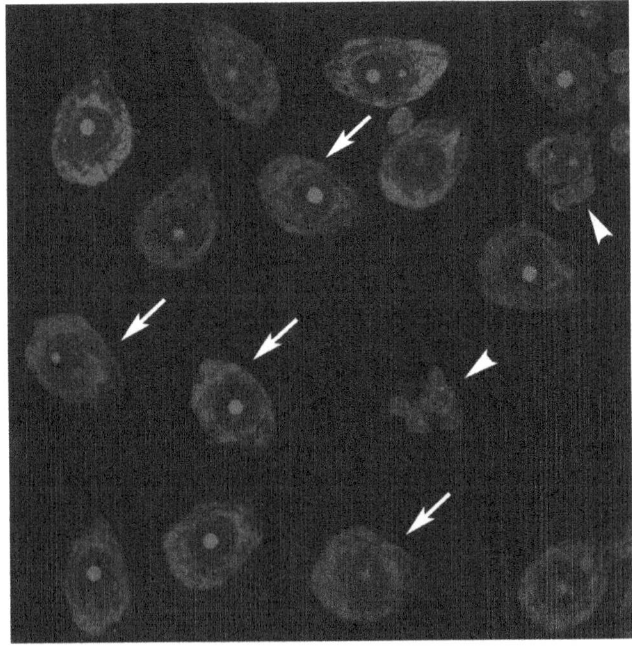

Fig. 1 Confocal microscopy image from "squash" preparation of Purkinje cells (PCs) isolated from the mouse cerebellar cortex, stained for PI. Observe the presence of PCs (*arrows*) and other cerebellar neural types (*arrowheads*). Note the well preservation of cell perikarya and the fine morphology of cell nucleus

6. Place the slides over dry ice until the samples are frozen (5 min are normally enough), and remove the coverslip by using a razor blade.

7. Process sequentially the cell samples in 96 % ethanol at 4 °C for 10 min and PBS at 4 °C. With this procedure, most neural cells will adhere to the slide.

8. Store samples at 4 °C until used.

3.3 Immuno-fluorescence Cell Staining

1. Incubate cell dissociates in 0.1 M glycine solution for 15 min using gentle shaking. *See* **Note 5**.

2. Permeabilize cells in PBS-Triton™ solution for 45 min using gentle shaking. *See* **Note 6**.

3. Rinse cell dissociates once in PBS-Tween™ solution briefly.

4. Incubate cell dissociates in primary antibody solution overnight at 4 °C in a humid chamber (Table 1). *See* **Notes 2, 7** and **8**.

5. Rinse cells with PBS-Tween™ for 5 min three times each, using gentle shaking.

6. Incubate cell dissociates with the adequate secondary antibody solution for 45 min in a humid dark chamber. *See* **Notes 3, 7** and **8**.

7. Rinse cells with PBS for 5 min three times each, using gentle shaking.
   Optional: PI counterstaining. *See* **Note 9**.

8. Place cell dissociates in a humid dark chamber and incubate them in PI stock solution diluted 1:2,000 in PBS for 15 min. *See* **Note 8**.

9. Rinse cells with PBS for 5 min three times at RT using gentle shaking.

10. Mount samples with an adequate antifade medium.

3.4 Imaging Nuclear Signs of Pre-neurodegeneration

1. Use a confocal microscope (*see* **Note 10**) and high magnification objectives (at least a 63× plan-apochromatic objective with a 1.4 numerical aperture).

2. Determine image acquisition parameters (filters, gain, offset, pinhole diameter, pixel depth, scanning resolution and zoom). *See* **Note 11**.

3. Use the parameters established in **step 2** and acquire single plane confocal images to analyze homogenously distributed staining or spots (*see* **Note 12**). Use *z*-stacks acquisition for volumetric measurements or analysis of heterogeneously distributed staining. *See* **Note 13**.

4. Process and quantitatively or qualitatively compare parameters of acquired images. *See* **Note 14**.

A list of nuclear signs of pre-neurodegeneration is shown in Table 2 (Figs. 2 and 3).

Table 1
List of primary antibodies

Antibody	Host sp	References	Dilution	Use
Anti-γH2AX	Rabbit	Millipore, Billerica, MA	1:300	DNA breaks detection
Anti-ATM pS1981	Mouse	Rockland Immunochemicals, Inc., Gilbertsville, PA	1:100	Identification of DNA breaks repair foci
Anti-53BP1	Rabbit	Novus Biologicals®, Littleton, CO	1:200	Detection of DNA breaks repair foci
Anti-H4K20me3	Rabbit	Millipore	1:100	Identification of transcriptional repressed chromatin
Anti-Acetylated H4	Mouse	Millipore	1:100	Detection of transcriptional active chromatin
Anti-pS2 RNA pol II	Mouse	Warren et al. [9]	1:100	Visualization of transcriptional active regions
Anti-B23	Mouse	Abcam, Cambridge, UK	1:200	Identification of nucleolar ribonucleoproteins
Anti-Fibrillarin	Mouse	Reimer et al. [10]	1:10	Localization of dense fibrillar component of nucleolus
Anti-Coilin	Rabbit	Bohmann et al. [11]	1:75	Coiled body detection

53BP1 p53-binding protein 1, *ATM pS1981* form of ataxia telangiectasia mutated protein kinase phosphorylated at serine 1981, *B23* nucleolar phosphoprotein B23, *γH2AX* phosphorylated form of the histone variant H2AX, *H4* histone 4, *H4-3Me-K20* histone H4 trimethylated at lysine 20, *pS2 RNA pol II* RNA polymerase II phosphorylated at serine 2

4   Notes

1. Although the perfusion can be performed using an adequate syringe, we strongly recommend the use of peristaltic pumps since a wide range of flow rates and back pressures (previously determined for each particular case), can be accommodated.

2. The dilution and incubation time should be determined for each particular antibody. Table 1 reports a list of antibodies and dilutions that we have previously used [2–4]. We highly recommend performing adequate negative and positive controls of the immunofluorescence protocol to verify the specificity and sensitivity of the technique. Classical negative controls are based on the omission of primary and/or secondary antibodies. Primary antibody specificity can be checked by using a primary antibody solution pre-incubated with a peptide containing the epitope that should be recognized (no staining should appear). As an example of a positive control, X-ray ionizing radiations

Table 2
Suggested primary antibody combinations

Antibody	Antibody/staining	Pre-neurodegeneration signs	References
Anti-γH2AX	PI	Bright and abundant γH2AX foci that can be also positive for PI	[3, 4]
	Anti-ATM pS1981 or Anti-53BP1	Low or none colocalization of pATM and 53BP1 with γH2AX indicating DNA repair miss function (Fig. 2)	[2, 4]
	Anti-H4K20me3	Microfoci of γH2AX colocalizing at the periphery of big H4K20me3 heterochromatin domains (Fig. 2)	[2]
Anti-H4K20me3	PI	Large perinucleolar masses and nuclear foci of H4K20me3	[4]
Anti-Acetylated H4	PI	Low intensity staining for Acetylated H4 in nucleoplasm and reduced number of bright foci	[4]
Anti-pS2 RNA pol II	PI	Low intensity staining for pS2 RNA pol II and reduced number of bright foci	[4]
Anti-B23	PI	B23 appears diffuse throughout the nucleoplasm and concentrated in segregated masses of granular component (Fig. 3).	[3]
Anti-Fibrillarin	PI	Segregated clusters of fibrillarin at the periphery of the nucleolus	[3, 4]
	Anti-Coilin	Coilin appears as a thin shell around large and small segregated masses of nucleolar fibrillarin (Fig. 3)	[3]
Anti-Coilin	PI	Reduced number of coiled bodies. Coilin appears as a thin shell around the nucleolus	[3, 4]

53BP1 p53-binding protein 1, *ATM pS1981* form of ataxia telangiectasia mutated protein kinase phosphorylated at serine 1981, *B23* nucleolar phosphoprotein B23, *γH2AX* phosphorylated form of the histone variant H2AX, *H4* histone 4, *H4-3Me-K20* histone H4 trimethylated at lysine 20, *PI* propidium iodide, *pS2 RNA pol II* RNA polymerase II phosphorylated at serine 2

can be used to induce DNA damage as a positive control for both DNA break and DNA repair markers [2].

3. Secondary antibody has to be raised against the constant region of the immunoglobulin of the species in which primary antibody was made. Primary and secondary antibodies combinations should be selected to avoid undesired cross-reactivity. Furthermore, pre-absorbed secondary antibodies are recommended if primary antibodies are made in related species like mouse and rat or sheep and goat. In any case, adequate controls should be made to be sure that there is no cross-reactivity. When using two primary antibodies, a simple

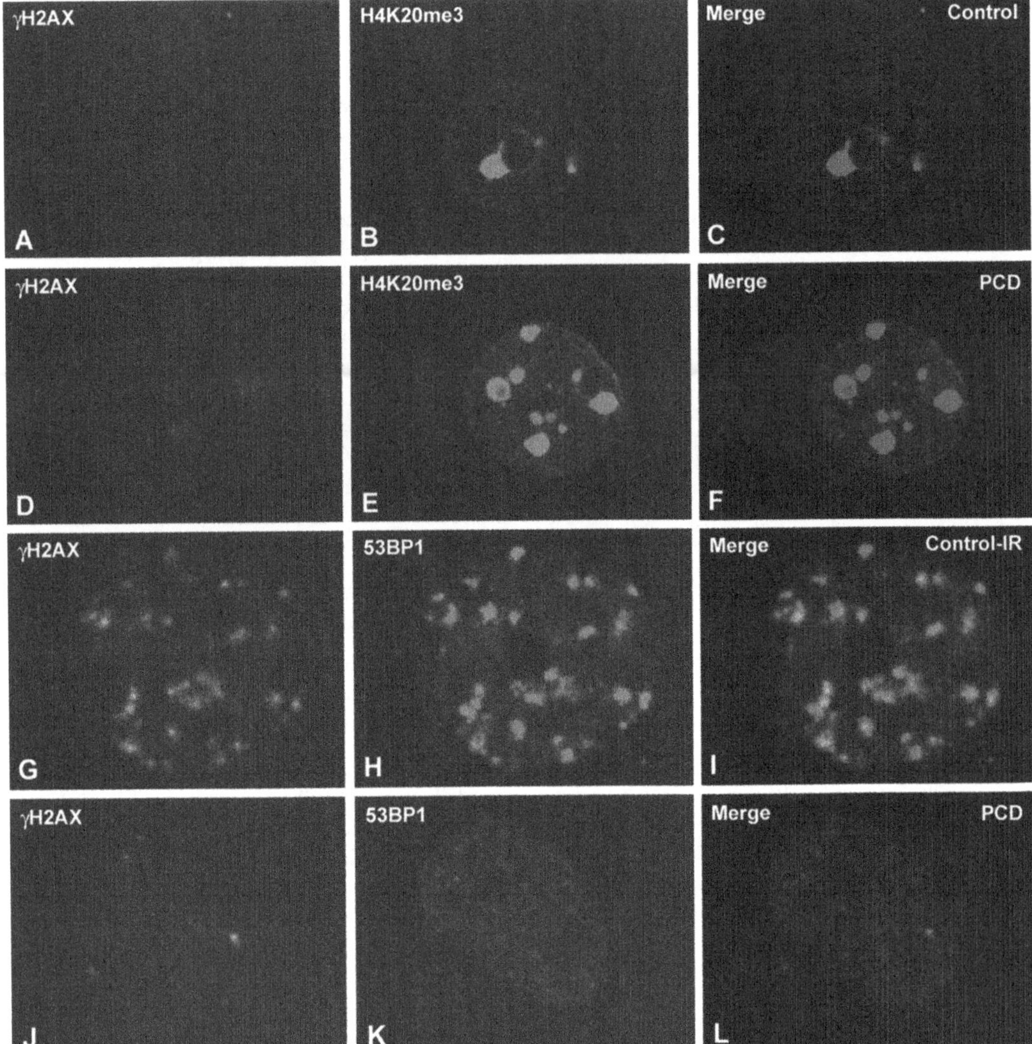

Fig. 2 Confocal microscopy image from "squash" preparations of Purkinje cells (PCs) isolated from cerebellum of control mice and Purkinje cell degeneration (PCD) mutants, co stained for γH2AX and H4K20me3 (**a–f**) or γH2AX and 53BP1 (**g–l**). (**a–c**) Two H4-3Me-K20-positive foci can be appreciated at the perinucleolar region of a control PC. (**d–f**) No nuclear signal of DNA damage was detected with the anti-γH2AX antibody in a PC isolated from PCD mutant. Note that when γH2AX-positive nuclear foci of DNA damage appear, an increase in the number and the size of H4K20me3-positive heterochromatin masses can be observed, then suggesting DNA damage-related genome silencing. (**g–i**) Double labeling for γH2AX and 53BP1 in dissociated PC from control mice exposed to ionizing radiation (IR). Note strong γH2AX immunofluorescent foci colocalizing with bright nuclear domains of 53BP1 as consequence of a successful DNA damage response. (**j–l**) Although 53BP1 was expressed in the nucleoplasm in PCD, it did not colocalize with either γH2AX-positive microfoci then revealing a fail in the DNA damage response and repair after IR

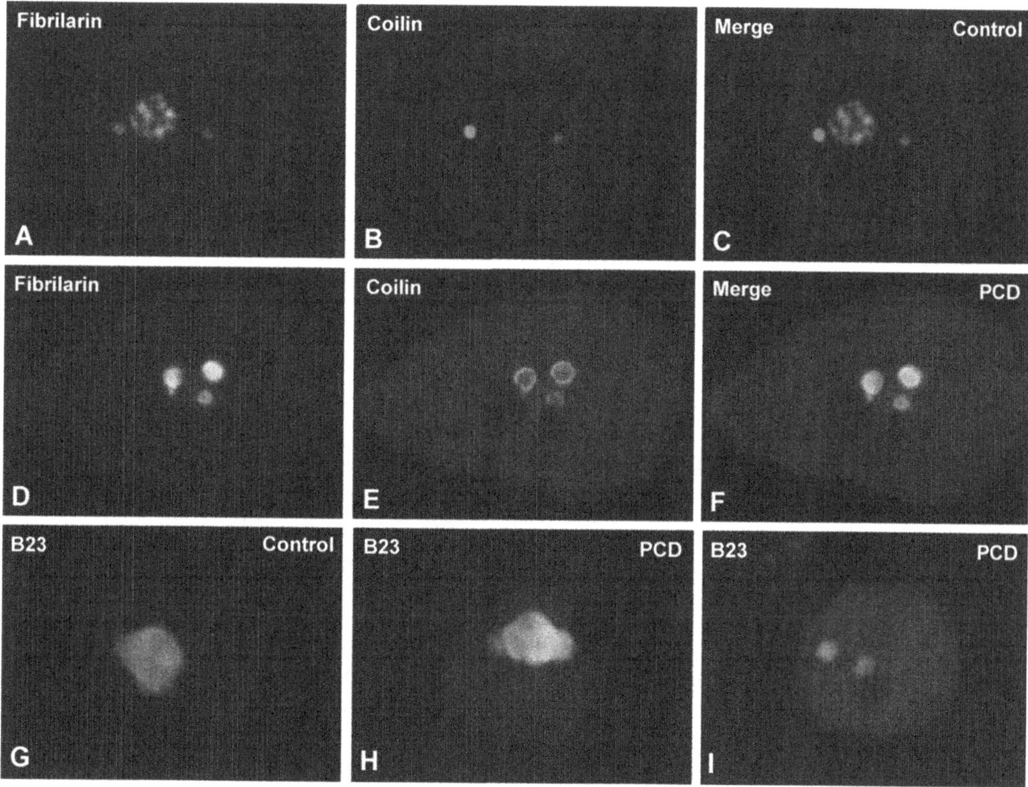

Fig. 3 (**a–f**) Confocal microscopy image from "squash" preparations of PCs costained for fibrillarin and coilin from control (**a–c**) and PCD mice (**d–f**). (**a–c**) A control PC shows Cajal bodies free in the nucleoplasm or attached to the nucleolus, in which both proteins colocalized. (**d–f**) PC mutation causes severe segregation of the nucleolar components, coilin appears relocalized as a thin ring surrounding the segregated masses of fibrillarin. (**g–i**) Immunostaining for B23 shows labeling of the nucleolus in a control PC (**g**), whereas PCs from mutant mice (**h–i**) show progressive segregation of the nucleolar B23 and an increment of its nucleoplasmic localization. *PC* Purkinje cell, *PCD* Purkinje cell degeneration

control consists in following the normal protocol but using just one primary antibody and the secondary antibody that recognizes the missing primary antibody. This control should be made for both primary antibodies. No primary antibody-related staining is expected if there is no cross-reactivity. Be also sure that your secondary antibodies do not recognize each other, we have found this a common mistake when using secondary antibodies made in different species. Importantly, excitation/emission wavelength ranges of secondary antibody fluorochromes should be separated enough to allow the specific discrimination of their fluorescent signals under the microscope in use.

4. The force of the mechanical percussion applied should be enough to separate the cell type/s of interest from the rest of neural types but maintaining the morphology and cytological

organization of cell perikarya. It depends of the cell type/s to isolate, the age of the animal and the fixation degree of the tissue. For example, it is easier to isolate neurons from young mice (2–3 postnatal weeks), which are located in well-known and recognizable anatomical regions, than isolating neurons from old mice (6 months onwards) with difficulty to identify their disposition.

5. This treatment is mainly aimed to reduce paraformaldehyde-related fluorescence and will not avoid other unspecific fluorescence issues. 1 % BSA (w/v) could be added to glycine solution (0.1 M in PBS) to block unspecific staining, while 0.01 % Tween™ 20 could be also used to softly permeabilize the cells and allow BSA penetration.

6. PBS-Triton™ solution can be also used as a washing medium to rinse the cells three times; we recommend 15 min each. Time and Triton™ X-100 dilution should be adjusted depending on the type of cell to be stained and type of epitope to be detected.

7. For multiple-immunofluorescence labelling sequential incubation with primary or secondary antibodies is recommended instead of co-incubation. Therefore, repeat **steps 4** and **5** for each primary antibody required and **steps 5** and **6** for each secondary antibody. Primary antibody combinations and their utility are listed in Table 2.

8. Incubations can be performed in different ways: (1) outlining cell dissociates with a grease pen and adding the adequate volume of primary antibody solution to cover all the cells or (2) adding the adequate amount of primary antibody solution and covering it with a small piece of Parafilm® (using it as a coverslip). This second strategy normally allows the use of smaller volumes of primary antibody solution.

9. We recommend the use of PI to counterstain cells. This step allows the visualization of cell morphology and nucleolus. Cells can be also counterstained with a fluorescent DNA intercalating agent (DAPI or Hoechst). The use of a counterstaining fluorescent dye will facilitate the focusing and identification of cells during image acquisition. Please consider that the excitation and emission wavelength ranges of counterstaining agents should be compatible with those of your secondary antibody fluorochromes.

10. If adequately performed, cell dissociates should appear as a monolayer of separated cells, allowing adequate visualization with an epifluorescence microscope. Therefore, a precise quantification of fluorescent spots or nuclear bodies per nucleus could be performed using an epifluorescence microscope and high resolution objectives.

11. This is especially important when intensity and area/volumetric measurements of stained regions are performed. Use the adequate bit depth intensity range for your analysis. Normally, 8-bit images with the adequate distribution of pixel intensities (from 0 to 255) are enough. However, 16-bit images could be required for small difference detection. We suggest the use of a grey scale lookup table with specific colours for saturated (e.g., red) and overexposed pixels (e.g., blue or green) to obtain images with the adequate intensity range while adjusting acquisition parameters. Our final image should contain few saturated and overexposed pixels. We also recommend the combination of high resolution parameters (at least $2{,}048 \times 2{,}048$ pixels) and zoom values from 1.5× to 2× for image acquisition.

12. When acquiring single plane images consider whether the staining to be analyzed is equally distributed through the nucleoplasm or concentrates in some regions by checking different planes. If staining or particles of interest accumulate in specific regions, try to obtain all images from planes situated at the same nuclear depth level (top, middle or bottom level) to perform adequate comparisons.

13. When acquiring multiple z-stacks we recommend scanning the entire volume of the nuclei during the acquisition (1 μm as maximum z-step size). If the whole volume of the nucleus is not covered please consider issues addressed in **Note 12**.

14. Tools and plugins from the free-access image processing package Fiji [5] are adequate to treat the images and perform manual and automated quantifications. We would like to highlight the following Fiji resources: cell counter plugin for manual quantification of particles, analyze particles tool for automated quantification and measurement of thresholded particles, find foci plugin [6] for segmentation of particles based on profile peak intensities (as a pre-step for automatic particle analysis), 3D object counter plugin [7] for automatic quantification and volumetric measurement of thresholded particles or stained regions, JACoP plugin [7] for staining colocalization/proximity analysis, or 3D Viewer plugin [8] for 3D rendering and visualization of image stacks.

References

1. Pena E, Berciano MT, Fernandez R et al (2001) Neuronal body size correlates with the number of nucleoli and Cajal bodies, and with the organization of the splicing machinery in rat trigeminal ganglion neurons. J Comp Neurol 430:250–263

2. Baltanás FC, Casafont I, Lafarga V et al (2011) Purkinje cell degeneration in pcd mice reveals large scale chromatin reorganization and gene silencing linked to defective dna repair. J Biol Chem 286:28287–28302

3. Baltanás FC, Casafont I, Weruaga E et al (2011) Nucleolar disruption and cajal body disassembly are nuclear hallmarks of DNA damage-induced neurodegeneration in purkinje cells. Brain Pathol 21:374–388

4. Valero J, Berciano MT, Weruaga E et al (2006) Pre-neurodegeneration of mitral cells in the pcd mutant mouse is associated with DNA damage, transcriptional repression, and reorganization of nuclear speckles and Cajal bodies. Mol Cell Neurosci 33:283–295

5. Schindelin J, Arganda-Carreras I, Frise E et al (2012) Fiji: an open-source platform for biological-image analysis. Nat Methods 9:676–682

6. Herbert A (2013) Find foci. MRC Genome Damage and Stability Centre: University of Sussex. http://www.sussex.ac.uk/gdsc/intranet/microscopy/imagej/findfoci

7. Bolte S, Cordelières FP (2006) A guided tour into subcellular colocalization analysis in light microscopy. J Microsc 224:213–232

8. Schmid B, Schindelin J, Cardona A et al (2010) A high-level 3D visualization API for Java and ImageJ. BMC Bioinformatics 11:274

9. Warren SL, Landolfi AS, Curtis C et al (1992) Cytostellin: a novel, highly conserved protein that undergoes continuous redistribution during the cell cycle. J Cell Sci 103:381–388

10. Reimer G, Pollard KM, Penning CA et al (1987) Monoclonal autoantibody from a (New Zealand black × New Zealand white) F1 mouse and some human scleroderma sera target an Mr 34,000 nucleolar protein of the U3 RNP particle. Arthritis Rheum 30:793–800

11. Bohmann K, Ferreira JA, Lamond AI (1995) Mutational analysis of p80 coilin indicates a functional interaction between coiled bodies and the nucleolus. J Cell Biol 131:817–831

Multi-parametric O₂ Imaging in Three-Dimensional Neural Cell Models with the Phosphorescent Probes

Ruslan I. Dmitriev and Dmitri B. Papkovsky

Abstract

Recent progress in bio-imaging has allowed detailed mechanistic studies of neural cell function in complex 3D tissue models including multicellular aggregates, neurospheres, excised brain slices, ganglia, and organoids. Molecular oxygen (O_2) is an important metabolite and an environmental parameter which determines the viability and physiological status of neural cells within tissue. Here we describe standard method for monitoring O_2 in 3D tissue models using phosphorescence lifetime imaging microscopy (PLIM) and cell-penetrating O_2-sensing probes. The O_2 probes can be multiplexed with many conventional fluorescence based live cell biomarkers and also end-point immunofluorescence staining. The multi-parametric O_2 imaging method is particularly useful for areas such as stem cell development and differentiation, hypoxia research, neurodegenerative disorders, regeneration of brain tissue, evaluation of new drugs, and development of novel tissue models.

Key words High-resolution microscopy, Imaging, Oxygen, Phosphorescence quenching, PLIM

1 Introduction

Neuronal cell death is associated with many common pathological and stress conditions such as stroke, neonatal hypoxia–ischemia, aging, and neurodegenerative disorders [1–3]. The critical roles of molecular oxygen (O_2), reactive oxygen and nitrogen species (ROS, RNS) in such processes have been appreciated for a long time [4]; however, newly emerged analytical and imaging methodologies have enabled to study them in greater detail. In particular, optical imaging by means of fluorescent probes and biosensors has revealed the interplay between mitochondrial function and cell necrosis, autophagy and apoptosis [1, 5, 6]. The interplay between O_2 dynamics and physiological parameters (ion and metabolite fluxes, distribution of specific cell types and death markers) has been studied in various neural cell models, including in vitro (cultured cells), ex vivo (explants, organoids), and in vivo (functional imaging of the brain), noninvasively and with high spatial resolution [7–9].

Laura Lossi and Adalberto Merighi (eds.), *Neuronal Cell Death: Methods and Protocols*, Methods in Molecular Biology, vol. 1254, DOI 10.1007/978-1-4939-2152-2_5, © Springer Science+Business Media New York 2015

The phosphorescent O_2-sensitive probes based on Pt(II)-porphyrins and some related structures have characteristic red/infrared emission, large Stokes shifts, high brightness ($\varepsilon \times q$), selective and optimal quenching by O_2 [10]. The unique photo-physical properties and long emission lifetimes (typically range 10–100 μs) of Pt(II)-porphyrins facilitate their multiplexing with many conventional fluorescent dyes including FITC, rhodamine, DAPI, fluorescent protein constructs (GFP, RFP and others) which have very different spectral characteristics and lifetimes in the picosecond–nanosecond range. *Quantitative* analysis of O_2 distribution and multi-parametric functional imaging of live tissue samples can be implemented on standard imaging platforms, including simple or ratiometric intensity measurements or fluorescence and phosphorescence lifetime-based imaging microscopy (FLIM, PLIM), under one or two-photon excitation. To enable high-performance imaging experiments on different platforms, with specific cell and tissue models and analytical tasks (e.g., semiquantitative vs. qualitative, point measurements or high-resolution O_2 imaging), a number of dedicated O_2-sensing probes have been developed which include small molecules, supramolecular and nanoparticle based structures. Some commonly used O_2 probes are presented in Table 1.

Here we describe a standard method which utilizes cell-permeable phosphorescent O_2 probes to perform high-resolution O_2 imaging in cultured multicellular aggregates (neurospheres) and excised brain slices [11, 12]. This imaging method allows to study development of neural progenitor (stem) cells, their proliferation, differentiation, drug effects and physiological processes involving neuronal cell death, at the level of individual cells.

Neurospheres are free-floating heterogeneous multicellular structures which consist of neural progenitor cells and differentiating neurons, astrocytes, oligodendrocytes and have a typical size of 50–500 μm. These ball-like clusters allow for proliferation of neural progenitor cells and manipulation by mitogen withdrawal to differentiate them into different neural cell types [13]. This model was used to study proliferation and differentiation capacity of human and murine neural stem cells in cell replacement therapies for neurodegenerative diseases [14–16].

Live brain slices isolated from embryonic and adult mice can be cultured for several days or weeks maintaining the intact cytoarchitecture of the brain [17]. This model finds use in neurotoxicology, inflammation, studies of neural cell development [18–21] and models of brain tumors such as glioblastoma [22]. In recent years even more complex and relevant models of the brain such as organoids have been developed [23].

Generation of consistent results with such models and imaging studies largely depends on precise control of tissue microenvironment throughout the experiment (carried out in vitro or ex vivo), particularly of its oxygenation state and other metabolic parameters.

Table 1
Phosphorescent O_2-sensitive probes tested in neural tissue imaging

Probe name	Probe description	Imaging mode, equipment	Suggested probes for multiplexing	Source, citation
PtP-C343	Cell-impermeable probe (supramolecular) for in vivo imaging of brain vasculature	Two-photon PLIM Ex.—840 nm Em.—676 nm	FITC-dextran—contrast agent for vasculature	[24]
MitoImage-NanO2	Cell-permeable probe (nanoparticle based) suitable for many cell types and multicellular spheroids	One-photon PLIM Ex.—405 nm, 534 nm Em.—650 nm	Blue, green, and red-emitting probes: DAPI, Hoechst 33342, propidium iodide; GFP, Calcein Green, CellTox Green, Rhodamine, Alexa Fluor (405–594 nm)	[12, 37]
Pt-Glc	Cell-permeable probe (small molecule) for use with neurospheres, brain slices, other 3D tissue models	One-photon PLIM Ex.—405 nm, 534 nm Em.—650 nm	Blue, green, and red-emitting probes: DAPI, Hoechst 33342, propidium iodide; GFP, Calcein Green, CellTox Green Rhodamine, Alexa Fluor (405–594 nm)	[11]
MitoImage-MM2	Cell-permeable probe (nanoparticle based) suitable for many cell types and multicellular spheroids. Can operate in different detection modalities	One- or two-photon, Ratiometric intensity, PLIM Ex. (one-photon)—405 nm or 534 nm Ex. (two-photon)—750–770 nm Em. 1–650 nm (O_2-sensitive) Em. 2–430 nm (O_2-insensitive reference)	Green and red-emitting probes: propidium iodide, GFP, Calcein Green, CellTox Green, Rhodamine, Alexa (488–594 nm)	[36]
NanO2-IR	Cell-permeable probe (nanoparticle based) suitable for many cell types and multicellular spheroids. Long-wave excitation and emission).	One-photon PLIM Ex.—590–620 nm Em.—760 nm	Blue, green, and red-emitting fluorescent cellular probes (e.g., DAPI, Hoechst 33342, GFP, Calcein Green, CellTox Green, Rhodamine, propidium iodide, Alexa (405–594 nm)	[38]

This can be effectively achieved by confocal (one-photon) or two-photon phosphorescence microscopy imaging [11]. The main prerequisites of this method are efficient staining of samples to achieve stable and uniform (laterally and in-depth) distribution of the O_2 probe allowing the analysis of individual cells and clusters of cells, without significant impact on cellular function. Optimized imaging conditions which ensure minimal probe photo-bleaching and sample photo-damage, and quantitative readout of O_2 concentration in 3D and over time (4D) are also critical. *See* **Note 1**.

Our protocol describes preparation and culturing of the neural tissue, staining with the O_2 probe, mapping of O_2 concentration, multiplexing with several other probes and analysis of imaging data. We mainly focus on PLIM (or FLIM in the microsecond time domain) as preferred method for O_2 imaging, which provides quantitative read-out of O_2 concentration $[O_2]$ and superior imaging performance [8, 9, 24–26]. Its main results are high-resolution 3D O_2 maps for samples of respiring neural tissue at different time points, correlated with relevant parameters produced by the other fluorescent probes or biomarkers. The preparation, culturing and staining of neurospheres and brain slices with O_2 probes are described in Subheadings 3.1 and 3.2, followed by acquisition and processing of imaging data in Subheading 3.3. This O_2 imaging methodology can be easily adapted for use with other models of neural tissue maintained in static cultures, perfusion cells or micro-fluidic devices, such as embryoid bodies, organoids, and spinal cord explants [15, 27]. Users who do not possess a PLIM system but have standard intensity-based imaging platforms can use the multimodal probe MM2 applying similar procedure with necessary modifications of image acquisition and processing [12].

2 Materials

Prepare all solutions using ultrapure water (Milli-Q grade, sterile-filtered 0.22 μm, 18 MΩ cm). Store all reagents at 4 °C, for no longer than 4 weeks (unless specified otherwise). To prevent contamination, perform all cell culture under laminar flow (class II Microbiological Safety Cabinet with HEPA filter) and aseptically. Wear gloves at all times and spray all used materials and surfaces with 70 % ethanol. Unless provided sterile, autoclave all glassware and plasticware and media prior to use (121 °C, 20 min).

2.1 Animals

Time-mated female Sprague-Dawley rats with embryonic days (E) 12–18 embryos or postnatal (P) 1–21 provided by relevant Biological Services facility. All the procedures must be performed under a license issued by relevant body and according to the legislation.

<table>
<tr><td>2.2 Microscope Set Up</td><td>

1. Laser-scanning microscope.

2. FLIM-PLIM imaging module, e.g., TCSPC DCS-120 confocal scanner (Becker & Hickl GmbH, Germany).

3. Pulsed diode laser, e.g., BDL-SMC (Becker & Hickl GmbH) 405 nm (pulse duration 40–70 ps, 0.25–1 mW power).

4. Emission filter (635–675 nm) for the O$_2$ probe.

5. Picosecond tunable 400–700 nm laser, e.g., SC400-4 Fianium, Southampton, UK or equivalent, to provide 488 nm excitation.

6. Emission filter (512–536 nm) for CellTox™ Green.

7. Live cell imaging incubator (controlled CO$_2$, temperature and O$_2$) chamber. Upright design with dipping water-immersion objective 20×–25× NA:1 (or higher) is recommended. *See* **Note 2**.

</td></tr>
<tr><td>2.3 Imaging</td><td>

1. O$_2$-sensitive probe Pt-Glc [11] *or* MitoImage-NanO2 (Luxcel Biosciences, Cork, Ireland). *See* **Note 3**.

2. Fluorescent probe(s) for multiplexing, e.g., CellTox™ Green Toxicity Kit (Promega, Madison, WI). *See* **Note 4**.
   Store all the probes in the dark and minimize exposure to light.

</td></tr>
<tr><td>2.4 Neurosphere Culture</td><td>

1. Phosphate buffered saline (PBS).

2. Acetic acid (glacial).

3. PBS-acetate: mix 50 mL of PBS with 300 μL of glacial acetic acid, filter-sterilize.

4. 7.5 % bovine serum albumin (BSA) in PBS.

5. DTT stock solution in PBS (100 mM): dissolve 0.31 g DTT in 20 mL of PBS, filter-sterilize and store at 4 °C.

6. Heparin stock solution in PBS: dissolve 50 mg of heparin in 1 mL PBS, filter-sterilize and store at 4 °C.

7. Epidermal growth factor (EGF).

8. Fibroblast growth factor (FGF).

9. EGF stock solution (20 μg/mL): take 10 mL PBS-acetate and mix with 133 μL of freshly prepared 7.5 % BSA in PBS. Add to EGF package to produce 200 μg/mL EGF. Dispense in 100 μL aliquots and store at –20 °C. Dilute one aliquot 1:10 with PBS-acetate to make 20 μg/mL EGF stock.

10. FGF stock solution (20 μg/mL): dissolve 25 μg FGF in 1.25 mL filter-sterilized PBS containing 0.1 % BSA, 1 mM DTT, 5 μg/mL heparin. Aliquot in 50 μL and store at –20 °C.

11. Growth medium: DMEM/F12 Ham with 2 mM L-Glutamine, 1 % penicillin-streptomycin, 2 % B27 supplement (Invitrogen™, Life Technologies™, Carlsbad, CA), 20 ng/mL FGF, 20 ng/mL EGF. Store at 4 °C for up to 1 week.

</td></tr>
</table>

12. Differentiation medium: DMEM/F12 Ham with 2 mM L-Glutamine, 1 % penicillin-streptomycin, 2 % B27 supplement (Invitrogen™, Life Technologies™), 1 % heat-inactivated fetal bovine serum (FBS). Store at 4 °C for up to 2 weeks.

13. Modified Hank's balanced salt solution (HBSS), with $NaHCO_3$, without phenol red, $CaCl_2$, and $MgSO_4$.

14. Trypsin inhibitor: dissolve 100 mg of soybean trypsin inhibitor (type I-S, 1 mg can inhibit 1–3 mg of trypsin, e.g., Sigma Chemical, St. Louis, MO) in 40 mL of sterile water, aliquot in 1.5 mL tubes (500 µL each) and store at –20 °C. For use, thaw one aliquot and add 1 mL HBSS.

15. Borate buffer 0.15 M: dissolve 0.927 g of boric acid (H_3BO_3) in 80 mL of sterile deionized H_2O, bring pH to 8.4 with NaOH, adjust volume to 100 mL.

16. Poly-D-lysine (PDL) stock solution: dissolve 5 mg PDL (Sigma Chemicals) in 5 mL borate buffer, sterilize with 0.22 µm filter, dispense in 1 mL aliquots and store at –20 °C.

17. 15 mL centrifuge tubes.

18. 1 mL pipettes.

19. Glass Pasteur pipettes (i.d. 0.5–1 mm, length 230 mm, Fisherbrand).

20. 70 µm, nylon cell strainer, e.g., BD Falcon™, Franklin Lakes, NJ.

21. Bench centrifuge.

22. 1.5 mL tubes.

23. Trypan Blue 0.4 % saline-phosphate solution (Sigma Chemicals).

24. Hemocytometer.

25. Cover slips, 22×22 mm, thickness No. 1.5.

26. 25 cm² flasks for suspension cells, sterile, vent cap.

27. Humidified CO_2/37 °C incubator. *Optionally* with O_2 control.

28. Microscopy mini-dishes. *See* **Note 5**.

2.4.1 Preparation of PDL-Coated Slices

Dilute the PDL stock 1:10 with borate buffer, apply 50–500 µL to the surface of microscopy mini-dishes. Incubate 12–16 h at room temperature, then collect the PDL solution, wash the surface three times with 100–500 µL of sterile deionized H_2O and dry under laminar flow hood. Store coated plasticware at 4 °C for up to 1 month. The PDL solution can be reused three times (stored at 4 °C).

2.5 Brain Slice Culture

1. Leibovitz (L15) medium.

2. Tissue slice medium (TSM): phenol red-free DMEM, supplemented with 25 % HBSS, 10 % FBS, 1 % penicillin-streptomycin, 10 mM glucose, 20 mM HEPES-Na, pH 7.2.

3. Porous membranes for immobilization of brain slices: Millicell® (0.4 μm, 30 mm diameter, EMD Millipore, Billerica, MA) or Alvetex™ (Reinnervate, Durham, UK).

4. Low melting agarose.

5. 5 % glucose in PBS.

6. Vibratome.

7. 35 mm petri dishes.

8. Electroporator with gold-plated electrodes and square wave pulses, e.g., BTX ECM 830 (Harvard Apparatus, Holliston, MA)—*optional* for plasmid DNA transfection.

3   Methods

3.1 Neurosphere Culture and Staining with Imaging Probes

1. Following standard protocol [12], dissect rat embryos, isolate neural tissue from the brain (e.g., 10–14 cortices from E18 embryos) and collect it in two 15 mL tubes kept on ice, each containing 5 mL of HBSS. *See* **Note 6**.

2. Distribute the tissue equally between two tubes, allow it to settle on ice (1–2 min), then carefully remove the supernatant and add to each tube 2.5 mL of HBSS containing 0.1 % trypsin. Gently flick the tubes two to five times and incubate 15–20 min at 37 °C.

3. Using a 1 mL pipette, carefully remove the supernatant leaving ~1 mL in each tube. Add to each tube 0.5 mL of trypsin inhibitor solution (0.83 mg/mL). Resuspend the sample, pass it ten times through a sterile glass Pasteur pipette, and then pool together.

4. Filter the suspension of cells through the cell strainer into a 50 mL tube.

5. Centrifuge the filtrate ($200 \times g$, 5 min, room temperature), then remove and discard the supernatant.

6. Using a 1 mL pipette, resuspend the pellet in 1 mL of growth medium and pass the suspension ten times through a glass Pasteur pipette. Add growth medium to a final volume of 10 mL.

7. Take 50 μL of cell suspension into a 1.5 mL tube, mix it with 50 μL of 0.4 % Trypan Blue solution, load in a hemocytometer and count viable cells.

8. Calculate the concentration of viable neural cells in the stock and immediately seed them in 25 cm² flasks for suspension cells using $3–6 \times 10^6$ cells and 5 mL of medium per flask. Seed a sufficient number of flasks, to include necessary controls in your experiments (e.g., for unstained samples). Place the flasks in a humidified CO_2/O_2 incubator set at 5 % CO_2, 21 % O_2, 37 °C. *See* **Note 6**.

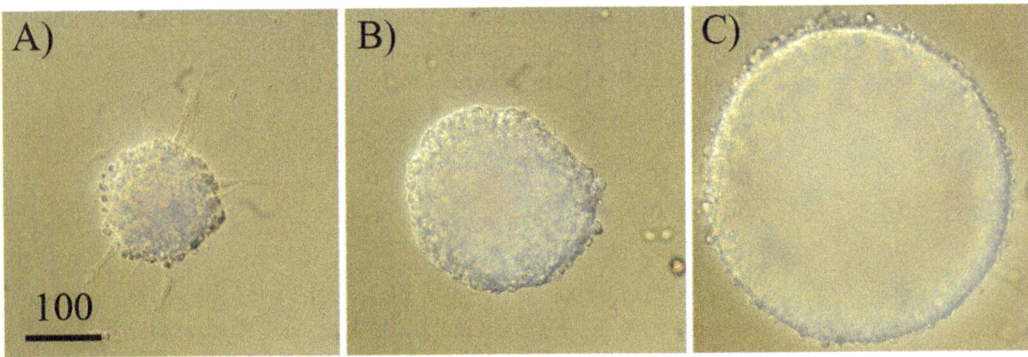

Fig. 1 Transmission light images of neurospheres from rat embryonic cortical tissue grown for 3–5 days at ambient (atmospheric) O_2. Sections (**a**, **b**, and **c**) indicate neurospheres grown for 1, 3, and 5 days in vitro, respectively. Scale bar in μm

9. Add fresh growth medium on days 2, 4, 6: let the flask stand for 15–20 min to settle the neurospheres, then carefully aspirate 2.5 mL of old medium, add 2.5 mL of fresh medium and return the flask back to the incubator. Monitor neurosphere formation during the culture by microscopic observation (Fig. 1).

10. While still in the culture, stain the neurospheres with Pt-Glc probe: add to the medium at 2 μM final concentration and incubate for 16 h in a CO_2 incubator at 37 °C (*see* **Note 7**). If required, stain the neurospheres with other imaging probes such as CellTox™ Green (0.2 %, 30 min), following corresponding protocols. *See* **Note 4**.

11. Using a 1 mL pipette, gently transfer the suspension of neurospheres to a 15 mL tube. Wait 2–3 min till they settle, then remove 4.5 mL of used medium and add fresh medium (1–10 mL of depending on the density).

12. Gently dispense 0.2 mL aliquots of neurosphere suspension to PDL-coated mini-dishes, allow them to attach for 30–60 min and proceed to imaging. *See* **Note 8**.

3.2 Brain Slice Culture and Staining with Imaging Probes

1. Using standard surgical procedures dissect sacrificed rat embryos or postnatal pups and extract brain tissue [28].

2. Embed brain tissue in low melting agarose (4 % in PBS with 5 mg/mL glucose). Mount on vibratome stage and prepare 300–400 μm thick slices in ice-cold L15 medium. Make sufficient number of slices, for control or treatment experiments. *See* **Note 9**.

3. Place each brain slice on a porous pre-wet scaffold membrane (Alvetex or Millipore) in a 35 mm petri dishes with TSM medium. Adjust the volume of medium to cover the tissue only slightly (Fig. 2).

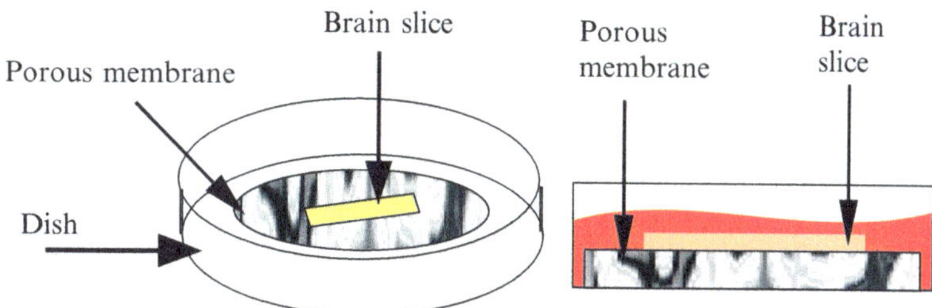

Fig. 2 A sample of rat brain slice cultured in a mini-dish with porous scaffold. Tissue section (300–400 μm thick) is adhered to the porous membrane and covered with 1–2 mm layer of medium

4. Incubate the slices in CO$_2$ incubator at ambient O$_2$ until they are ready for O$_2$ analysis (e.g., 3 h—5 days). Add fresh medium as required. Optional: transfect the tissue with relevant DNA plasmid encoding cell marker by electroporation, 16–48 h prior to analysis. *See* **Note 4**.

5. Add 5 μM Pt-Glc and 0.2 % CellTox™ Green probes to the medium and incubate at 37 °C for 2–3 h to stain tissue. Apply other imaging fluorescent probes, as required. *See* **Note 4**.

6. Change the medium to fresh prior to imaging.

3.3 Imaging and Data Processing

1. Turn on the microscope, confocal scanner, lasers, incubator system, computer and other electronic modules. Set up the required temperature and gas levels (e.g., 37 °C, 19 % O$_2$, 5 % CO$_2$) in sample compartment. Allow the system to warm up for 30 min. Prepare system control software and select the appropriate filter cubes for the probe(s) used.

2. Take one mini-dish with sample (neurospheres adhered on PDL surface or a brain slice immobilized on membrane) and place it in the incubator on microscope stage. Select and bring the objective to working position for imaging. Allow the sample to equilibrate for 5–45 min (optimal time depends on the sample, incubator and O$_2$ levels used, *see* **Note 10**).

3. Preview the sample in transmission light mode. Select region of interest (ROI) for imaging, adjust the focus and Z-axis positions. Record transmission light image, if necessary.

4. Select the desired probe and make preview scan(s) in fluorescence intensity mode under recommended image acquisition settings (e.g., exc/em. 405/650 nm for Pt-Glc, or 488/520 nm for CellTox™ Green). Review the results and, if necessary, adjust the acquisition settings for intensity (S:N > 5) or PLIM imaging. For DCS-120 confocal scanner this is achieved by changing the ND filter and pinhole size. *See* **Note 11**.

5. Scan one or a number of Z-stacks (as required). Once image acquisition is complete, save the data. For PLIM data, open

the file in SPCImage software and check the decay function to ensure that the number of collected photons is high enough (normally >100 counts in first bins) for reliable decay fitting and lifetime calculation.

6. Repeat **steps 4** and **5** for CellTox™ Green or other fluorescent probes used for multi-parametric imaging, using the same X,Y,Z coordinates. These probes usually require collection of intensity images.

7. To ensure consistency and reproducibility of imaging data, repeat the measurements at a different time or moving to a different spheroid or ROI. *See* **Note 12**. Change conditions in analyzed sample (e.g., apply drug stimulation or a different level of atmospheric O_2). Optionally, after the live cell imaging experiments are completed, fix the tissue and perform immunofluorescence staining with relevant antibodies.

8. Process collected CellTox™ Green intensity image(s) in an ordinary fashion, using a built-in image analysis software or freeware such as ImageJ. Count the number of nuclei positively stained with CellTox™ Green probe (i.e., dead cells) and compare with control conditions (e.g., tissue unstained with O_2 probe, *see* **Note 14**). Export processed imaging data as ASCII (photon counts) and TIF files.

9. Using SPCImage software, open one data file generated with the O_2 probe (2D PLIM image). Fit the phosphorescence decays and calculate lifetime values for each pixel in measured ROI to generate 2D lifetime image. Export the intensity (photon counts) and lifetime images as ASCII and TIF files. A screenshot of PLIM image opened in SPCImage software with the main parameters for FLIM-TCSPC data processing is shown in Fig. 3. Using a calibration function for the O_2 probe provided by vendor or generated in a separate experiment (*see* **Note 13**), convert the lifetime matrices (ASCII files) into O_2 concentration maps.

10. (Optional) apply batch processing to the other PLIM and fluorescent images (Z-stacks or time-lapse images). Using a suitable software (e.g., MS Excel, ImageJ), reconstruct 3D images of intensity, lifetime or O_2 concentration.

11. Correlate O_2 concentrations (or lifetime values) in selected ROIs with the number of dead cells. Representative imaging data with spheroids and brain slices are shown in Fig. 4.

---

4 Notes

1. Modern imaging platforms allow scanning of complex biological samples, however many traditional fluorescent and phosphorescent probes have poor penetration into tissue (deeper

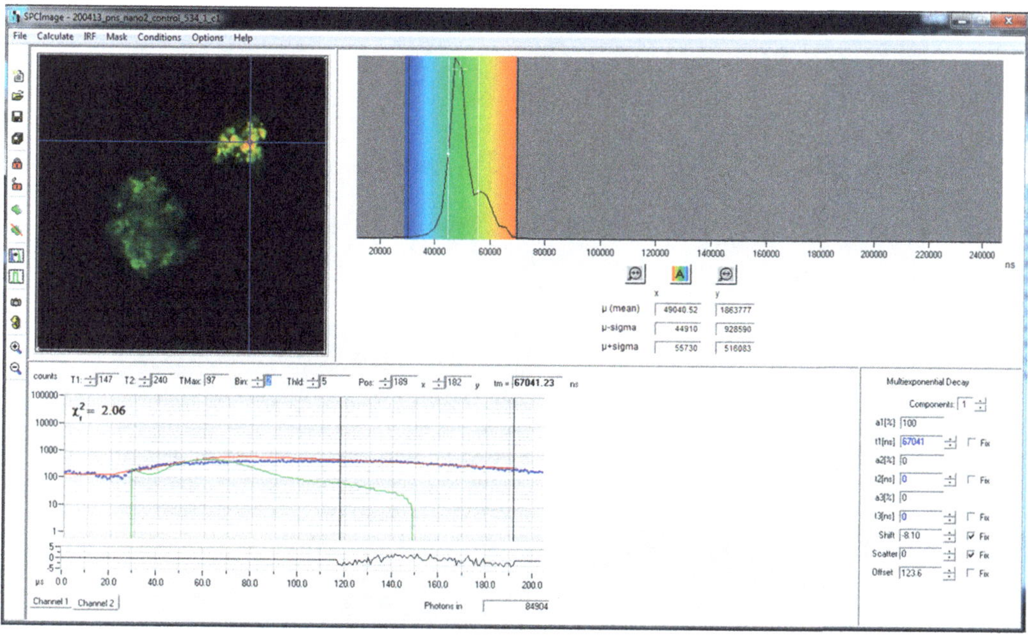

Fig. 3 Screenshot of data processing and decay fitting in SPCImage software (Becker & Hickl GmbH). *Top left window*: PLIM images of neurospheres stained with NanO2 probe produced by fitting. *Bottom left*: the distribution of photons (*blue dots*) and mono-exponential decay fit (*red*). Parameters T_1, T_2, and Shift allow optimization of fitting. *Top right*: lifetime distribution for the whole image (Color figure online)

than 50 μm). The O$_2$ imaging probes described here show efficient penetration and staining of neural tissue. For the small molecule Pt-Glc probe [11] this is achieved in 3–12 h, for the anionic nanoparticles in >2 h, while for the cationic nanoparticle probe requires continuous presence during the culture [12]. Small molecule fluorescent probes such as Hoechst 33342 and TMRM also demonstrate satisfactory staining of tissue, while genetically encoded DNA and RNA constructs (e.g., plasmids) require an electroporation step for transfection [29]. Careful testing and optimization of staining and transfection conditions are highly recommended for each probe used. Different photophysical properties may result in different spatial and temporal resolution (e.g., slow scanning rates for phosphorescent probes). Possible toxic effects of the probe must be evaluated prior to starting complex physiological experiments: concentrations which provide sufficient staining and minimal toxicity should be used (typically <50 μg/mL for nanoparticle and <10 μM for small molecule O$_2$ probes). For multiplexed detection, spectral cross-talk between the different probes (i.e., emission and excitation wavelengths) should be also kept in mind.

2. Laser-scanning imaging systems and probes with long wave excitation and emission (>600 nm) provide deep light penetration into tissue (200–500 micrometers), low photo-damage

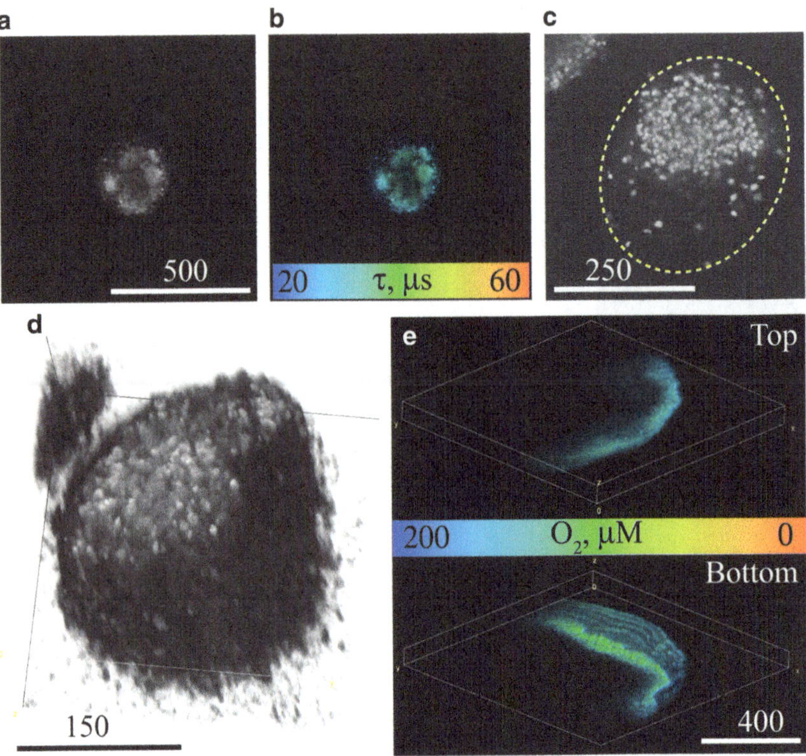

Fig. 4 Representative images of neurospheres and brain slices. (**a**, **b**) Optical sections of a neurosphere stained with Pt-Glc probe (5 µM, 16 h) in intensity (**a**) and lifetime (**b**) scales. (**c**) Optical section of a neurosphere exposed to CellTox™ Green (0.2 %, 30 min). Positively stained nuclei indicate dead cells and *dashed yellow line*—border of the neurosphere. (**d**) 3D reconstruction of the distribution of Pt-Glc probe (intensity image) within live neurosphere attached to a mini-dish. Total height = 38 µm (*Z*). (**e**) 3D reconstruction of O_2 distribution across the stained region (125 µm depth of scanning from the surface) of cultured rat brain cortical slice (P7, 2 DIV). *Top view* reveals higher oxygenation at the surface than inside the tissue (*bottom*). 3D reconstructions were created in Fiji software (Volume Viewer plugin). Scale bar in µm (Color figure online)

and high spatial resolution. Two-photon or light sheet microscopy systems [15, 30–32] coupled with FLIM-PLIM hardware and software [26] are generally preferred. The microscope should be spectrally compatible with the O_2 probe and allow for multiplexing with relevant fluorescent probe. For PtPFPP-based probes such as Pt-Glc, MitoImage™ NanO2, and MM2 [11, 12, 33], one-photon excitation at 390–405 nm (one-photon ps/ns laser) and 640–660 nm emission filter are required. Multimodal MM2 probe is also excitable at 750–770 nm (two-photon femtosecond laser), and for ratiometric intensity imaging it also requires a second emission filter at 420–450 nm (Table 1). For other probes appropriate lasers and filters should be provided. The preferred mode is phosphorescence lifetimes imaging (PLIM covering the range 10–100 µs). Alternatively, dual channel ratiometric intensity scanning mode can be used. Incubator system which controls CO_2, O_2, humidity, and temperature in the microscope sample compartment, and motorized Z-stage for 3D scanning are necessary.

We use an upright microscope AxioExaminer Z1 (Carl Zeiss, Oberkochen, Germany) with a 20× water immersion dipping objective (NA:1), motorized heated stage and connected to a confocal scanner DCS-120 (Becker & Hickl GmbH) with two excitation and two emission channels. A 405 nm picosecond diode laser (Becker & Hickl GmbH) and a picosecond supercontinuum 400–650 nm laser SC400-4 (Fianium) connected to the DCS-120 scanner with an optical fiber are used for excitation (488 nm for CellTox™ Green). Emission is detected with an R10467U-40 photon counting detector (Hamamatsu Photonics, Hamamatsu City, Japan) connected to the scanner and TCSPC hardware (Becker & Hickl GmbH). Long-pass excitation (>435 nm) and band-pass emission (635–675 nm) filters are used to detect the O_2 probes. MicroToolBox software (Carl Zeiss, v. 2011) is used to control the microscope. Image acquisition and data processing are performed with the SPCM software (Becker & Hickl GmbH). The two-photon microscopy set-up for MM2 O_2 probe is described in [12].

3. The list of phosphorescent O_2 probes is growing (>2 structures each year), some of them are available commercially. Apart from Table 1, we refer the reader to most recent publication and reviews on this topic, e.g., [8, 33].

4. Pt-Glc and MitoImage-NanO2 probes [11] can be multiplexed with blue (405/420–460 nm ex/em), green, and red fluorescent dyes (470–600/510–630 nm ex/em) (Table 1). Lifetime multiplexing can also be used in FLIM-PLIM, whereby emission decay of the O_2 probe (20–60 μs) is separated from fluorescent dye (1–10 ns). Many common fluorescent indicators (BCECF, GFP for pH, Fluo-4, Calcein Green for Ca^{2+}, TMRM, JC-1 for membrane potential), cell proliferation and viability markers (Click-It EdU, CellTox™ Green, propidium iodide, Image-IT DEAD Green, TUNEL assay), and live cell markers (e.g., cholera toxin, staining predominantly neuronal cells or sulforhodamine 101) can be used for multiplexed imaging. After live cell imaging of O_2, antibody-aided visualization of cell markers is also possible. However, locating the same regions of tissue after immunofluorescence staining can be difficult. Prior to O_2 imaging neural tissue can also be transfected with DNA plasmids expressing fluorescently tagged markers which can then be identified and correlated with observed O_2 distribution. The efficiency of tissue transfection is typically quite low (<10 %), so only populations of the cells expressing specific marker will be identified.

5. Different microscopy mini-dishes can be used, including plastic standard 35 mm petri mini-dishes or gridded dishes (e.g., Ibidi GmbH, Munich, Germany) which allow tracking of individual spheroids and repetitive measurements. Inverted microscopes require glass-bottom mini-dishes of specific thickness (e.g., MatTek, Ashland, MA).

6. This procedure usually takes 3–4 h and produces 3–4×10^7 viable cells from cortices of 10–14 E18 rat embryos. If cell viability is below 70–80 %, shorten trypsinization and processing times. The free-floating culturing method described here often produced spheroids of different size. Plasticware and petri dishes with low cell attachment will facilitate formation of spheroids. More advanced methods of spheroid formation, such as hanging drop system [34], can be used instead. Neurospheres can also be cultured under low O_2 atmosphere (e.g., 3 %).

7. For Pt-Glc probe normal staining time is 6–16 h for neurospheres and 2–3 h for brain slices. Depending on the sample staining efficiency, the concentration can be increased to 5 μM. Other probes such as NanO2 or MM2 require longer incubations or continuous staining [12]. Staining efficiency also depends on the number of viable cells and size of spheroids. Spheroid core can display lower staining due to cell necrosis (can be confirmed with dead cell stain CellTox™ Green). After loading with the O_2 probe, spheroids produce stable signals and can be imaged for at least 2–4 days.

8. We recommend seeding of 10–20 spheroids per dish; however this depends on particular analytical task. For stable focusing and imaging, neurospheres and brain slices should be adhered to the surface. Therefore PDL-coated dishes are used for spheroids and porous membrane scaffold for brain slices. Dishes with gridded bottom allow tracking and repetitive measurements with the same spheroid, particularly when O_2 analysis is combined with immunofluorescence staining, drug treatment or sample manipulation. If spheroids tend to detach and move during imaging or washing procedures, porous membrane scaffold should be used (though it is not transparent for transmission light) or samples can be incubated in differentiation media. In this case, collect spheroids in 15 mL tube, wash with HBSS, resuspend neurospheres in differentiation medium, seed on 0.2 mL on PDL-coated mini-dishes, incubate for 2–3 h in CO_2/O_2 incubator and monitor their attachment, then proceed to imaging.

9. Embedding slices in agarose makes them easier to handle and is not always necessary.

10. Effective control of sample temperature and pH, and atmospheric O_2, CO_2 and humidity in the chamber is very important, [35].

Normally, temperature is set to 37 °C and O$_2$ is ambient (~20 %), however this depends on the experimental task. Minimize the time the sample spends outside the incubator, avoid large fluctuations of sample parameters. For imaging use phenol red-free media with 5–20 mM HEPES, pH 7.2–7.4. During long-term experiments (>1–2 h) monitor evaporation of media from the sample (critical for water immersion objective) and use larger volume (2 mL per mini-dish).

11. For imaging tissue samples stained with Pt-Glc probe, typical PLIM settings on DCS-120 system (*see* **Note 2**) are: excitation—405 nm laser, emission filter—635–675 nm. Software parameters: PS FLIM, MCS FLIM, trigger marker 0, Offset -9 %, 32 MT units (204.8 µs time range), routing channel *X2*. $T_{pxl} = 201.5$ µs, $T_{line} = 51.83$ ms, $T_{frame} = 13.27$ s. Collection time—120 s, Stop T = ON. Multiplexing—PXL, steps. Depending on the signals collected, multiplexing parameter (percent of time when the laser is ON during PLIM scanning) can be optimized.

12. Import imaging data file into SPCImage software using "Import" function in "File" menu (Fig. 3). If more than one detector is installed, select relevant spectral channel. Software automatically selects the brightest pixel and fit the decay, however binning, T1, T2 and Shift parameters must be adjusted manually to perform the fitting and calculation of lifetime values more accurately. Offset value can also be preset, if required. After this verify the fitting by moving the cursor to another pixel with a different intensity. Select ROI or the whole image and perform lifetime calculation.

13. For PLIM experiments with the same probe and conditions, published O$_2$ calibrations can be used:
NanO2, 37 °C: $[O_2, \mu M] = 9{,}274.667 \times \exp(-\tau/7.27143)$; [12].
    Pt-Glc, 37 °C: $[O_2, \mu M] = 1{,}768.574 \times \exp(-\tau/9.72584)$; [11].
    MM2, 37 °C: $[O_2, \mu M] = 18{,}861 \times \exp(-\tau/5.79) + 9.62$; [36].
    Where τ is expressed in µs.
    For the other probes (e.g., pH, Ca^{2+}) and imaging modalities (intensity based or ratiometric O$_2$ imaging), calibration may be influenced by optics, image acquisition settings and conditions used. It is therefore recommended to calibrate your own system. For O$_2$ probes typical calibration experiment involves staining a small spheroid or tissue slice, blocking its respiration with a mitochondrial inhibitor (e.g., 5–10 µM antimycin A, 5–30 min). The sample is then exposed to known levels of atmospheric O$_2$ (e.g., 20, 15, 10, 5, 3, 1, 0 % O$_2$) and after equilibration the sample is measured and mean lifetime (or intensity ratio) values are calculated for each concentration. These experimental points are then used to establish O$_2$ cali-

bration (analytical function $[O_2] = f(\tau)$ [8]. Calibrations should be performed as 2–3 independent experiments. As $[O_2]$ decreases, phosphorescence lifetime and intensity signals of the probe increase significantly. Therefore signal acquisition settings should be optimized such that at all O_2 concentrations sufficient number of photons is collected: not too high to saturate the detector and not too low to reliably see and fit the decay [26].

14. Due to nonideal isolation procedure, ex vivo culturing and probe staining, multicellular spheroids and tissue slices contain a significant number of dead cells (stained positively with CellTox™ Green or propidium iodide). This should be assessed by comparing processed sample and untreated control, particularly after staining with probes, treatment with drugs or imaging session. In addition, viable cell markers can be employed, such as staining of healthy mitochondria (TMRM) or endocytosis tracers (Cholera toxin, Transferrin, LysoTracker).

Acknowledgments

This work was supported by the Science Foundation Ireland, grants 12/RC/2276 and 12/TIDA/B2413, and the EC FP7 Program, grant NanoBio4Trans (No. 304842-2). We thank Dr. J. M.P. Pakan and Dr. Y.M. Nolan (Department of Anatomy and Neuroscience, University College Cork) for the help in preparation of brain slices and neurosphere cultures.

References

1. Northington FJ, Chavez-Valdez R, Martin LJ (2011) Neuronal cell death in neonatal hypoxia-ischemia. Ann Neurol 69:743–758

2. Green DR, Galluzzi L, Kroemer G (2011) Mitochondria and the autophagy–inflammation–cell death axis in organismal aging. Science 333:1109–1112

3. Sims NR, Muyderman H (2010) Mitochondria, oxidative metabolism and cell death in stroke. Biochim Biophys Acta 1802:80–91

4. Kalyanaraman B, Darley-Usmar V, Davies KJA et al (2012) Measuring reactive oxygen and nitrogen species with fluorescent probes: challenges and limitations. Free Radic Biol Med 52:1–6

5. Tait SW, Green DR (2010) Mitochondria and cell death: outer membrane permeabilization and beyond. Nat Rev Mol Cell Biol 11:621–632

6. Jespersen SN, Østergaard L (2011) The roles of cerebral blood flow, capillary transit time heterogeneity, and oxygen tension in brain oxygenation and metabolism. JCBFM 32:264–277

7. Devor A, Sakadžić S, Srinivasan VJ et al (2012) Frontiers in optical imaging of cerebral blood flow and metabolism. JCBFM 32:1259–1276

8. Papkovsky DB, Dmitriev RI (2013) Biological detection by optical oxygen sensing. Chem Soc Rev 42:8700–8732

9. Lecoq J, Roussakis A, Parpaleix E et al (2011) Simultaneous two-photon imaging of oxygen and blood flow in deep cerebral vessels. Nat Med 17:893–898

10. Quaranta M, Borisov SM, Klimant I (2012) Indicators for optical oxygen sensors. Bioanal Rev 4:115–157

11. Dmitriev RI, Kondrashina AV, Koren K et al (2014) Small molecule phosphorescent probes for O2 imaging in 3D tissue models. Biomater Sci 2:853–866

12. Dmitriev RI, Zhdanov AV, Nolan YM et al (2013) Imaging of neurosphere oxygenation with phosphorescent probes. Biomaterials 34:9307–9317

13. Reynolds BA, Weiss S (1992) Generation of neurons and astrocytes from isolated cells of the adult mammalian central nervous system. Science 255:1707–1710

14. Monni E, Congiu T, Massa D et al (2011) Human neurospheres: from stained sections to three-dimensional assembly. Translational Neurosci 2:43–48

15. Pampaloni F, Reynaud EG, Stelzer EH (2007) The third dimension bridges the gap between cell culture and live tissue. Nat Rev Mol Cell Biol 8:839–845

16. Gil-Perotín S, Duran-Moreno M, Cebrián-Silla A et al (2013) Adult neural stem cells from the subventricular zone: a review of the neurosphere assay. Anat Rec 296:1435–1452

17. Gähwiler BH, Capogna M, Debanne D et al (1997) Organotypic slice cultures: a technique has come of age. Trends Neurosci 20:471–477

18. Ohnishi T, Matsumura H, Izumoto S et al (1998) A novel model of glioma cell invasion using organotypic brain slice culture. Cancer Res 58:2935–2940

19. Zimmer J, Kristensen BW, Jakobsen B et al (2000) Excitatory amino acid neurotoxicity and modulation of glutamate receptor expression in organotypic brain slice cultures. Amino Acids 19:7–21

20. Dionne KR, Smith Leser J, Lorenzen KA et al (2011) A brain slice culture model of viral encephalitis reveals an innate CNS cytokine response profile and the therapeutic potential of caspase inhibition. Exp Neurol 228:222–231

21. Su T, Paradiso B, Long Y-S et al (2011) Evaluation of cell damage in organotypic hippocampal slice culture from adult mouse: a potential model system to study neuroprotection. Brain Res 1385:68–76

22. McCord AM, Jamal M, Shankavarum UT et al (2009) Physiologic oxygen concentration enhances the stem-like properties of CD133+ human glioblastoma cells in vitro. Mol Cancer Res 7:489–497

23. Lancaster MA, Renner M, Martin C-A et al (2013) Cerebral organoids model human brain development and microcephaly. Nature 501:373–379

24. Finikova OS, Lebedev AY, Aprelev A et al (2008) Oxygen microscopy by two-photon-excited phosphorescence. Chem Phys Chem 9:1673–1679

25. Kazmi SMS, Salvaggio AJ, Estrada AD et al (2013) Three-dimensional mapping of oxygen tension in cortical arterioles before and after occlusion. Biomed Opt Express 4:1061–1073

26. Becker W (2012) Fluorescence lifetime imaging – techniques and applications. J Microsc 247:119–136

27. Bershteyn M, Kriegstein AR (2013) Cerebral organoids in a dish: progress and prospects. Cell 155:19–20

28. Stoppini L, Buchs P-A, Muller D (1991) A simple method for organotypic cultures of nervous tissue. J Neurosci Meth 37:173–182

29. Barry DS, Pakan JMP, O'Keeffe GW et al (2013) The spatial and temporal arrangement of the radial glial scaffold suggests a role in axon tract formation in the developing spinal cord. J Anat 222:203–213

30. Pampaloni F, Ansari N, Stelzer EH (2013) High-resolution deep imaging of live cellular spheroids with light-sheet-based fluorescence microscopy. Cell Tiss Res 352:161–167

31. Wilt BA, Burns LD, Wei Ho ET et al (2009) Advances in light microscopy for neuroscience. Annu Rev Neurosci 32:435–506

32. Graf BW, Boppart SA (2010) Imaging and analysis of three-dimensional cell culture models. In: Live cell imaging, vol 591, Methods in molecular biology. Springer, New York, pp 211–227

33. Dmitriev RI, Papkovsky DB (2012) Optical probes and techniques for O$_2$ measurement in live cells and tissue. Cell Mol Life Sci 69:2025–2039

34. Mehta G, Hsiao AY, Takayama S (2012) Opportunities and challenges for use of tumor spheroids as models to test drug delivery and efficacy. J Control Release 164:192–204

35. Frigault MM, Lacoste J, Swift JL et al (2009) Live-cell microscopy–tips and tools. J Cell Sci 122:753–767

36. Kondrashina AV, Dmitriev R, Borisov SM et al (2012) A phosphorescent nanoparticle-based probe for sensing and imaging of (intra)cellular oxygen in multiple detection modalities. Adv Funct Mater 22:4931–4939

37. Fercher A, Borisov SM, Zhdanov A et al (2011) Intracellular O$_2$ sensing probe based on cell-penetrating phosphorescent nanoparticles. ACS Nano 5:5499–5508

38. Tsytsarev V, Arakawa H, Borisov SM et al (2013) In vivo imaging of brain metabolism activity using a phosphorescent oxygen-sensitive probe. J Neurosci Meth 216:146–151

Chapter 6

Calcium Imaging in Neuron Cell Death

María Calvo, Carlos Villalobos, and Lucía Núñez

Abstract

Intracellular Ca2+ is involved in control of a large variety of cell functions including apoptosis and neuron cell death. For example, intracellular Ca2+ overload is critical in neuron cell death induced by excitotoxicity. Thus, single cell monitoring of intracellular Ca2+ concentration ([Ca2+]cyt) in neurons concurrently with apoptosis and neuron cell death is widely required.

Procedures for culture and preparation of primary cultures of hippocampal rat neurons and fluorescence imaging of cytosolic Ca2+ concentration in Fura2/AM-loaded neurons are described. We also describe a method for apoptosis detection by immunofluorescence imaging. Finally, a simple method for concurrent measurements of [Ca2+]cyt and apoptosis in the same neurons is described.

Key words Calcium imaging, Neuron death, Apoptosis, Hippocampal neurons, Fura2/AM

1 Introduction

Intracellular free calcium concentration ([Ca2+]cyt) plays a pivotal role in control of a large variety of cell and physiological functions in most cells and tissues from the very short term (neurotransmitter release, muscle contraction) to the long-term scale (gene expression, cell proliferation). In general, an increase in [Ca2+]cyt triggers cell activation. However, if the rise in [Ca2+]cyt is large and/or sustained enough, it promotes rather cell death in many cell types including neurons [1]. Examples include overly activation of ionotropic glutamatergic receptors during excitotoxicity, particularly N-methyl D-aspartate (NMDA) [2, 3] as well as Ca2+ overload induced by amyloid β oligomers in Alzheimer's disease [4]. These events may be followed by a delayed [Ca2+]cyt increase long after stimuli removal and lead to calpain activation, turning on a proteolytic cascade culminating in neuron cell death [5]. This process is usually limited by endogenous buffers including proteins like calbindin and organelles like mitochondria. However, if mitochondria take up too much Ca2+, they favor the delayed rise in [Ca2+]cyt [6] as well as the opening of the so-called mitochondrial

Laura Lossi and Adalberto Merighi (eds.), *Neuronal Cell Death: Methods and Protocols*, Methods in Molecular Biology, vol. 1254, DOI 10.1007/978-1-4939-2152-2_6, © Springer Science+Business Media New York 2015

permeability transition pore, a not well understood phenomenon that precedes release of cytochrome c and other pro-apoptotic factors culminating in apoptosis [7]. Therefore, changes in neuronal [Ca2+]cyt are critical in neuron apoptosis and cell death. In the present chapter, a detailed description of methods for primary culture of rat hippocampal neurons is provided together with means for imaging [Ca2+]cyt rises in neurons and apoptosis formation in the same cultures. Finally, a concurrent procedure is presented for monitoring both Ca2+ levels and apoptosis in the same single neurons.

2 Materials

2.1 Primary Culture of Rat Hippocampal Neurons

1. Neonatal Wistar rats. *See* **Note 1**.

2. HEPES-buffered saline (HBS): 145 mM NaCl, 5 mM KCl, 1 mM MgCl$_2$, 1 mM CaCl$_2$, 10 mM glucose, 10 mM sodium-HEPES, pH 7.4. Use double distilled water for preparation.

3. 4 % bovine serum albumin (BSA) in Hank's balanced salt solution (HBSS) without Ca2+ and Mg2+ (Gibco®, Life Technologies™, Gaithersburg, MD): dissolve 4 g of BSA in 100 mL HBSS.

4. Hank's medium + 0.6 % BSA: mix 85 mL HBSS without Ca2+ and Mg2+ (Gibco®, Life Technologies™) with 15 mL 4 % BSA in HBSS.

5. Papain (Worthington, Lakewood, NJ): Prepare cell dissociation solution diluting the enzyme at 20 U/mL in Hank's + 0.6 % BSA.

6. Deoxyribonuclease I (DNase I) from bovine pancreas (Sigma Chemicals, St. Louis, MO): Dissolve DNase in Hank's medium + 0.6 % BSA to a final concentration of 1 mg/mL.

7. Neurobasal® Culture Medium (Gibco®) with 10 % fetal bovine serum (FBS, Lonza, Basel, Switzerland), 2 % B27 (Gibco®), 1 µg/mL gentamicin (50 mg/mL, Gibco®) and 2 mM L-glutamine (Gibco®).

8. 12 mm glass coverslips (Marienfeld GmbH & Co. KG, Lauda-Königshofen, Germany).

9. Poly-D-lysine (Becton Dickinson, Franklin Lakes, NJ).

10. 4-Well multidish plaques for 12 mm glass coverslips (Nunc, Rochester, NY).

11. 5 and 10 mL sterile pipettes (Fisher Scientific, Lough-borough, UK).

12. HeraCell 150 Incubator (Thermo Scientific, Waltham, MA).

13. Centrifuge.

14. Neubauer counting chamber.

2.2 Calcium Imaging

1. HEPES-buffered saline (HBS): 145 mM NaCl, 5 mM KCl, 1 Mm $MgCl_2$, 1 Mm $CaCl_2$, 10 Mm glucose, 10 mM sodium-HEPES, pH 7.4.

2. Fura2/AM (Invitrogen™, Life Technologies™, Gaithersburg, MD): 2 mM stock solution in DMSO. Store at – 20 °C.

3. Inverted fluorescence microscope *or* confocal microscope. *See* **Note 2**.

4. Cell perfusion system for living cells. *See* **Note 3**.

5. Light-proof box covering the imaging setup (about $100 \times 100 \times 100$ cm): Also available from commercial sources, e.g., Hamamatsu Photonics.

2.3 Excitotoxicity and Apoptosis Assays

1. Mg2+-free, HEPES buffered saline: 146 mM NaCl, 5 mM KCl, 1 mM $CaCl_2$, 10 mM glucose, 10 mM sodium-HEPES, pH 7.42.

2. *N*-methyl-D-aspartic acid (NMDA), e.g., Sigma Chemicals.

3. Glycine.

4. FITC-conjugated Annexin V (Becton Dickinson).

5. Annexin V binding buffer: 140 mM NaCl, 2.5, $CaCl_2$, 10 mM sodium-HEPES, pH 7.4.

6. Fluorescence microscope, e.g., Nikon Eclipse TS100 microscope (objective 40×) coupled with fluorescence filters (Nikon, Tokyo, Japan).

3   Methods

3.1 Primary Rat Hippocampal Neuron Cell Culture

1. Sterilize 12 mm glass coverslips in ethanol for 24 h, and allow drying under sterile conditions.

2. Cover one side of the coverslip overnight with approximately 200 μL of 1 mg/mL poly-D-lysine solution.

3. On the next day wash coverslips with double distilled sterile water every 15 min for 90 min under sterile conditions.

4. Place coverslips in a 4-well multidish plaque filled with 500 μL of Neurobasal® Culture Medium, and maintain in a humidified 37 °C incubator with 5 % CO_2 until use. Treated coverslips can be used for up to 1 week.

5. Obtain hippocampal neurons from two newborn Wistar rat pups [4, 9]. Kill newborn rat pups by decapitation, and quickly wash the head in sterile HBS medium. Then, open the skull to extract the brain, and wash it quickly with sterile HBS medium before dissecting the hippocampus.

6. Make a diagonal cut with a scalpel in each hemisphere, and transfer the top side of each hemisphere to a Petri dish containing

sterile HBS medium. With the help of a magnifying glass, carefully remove the meninges and separate the hippocampi from the cortex.

7. Wash hippocampal tissue with sterile Hank's medium without Ca2+ or Mg2+ + 0.6 % BSA and then cut it into small pieces of about 2×2 mm.

8. Transfer small hippocampal pieces to a 10 mL centrifuge tube containing 10 mL of pre-filtered papain solution. Close the tube and put inside the 37 °C incubator. After 15 min incubation, add 500 µL of DNase I solution (50 µg/mL final concentration) and further incubate at 37 °C for another 15 min with occasional, gentle shaking.

9. Wash tissue fragments three times with fresh Neurobasal® Culture Medium. To do so, allow pieces to go to the bottom of the tube (by gravity) and remove medium. Then add 5 mL sterile Neurobasal® Culture Medium and gently shake the suspension. Repeat this procedure two more times.

10. Disperse tissue fragments into a cell suspension using a 5 mL plastic pipette. Specifically, after the last wash, add 3 mL Neurobasal® Culture Medium and disperse pieces by passing them 10–12 times through a 5 mL plastic pipette. Allow small pieces to set by gravity and collect the 3 mL medium in another 10 mL centrifuge tube. Repeat this procedure three times.

11. Centrifuge the cell suspension gently ($160 \times g$, 5 min). Remove the supernatant and suspend the cell pellet is in 1,000 µL Neurobasal® Culture Medium.

12. Estimate cell density using a Neubauer counting chamber. Add 30×10^3 cells (around 50–70 µL cell suspension) to each well previously filled with poly-D-lysine coated coverslips in 500 µL Neurobasal® Culture Medium.

13. Maintain hippocampal cells in culture for 14–16 days in vitro (DIV, *see* **Note 4**) in a humidified incubator at 37 °C and 5 % CO_2 (*see* **Note 5**) before experiments. Figure 1 shows bright field pictures of rat hippocampal neurons cultured for 2, 8 and >14 DIV as well as the typing of glia and neurons by specific immunofluorescence (*see* **Note 6**) for details on identification of glia and neurons by double immunofluorescence microscopy.

3.2 Fluorescence Imaging of Cytosolic Ca2+

1. Wash coverslips containing cultured hippocampal cells with HBS medium twice and incubate in the same medium containing 4 µM Fura2/AM for 60 min at room temperature and in the dark.

2. Place coverslips on a thermostatic platform for open 12 mm glass coverslips (total volume about 500 µL) on the stage of the inverted microscope.

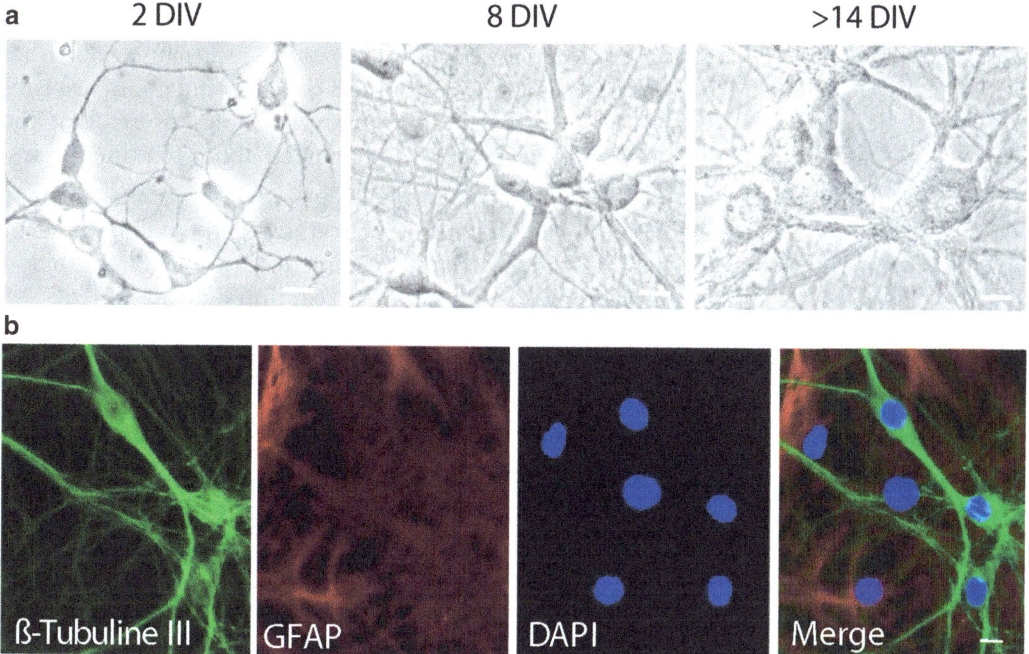

Fig. 1 Primary rat hippocampal neurons in culture. (**a**) Representative microphotographs of neurons as they grow and develop in culture. It can be appreciated that processes are not completely developed until around 8 days in vitro (DIV). Bars represent 10 μm. (**b**) Images of a double immunocytochemistry on a mixed primary culture of hippocampal cells, where neurons labeled with β-tubulin III (*green*) and glia labeled with GFAP (*red*) are appreciated. Nuclei are stained with DAPI. *Bottom right* image shows the merge of the above fluorescent images. Bar represents 10 μm

3. Continuously perfuse cells in the platform containing coverslip with heated (37 °C) HBS medium at a rate of about 5–10 mL/min. *See* **Notes 3** and **7**.

4. Found focal plane before searching for a representative microscopic field that should contain ideally at least 5–10 neurons and glial cells for imaging (Fig. 1).

5. Once the microscopic field is selected, insert excitation filters for Fura2/AM imaging (340 and 380 nm) and test fluorescence emission at 520 nm using standard conditions for Ca2+ imaging: camera exposure time of about 100 ms for each wavelength, camera gain (about 50 %), offset (about 10 %) and binning (2×2). These conditions allow having good quality signal to noise ratio. Change settings as required according to the quality of the signal. *See* **Note 8**.

6. Epi-illuminate cells alternately at 340 and 380 nm using band pass filters located on the excitation filter wheel. Light emitted above 520 nm at both excitation lights is filtered by the Fura-2 dichroic mirror, collected every 5–10 s with a 40×, 1.4 NA, oil objective and recorded using the CCD camera.

7. Capture a background image at both excitation wavelengths with the shutter closed. During recordings, perfuse cells either with heated (37 °C) control HBS or HBS containing test substances at a flow of 5 mL/min. This flow ensures that the medium bathing the cells is exchanged about ten times in less than 1 min. *See* **Note** 7.

8. At the end of the recording period, store in the computer for further analysis the complete sequences of images emitted at 520 nm after 340 and 380 nm excitation light.

9. Use the Aquacosmos software, to subtract background images and calculate the pixel by pixel ratio in the resulting images to obtain a sequence of ratio images. Code ratio images in pseudocolor to better appreciate changes in Ca2+ concentrations. Perfusion of hippocampal neurons with agonists such as 100 μM NMDA or high K+ (145 mM) induces rather dramatic increases in these ratios (Fig. 2).

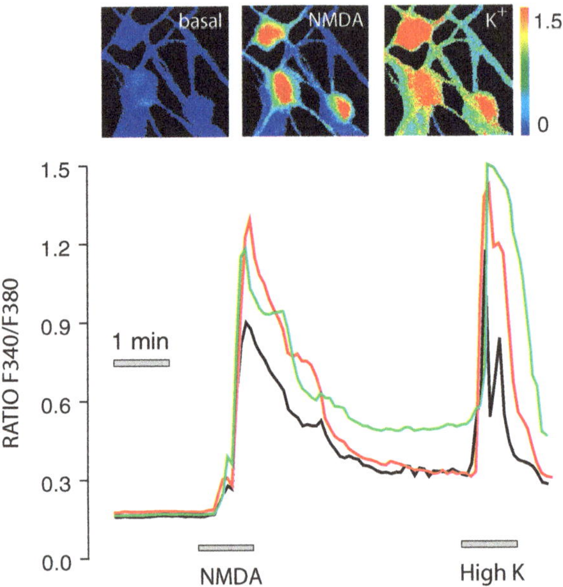

Fig. 2 NMDA and high-K+ medium increased cytosolic [Ca2+] in neurons. Hippocampal cells were cultured, loaded with Fura2/AM and subjected to Fura2 imaging. Pictures show pseudocolor images (Ratio F340/F380) of hippocampal neurons before (basal, *left*) and after stimulation with either 100 μM NMDA (NMDA, middle) or high K+ (K+, *right*). Warmer colors reflect increased cytosolic Ca2+ concentration (pseudocolor scale is shown at *right*). Recordings show cytosolic Ca2+ concentrations (Ratio F340/F380) averaged for a region of interest (ROI) corresponding to each individual neuron and taken every 5 s in three individual cells. Perfusion of HBS medium containing 100 μM NMDA or 145 mM K+ solutions increased cytosolic Ca2+ concentrations in all three neurons

10. For quantitative analysis of individual cells, draw regions of interest (ROIs) on individual cells and average all ratio values corresponding to all pixels within each ROI for each image resulting in a recording of ratio values for individual ROIs (cells).

11. Recorded ratio values can be converted into Ca2+ concentration values using the algorithm developed by Grynkiewicz et al. [10]. *See* **Note 9**.

12. For analysis of differences, express changes in fluorescence ratio as area under curve (AUC). Perform calculation of AUC using Origin Lab 7.0. Curves are defined as the period between which fluorescence ratio significantly exceeded and returned to the basal level following a stimulus.

3.3 Excitotoxicity and Apoptosis Assessment

1. Wash primary rat hippocampal neurons in culture with HBS and then treat them for 1 h with or without NMDA (100 µM) in Mg2+-free, HBS pH 7.42 supplemented with 10 µM glycine. During this time keep cells at 37 °C.

2. After NMDA treatment, wash coverslips containing hippocampal neurons in HBS and then return them to the original Neurobasal® Culture Medium and culture for 24 h in the incubator at 37 °C and 5 % CO_2.

3. Wash cells once with PBS.

4. Test for apoptosis by incubating cells for 10 min with Annexin V diluted 1:20 in Annexin V binding buffer.

5. Assess staining (apoptosis) by fluorescence microscopy using a 40× objective.

6. For quantitative analysis, evaluate the percentage of apoptotic cells by counting the number of Annexin V-positive cells (green fluorescent channel), then dividing by the total number of cells in the field and multiplying by 100. Choose at least four microscopic fields randomly for a total of at least 60 neurons for each coverslip. Score two coverslips per condition in each experiment (Fig. 3).

3.4 Combination of Cytosolic Ca2+ Imaging and Annexin V Staining

This procedure is intended to combine the two methods described above (Subheadings 3.2 and 3.3) concurrently on the very same cells. The method, although time consuming, may allow correlating Ca2+ changes with susceptibility to apoptosis in the same cells.

1. Incubate coverslips containing cultured hippocampal cells in HBS medium containing 4 µM Fura2/AM for 60 min and then subject to calcium imaging as described before (*see* Subheading 3.2).

2. Perfuse hippocampal neurons with apoptosis-inducing agonists such as NMDA and record rises in [Ca2+]cyt in real time.

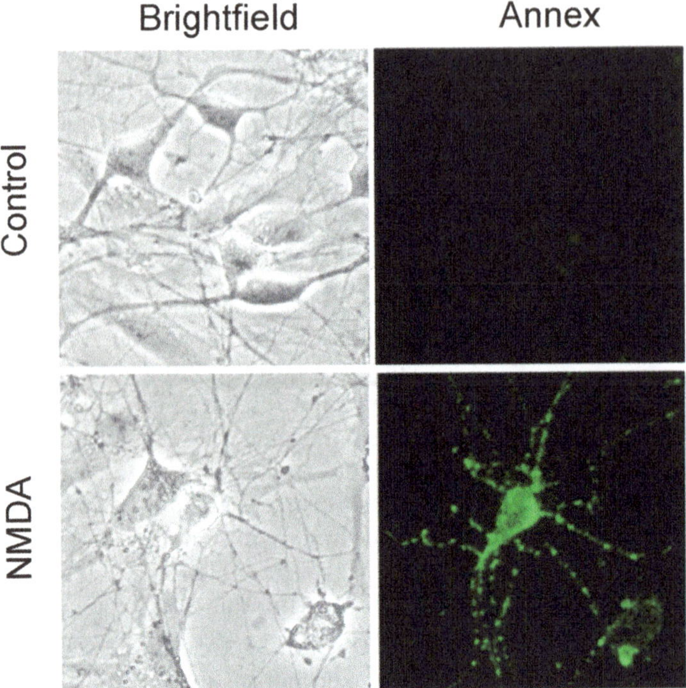

Fig. 3 NMDA induces apoptosis in neurons. Representative microphotographs of cultured hippocampal neurons in control conditions or treated with 100 μM NMDA. Hippocampal neuron cells were treated for 1 h in absence (control) or presence of NMDA (100 μM), and apoptosis was assessed 24 h later by staining with Annexin V. Microphotographs show phase contrast images (*left*) and the green fluorescence staining (Annexin V, *right*) of apoptotic cell bodies. Bars represent 10 μm

Duration of the stimulation depends on the stimulus. For NMDA a 60–70 min period is typically used. During that period, cut off for 10 min excitation light to limit cell damage by UV light (*see* straight line gaps in Fig. 4).

3. At the end of the recording period, store the sequences of images emitted at 520 nm after 340 and 380 nm excitation light in the computer. Analyze them later using the Aquacosmos software.

4. Wash cells with prewarmed HBS at 37 °C during 10 min to remove the stimulus. Then keep in the same medium at room temperature for several hours before assessing apoptosis in the same microscopic field. *See* **Note 10**.

5. With extreme care to avoid moving the microscopic field, incubate cells for 10 min in the platform with Annexin V diluted 1:20 in Annexing V binding buffer 1× (*see* Subheading 3.3). Carry out addition of Annexing V by emptying carefully the chamber and by adding drops of Annexin V solution to the coverslip.

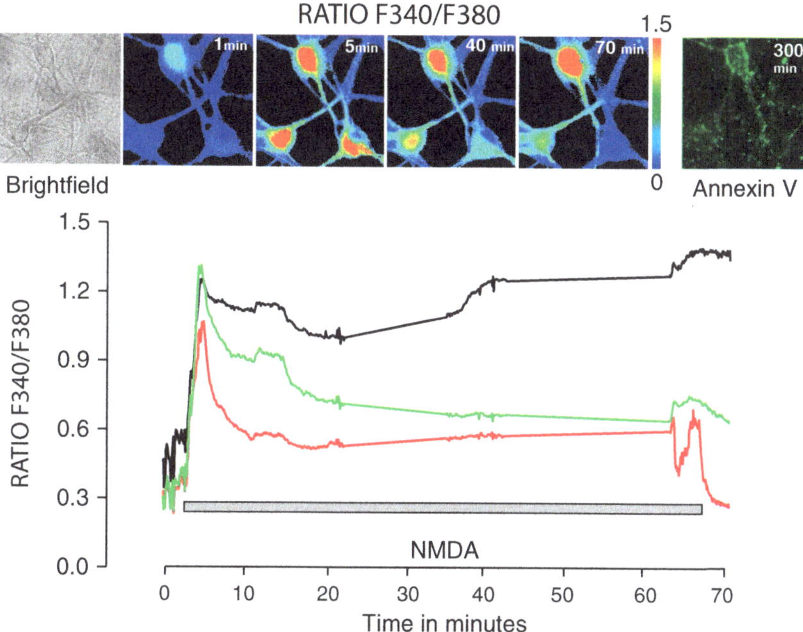

Fig. 4 Concurrent imaging of NMDA-induced cytosolic Ca2+ increases and apoptosis in the same neurons. Pictures show a bright field (*left*), pseudocolor images (Ratio F340/F380) before (basal) and at different times after stimulation with 100 μM NMDA. Annexin V staining of the same cells after several hours of treatment is shown at right. Picture shows the time from the beginning of the stimulus presentation to apoptosis assessment, in this particular case 300 min (5 h). Recordings show cytosolic Ca2+ concentrations (Ratio F340/F380). Perfusing cells with a solution containing 100 μM NMDA for 60 min increased [Ca2+]cyt in all three neurons and induced apoptosis in some of them. During recording, excitation light is cut off for 10 min periods (*straight lines* in records) to limit cell damage by UV light

6. Wash Annexin V with HBS solution and assess staining by fluorescence in the same microscopic field using a 40× objective (Fig. 4).

7. If required, correlate analysis of Ca2+ values with apoptosis induction at a given time in the same ROIs (cells).

4 Notes

1. Animals must be obtained from an approved facility and protocols must be approved by ad hoc institutional body. In our case rats were obtained from the Valladolid University animal facility and handled under protocols approved by the Valladolid University animal housing facility in agreement with the European Convention 123/Council of Europe and Directive 86/609/EEC.

2. The set-up in use is made of a Zeiss Axiovert S100 TV inverted microscope (Carl Zeiss Inc., Gottingen, Germany) equipped with a Zeiss Fluar 40×, 1.3 NA oil objective, a Xenon XBO75 fluorescence excitation lamp or a XCITE illumination system (EXFO, Ontario, Canada), an excitation filter wheel (Sutter Instrument Company, Novato, CA) with band pass filters for Fura2 excitation (340 and 380 nm) and a Fura2 dichroic mirror. Attached in the lateral port of the microscope is a Hamamatsu Orca ER Digital Camera (Hamamatsu Photonics, Hamamatsu, Japan). Camera capturing and filter wheels are handled by Aquacosmos software (Hamamatsu Photonics). A schematic of the imaging setup has been described in detail in a recent chapter [8].

3. The cell perfusion system is mounted in a PH-3 thermostatic platform for open 12 mm glass coverslips using an 8-lines gravity-driven perfusion system equipped with pinch valves (VC-8 valve controller) and solutions heated using a SH-27B inline heating system. All the above components are from Warner Instruments, Hamden, CT.

4. Hippocampal neurons require several days in vitro for establishment of neural connections and proper responsiveness to glutamate receptor agonists. Accordingly, to test for excitotoxicity it is required to culture hippocampal cells for at least 2 weeks (>14 DIV). Neuron cells cultured for less than 2 weeks may show resistance to apoptosis [11].

5. An incubator set to hold a 5 % of CO_2 is to be used when Neurobasal® Culture Medium is employed because it contains 24 mM $NaHCO_3$. In case that other culture mediums containing 44 mM (instead 24 mM) $NaHCO_3$ are used (as DMEM for instance), the incubator should be set to 10 % CO_2 to keep pH in physiological conditions (pH 7.4).

6. In order to identify glia and neurons in the culture, double immunofluorescence can be used. To accomplish this treat cells in the glass coverslip with 4 % paraformaldehyde (PFA) prepared in a phosphate buffer solution (PBS) for 20 min. Wash the coverslips three times with PBS. Treat them with 0.1 % Triton X-100 solution for 10 min and then wash again coverslips three times with PBS. Incubate next the cells in PBS containing 20 % goat serum for 20 min to suppress non-specific binding of antibody. Then, incubate coverslips with primary antibodies prepared in PBS containing 10 % goat serum at 1:300 dilution for 1 h. To identify neurons and glia, mouse anti-β-tubulin III and rabbit anti-glial fibrillary protein (GFAP) can be used. After 1 h incubation, wash coverslips three times with PBS before incubating cells with anti-mouse IgG FITC labeled secondary antibody (dark room), which

emits green fluorescence, and anti-rabbit IgG Alexa Fluor 594 labeled secondary antibody, which emits red fluorescence. These antibodies had also been prepared in PBS containing 10 % goat serum at a 1:300 dilution. After 45 min incubation, wash coverslips three times with PBS and incubate fixed cells for 5 min with DAPI, prepared 1:5,000 in PBS containing 10 % goat serum. Finally, place coverslips on a slide with mounting medium (50 % glycerol in PBS) and capture images in a fluorescence microscope. It is possible to carry out this procedure in the same coverslip used for Ca^{2+} imaging. In that case, do not remove coverslip from the recording chamber and add and remove solutions carefully in order to avoid losing the microscopic field. In this case, it is possible to use the same microscope for calcium imaging and immunofluorescence in exactly the same single cells.

7. Perfusion is helpful during imaging not only to quickly and easily add and remove test solutions to and from the cell chamber, but also to keep physiological conditions (37 °C or even CO_2 if required) using the in-line heating system. This is particularly true for long-term experiments where water evaporation concentrates solutes modifying physiological conditions. In the case one needs to test very expensive compounds such as, for instance, Aβ species, perfusion may be cost limiting. In those cases were perfusion is not available or convenient, solutions can be provided by carefully adding a drop of 2× test-containing solution, to a half-filled chamber and making a quick mix of solutions with the tip of the pipette. In this case, it is best to keep the heating system off and make the experiment at room temperature to avoid changes in the saline composition due to evaporation.

8. Fluorescence settings have to be adjusted in each imaging setup depending on multiple factors including excitation lamp intensity in the UV range (340 and 380 nm), quality of the lens, dichroic mirror, and quality of the camera. Usually Fura2/AM-loaded cells do not emit much light when excited at 340 nm. In those cases it may be possible to use different setting for each wavelength. In general, glass coverslips are required to have good light transmission in the UV. Avoid plastics that absorb UV light. If not enough signal is achieved, it is possible to increase gain, decrease offset, increase the binning to 4×4 pixels and also the Fura2/AM loading time.

9. Binding of Ca2+ to Fura2 changes Fura2 fluorescence according to the low of mass action [10]:

$$\left[Ca^{2+}\right] = K_d \left(F_{max} - F\right) / \left(F - F_{min}\right);$$

where K_d, is the dissociation constant of the Fura2-Ca2+ complex (224 nM at 37 °C); F is the fluorescence emission for each [Ca2+]; F_{max}, is the fluorescence emission when Fura2 is saturated with Ca2+ and Fmin is fluorescence emission when Fura2 is free of Ca2+. If we apply the above algorithm to both wavelengths and do the ratio, then we obtain the following algorithm:

$$\left[Ca^{2+} \right] = K_d \beta \left(R_{max} - R \right) / \left(R - R_{min} \right);$$

where R is the ratio of fluorescence recordings obtained after exciting at 340 and 380 nm for a given [Ca2+]; R_{max}, is the same ratio when Fura2 is saturated with Ca2+ and R_{min} is the same ratio when Fura2 is free of Ca2+. Finally, β is the ratio of F_{max} / F_{min} at 380 nm. This algorithm allows estimation of [Ca2+] knowing the R values at any point in time. R_{max}, R_{min} and β values can be determined experimentally using Fura2/AM solutions in the presence of saturating concentrations of Ca2+ (HBS containing 1 mM Ca2+) and in the absence of Ca2+ free medium (HBS without added Ca2+ and containing EGTA 5 mM).

10. Usually, this kind of cell death assays are performed several hours after treatment because the stimulus induce a delayed cell death [6], and it involves events during or shortly after the acute phase of stimulation that would detonate the death sequence. During this time cells are kept in HEPES-containing buffer at room temperature to limit water evaporation and changes in saline. Alternatively, the whole experiment could be carried out in an incubator attached on the microscope's platform and carried out in $NaHCO_3$ buffered medium and at 37 °C. There are commercially available incubators for inverted microscopes from Carl Zeiss Inc., Gottingen, Germany.

Acknowledgments

This work was supported by Ministerio de Economía y Competitividad (Spain) grants BFU2009-08967 and BFU2012-37146 and Regional Government Junta de Castilla y León (Spain) grants BIO103/VA45/11, VA145U13, and BIO/VA33/13. María Calvo was supported by a predoctoral fellowship from Regional Government Junta de Castilla y León (Spain) and the European Social Fund.

References

1. Berridge MJ (2012) Calcium signalling remodelling and disease. Biochem Soc Trans 40:297–309
2. Sattler R, Tymianski M (2001) Molecular mechanisms of glutamate receptor-mediated excitotoxic neuronal cell death. Mol Neurobiol 24:107–129
3. Liu Y, Wong TP, Aarts M et al (2007) NMDA receptor subunits have differential roles in mediating excitotoxic neuronal death both in vitro and in vivo. J Neurosci 27:2846–2857
4. Sanz-Blasco S, Valero RA, Rodríguez-Crespo I et al (2008) Mitochondrial Ca2+ overload underlies Aβ oligomers neurotoxicity providing an unexpected mechanism of neuroprotection by NSAIDs. PLoS One 3:e2718
5. Brustovetsky T, Bolshakov A, Brustovetsky N (2010) Calpain activation and Na+/Ca2+ exchanger degradation occur downstream of calcium deregulation in hippocampal neurons exposed to excitotoxic glutamate. J Neurosci Res 88:1317–1328
6. Pivovarova NB, Nguyen HV, Winters CA et al (2004) Excitotoxic calcium overload in a subpopulation of mitochondria triggers delayed death in hippocampal neurons. J Neurosci 24:5611–5622
7. Jordán J, Ceña V, Prehn JH (2003) Mitochondrial control of neuron death and its role in neurodegenerative disorders. J Physiol Biochem 59:129–141
8. Villalobos C, Caballero E, Sanz-Blasco S et al (2012) Study of neurotoxic intracellular calcium signalling triggered by amyloids. Meth Mol Biol 849:289–302
9. Pérez-Otaño I, Luján R, Tavalin SJ et al (2006) Endocytosis and synaptic removal of NR3A-containing NMDA receptors by PACSIN1/syndapin1. Nat Neurosci 9:611–621
10. Grynkiewicz G, Poenie M, Tsien RY (1985) A new generation of Ca2+ indicators with greatly improved fluorescence properties. J Biol Chem 260:3440–3450
11. Zhou M, Baudry M (2006) Developmental changes in NMDA neurotoxicity reflect developmental changes in subunit composition of NMDA receptors. J Neurosci 26:2956–2963

Chapter 7

Monitoring Mitochondrial Membranes Permeability in Live Neurons and Mitochondrial Swelling Through Electron Microscopy Analysis

Macarena S. Arrázola and Nibaldo C. Inestrosa

Abstract

Maintenance of mitochondrial membrane integrity is essential for mitochondrial function and neuronal viability. Apoptotic stimulus or calcium overload leads to mitochondrial permeability transition pore (mPTP) opening and induces mitochondrial *swelling*, a common feature of mitochondrial membrane permeabilization. The first phenomenon can be evaluated in cells loaded with the dye calcein-AM quenched by cobalt, and mitochondrial *swelling* can be detected by electron microscopy through the analysis of mitochondrial membrane integrity. Here, we describe a live cell imaging assay to detect mitochondrial permeability transition and the development of a detailed analysis of morphological and ultrastructural changes that mitochondria undergo during this process.

Key words Mitochondrial permeability transition, Mitochondrial swelling, Live cell imaging, Mitochondrial membrane permeabilization, Electron microscopy, Mitochondrial calcein, Cobalt, Mitochondrial ultrastructure

1 Introduction

Mitochondria are membrane-enclosed organelles present in all cellular types, including neurons, where they participate in several vital processes, such as cellular bioenergetics, metabolism, and calcium homeostasis [1, 2]. Calcium regulation is essential for the synaptic function of a neuron [3] and therefore for its viability [4]. Calcium ions influx into neurons is regulated by different cellular compartments, including the endoplasmic reticulum and mitochondria [5]. Calcium overload into mitochondria has been described as an inductor of the mitochondrial permeability transition [6], which is a phenomenon characterized by the formation of large conductance permeability transition pores (mPTP) that make the mitochondrial membrane abruptly permeable to solutes up to 1,500 Da, resulting in the loss of mitochondrial membrane potential [7]. The permeabilization of

Laura Lossi and Adalberto Merighi (eds.), *Neuronal Cell Death: Methods and Protocols*, Methods in Molecular Biology, vol. 1254, DOI 10.1007/978-1-4939-2152-2_7, © Springer Science+Business Media New York 2015

the mitochondrial inner membrane induces morphological changes in this organelle, including increased volume, a phenomenon known as swelling, membrane disruption, cristae disorganization, and the release of pro-apoptotic factors to the cytoplasm, to finally activate neuronal death cascades [8, 9]. For this reason, the maintenance of the structural integrity and function of mitochondria is crucial for their proper function and, therefore, for neuronal viability.

Mitochondrial swelling can be detected in vitro with a spectrophotometric assay from isolated mitochondria, as previously described [10]; however, the amount of material required to obtain high quality purification is a limiting step when the experimental model in use is a primary culture of neurons. For this reason, we describe here a direct live cell imaging technique to monitor the mPTP opening, and the development of a detailed data analysis. This assay was developed by Petronilli et al. in 1998 [11] and, later, was validated in a neuronal model [12]. The assay is based on the loading of living neurons with a calcein-AM/cobalt mix to quench the cytosolic signal of calcein, but leaving intact the mitochondrial stain. The exposure of neurons to mPTP inductors, such as the ionophore ionomycin [13], or to calcium overload results in the decay of the mitochondrial calcein fluorescence, indicating mitochondrial permeability transition. Mitochondrial swelling can be also detected through the study of certain ultrastructural parameters by transmission electron microscopy [14]. We describe here a detailed analysis to evaluate morphological changes in mitochondria exposed to permeability transition.

2   Materials

2.1 *Mitochondrial Permeability Transition by Live Cell Imaging*

1. Glass coverslips of 25 mm diameter (Marienfeld GmbH & Co. KG, Lauda-Königshofen, Germany).

2. 1× Neurobasal Medium (Gibco®, Life Technologies™, Carlsbad, CA).

3. B27 supplement (Gibco®, Life Technologies™).

4. DMSO (Sigma Chemicals, St. Louis, MO).

5. Stock probes: 1 mM Calcein-AM, 10 µM MitoTracker Orange CMTMRos, and 1 mM Hoechst 33342 (all from Molecular Probes®, Life Technologies™, Carlsbad, CA) in DMSO. Store at −20 °C protected from light.

6. 1 mM Cobalt (II) chloride hexahydrate ($CoCl_2 \times 6H_2O$) stock solution: To prepare 5 mL dissolve 1.1897 g in 5 mL of sterile distilled water. Store at −20 °C protected from light.

7. Tyrode buffer: 135 mM NaCl, 5 mM KCl, 1.8 mM $CaCl_2$, 1 mM $MgCl_2$, 10 mM HEPES, 5.6 mM glucose, pH to 7.3

with concentrated HCl. To prepare 0.5 L weigh 3.944 g NaCl, 0.186 g KCl, 0.132 g $CaCl_2$, 1.19 g HEPES, and 0.280 g glucose and transfer to a 0.5 L glass beaker containing 300 mL distilled water. Mix and then add 500 µL 1 M $MgCl_2$. Adjust pH with HCl and make up to 0.5 L. Make aliquots of 50 mL in Falcon tubes or glass bottles and store at −20 °C. Defrost 1 aliquot of Tyrode buffer for the assay, use it along the day and then discard. For calcium free experiments do not add $CaCl_2$ to the preparation. *See* **Note 1**.

8. Confocal microscope with appropriate laser sources (*see* below).

2.2 Mitochondrial Swelling Detection from Electron Microscopy Analysis

1. Cold artificial cerebrospinal fluid (ACSF): 124 mM NaCl, 2.69 mM KCl, 1.25 mM KH_2PO_4, 1.3 mM $MgSO_4$, 2.6 mM $NaHCO_3$, 10 mM glucose, 2.5 mM $CaCl_2$. To prepare a 10× stock solution of ACSF, weigh 72.466 g NaCl, 2.006 g KCl, 1.701 g KH_2PO_4, 1.565 g $MgSO_4$ or 3.203 g $MgSO_4 \times 7H_2O$, 21.843 g $NaHCO_3$, and 19.817 g glucose. Mix and make up to 1 L. To prepare 1× ACSF solution, dilute 100 mL of the 10× stock solution in 900 mL of distilled water. Oxygenate the solution by bubbling with an O_2/CO_2 mix for 10 min and adjust pH to 7.4 with NaOH/HCl 1 N. Add 0.368 g $CaCl_2 \times 2H_2O$ and make up to 1 L. Check again the pH and if it is necessary readjust to 7.4.

2. Cell strainer 70 µm nylon (BD Biosciences, Bedford, MA).

3. 6-well plates for cell culture.

4. Silicone tubes.

5. 0.1 M cacodylate buffer, pH 7.2 with HCl: Dissolve 2.14 g of sodium cacodylate in 80 mL of distilled water. Adjust pH to 7.2 and then make up to 100 mL with distilled water.

6. Fixative solution: 2.5 % glutaraldehyde (e.g., Sigma Chemicals): Dilute 1 mL of 25 % glutaraldehyde stock solution into 9 mL of 0.1 M cacodylate buffer, pH 7.2.

7. 1 % osmium tetroxide (OsO_4): Dilute 25 mL of a 4 % osmium tetroxide stock solution (Sigma Chemicals) with 75 mL of distilled water. Store at 4 °C.

8. 1 % uranyl acetate [$UO_2 (OCOCH_3)_2 \times 2H_2O$]: Weigh 10 mg uranyl acetate and dissolve in 1 mL distilled water.

9. Epon resin (e.g., EMbed-812, Electron Microscopy Sciences, Hatfield, PA).

10. 4 % uranyl acetate [$UO_2 (OCOCH_3)_2 \times 2H_2O$]: Weigh 0.4 g of uranyl acetate and dissolve in 10 mL pure methanol.

11. Lead citrate staining solution: mix 1.33 g $Pb(NO_3)_2$, 1.76 g $Na_3(C_6H_5O_7) \times 2H_2O$ and 30 mL distilled water in a 50 mL volumetric flask. Shake vigorously for 1 min and allow standing with intermittent shaking. After 30 min add 8.0 mL 1 N

NaOH and make up to 50 mL with distilled water. Mix until lead citrate dissolves. The final pH is 12.0.

12. Cooper grids for electron microscopy (e.g., Ted Pella Inc., Redding, CA).

13. Biosafety cabinet.

14. CO_2 incubator.

15. Vibrating microtome.

16. Ultramicrotome.

17. Transmission electron microscope.

---

3   Methods

For live cell imaging, all procedures are carried out at 37 °C and under sterilized conditions unless otherwise specified.

3.1   Mitochondrial Permeability Transition by Live Cell Imaging

1. Seed neurons on glass coverslips of 25 mm diameter in a 35 mm dish or in the available format (*see* **Note 2**). Neurons are maintained in Neurobasal medium supplemented with B27 at 37 °C in an incubator with 5 % CO_2.

2. For live cell imaging, load the cells with the fluorescent probes (*see* **Note 3**). In this step work with the cells in a biosafety cabinet protected from light. Before removing the cells from the incubator, prepare the loading mix: add 1 μL of 1 mM calcein-AM stock solution, 1 μL of 1 M $CoCl_2$, and 5 μL of 10 μM MitoTracker Orange in 993 μL of Neurobasal medium to make 1 mL. Final mix concentration is 1 μM calcein-AM, 1 mM $CoCl_2$, 50 nM MitoTracker Orange (*see* **Note 4**). Hoechst dye can also be added to the loading mix to evaluate neuronal viability if necessary. *See* **Note 5**.

3. Wash cells twice with 1 mL Neurobasal medium without supplements (*see* **Note 6**). For 25 mm covers add 1 mL of loading mix to the dish and incubate neurons at 37 °C for 30 min, protected from light.

4. Remove the loaded neurons from the incubator and wash twice with 1 mL of Neurobasal medium, working in a biosafety cabinet. Change medium to Tyrode buffer and incubate cells for 10 min at 37 °C to equilibrate probes.

5. Transfer the cover to the microscope room and assemble the system as shown in Fig. 1a, b. Use little amounts of solid paraffin between both sides of the cover to fix it at the microscope system. *See* **Note 7**.

6. Set microscope parameters to start the mPTP live cell assay. For a spinning disk confocal microscope, use Texas Red filter to detect MitoTracker fluorescence, FITC for calcein-AM, and

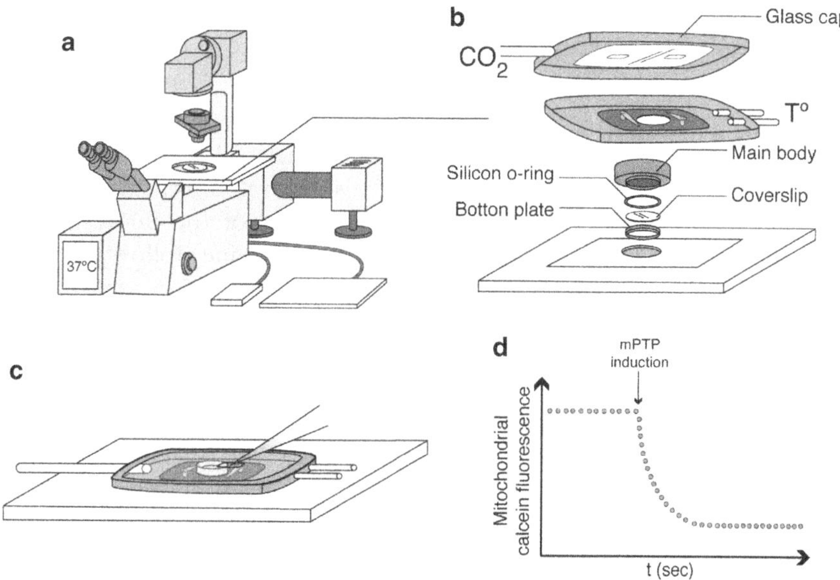

Fig. 1 Sample assembly for live cell imaging assay. (**a**) Microscope system equipped for live cell assays. (**b**) Neurons seeded on a glass coverslip mounted in the microscope chamber with CO_2 and temperature regulation. The glass cap is used to maintain the gas pressure and the temperature at 37 °C. (**c**) Closed chamber ready to use and for the stimulation of neurons. (**d**) Graphical representation of a common record obtained along the experiment after mPTP induction, i.e., with ionomycin. Mitochondrial calcein fluorescence decay is usually observed under mitochondrial permeability transition conditions

UV filter for Hoechst, or set the adequate excitation/emission wavelengths depending on the microscope used (*see* **Note 8**). Water immersion objectives are recommended for live cell imaging assays. Set temperature to 37 °C and the gas pressure to ensure a constant flux of 5 % CO_2 in the culture chamber of the microscope.

7. Set the time of the experiment and the interval between digital images. It is recommened to measure fluorescence changes for 15 min at least, with an image acquisition time of 10–15 s between frames. Measure a baseline for 3 min and then add the stimulus (Fig. 1c) (*see* **Note 9**). Finally, check the focus and start the experiment. It is recommended to perform positive and negative controls of the mitochondrial permeability transition pore opening to check the efficiency of the assay. *See* **Note 10**.

8. For image analysis use the NIH ImageJ software and install the "Delta F" plugin. Follow the sequential steps to analyze fluorescence changes on the images: Open the image with Image J > Image > Stacks > Z project, type the number of images corresponding to the basal line, i.e., a 3-min baseline corresponds to 180 s, if images were taken every 15 s, and then the

first 12 images correspond to the baseline; so type 1 in the start slice box and 12 in the stop slice > OK. This step creates an average image of the baseline.

9. Run "Delta F" plugin to determine fluorescence changes from the image stacks: Plugins > Stacks>T-functions > Delta F. A new image called "Delta F" is generated from this step. To obtain the value of $\Delta F/Fo$ (variation of the fluorescence between each image compared to the baseline) follow the next steps on Image J: Process > Image calculator, on Image 1 select the Delta F image created in the previous step; operation: Divide; and Image 2 is the average image created first > Yes. A new image is created called "Result of Delta F". From this image select a neurite with the polygon selection tool and perform the analysis: Images > Stacks > Plot z-axis Profile. A "Results" window is displayed with the mean fluorescence values corresponding to each time point. Plot the data to obtain fluorescence changes along the live cell imaging assay, as shown Fig. 1d. Repeat this step with each neurite selected until complete around 8–10 neurites per neuron.

10. Fluorescence analysis can be done with the mitochondrial calcein images to detect mPTP opening and also with those obtained from the MitoTracker loading, to detect mitochondrial membrane potential changes in response to the same stimulus (Fig. 2).

3.2 Mitochondrial Swelling Detection from Electron Microscopy Analysis

1. Prepare brain slices from rat or mouse brain [15, 16]. Cut transverse slices of 400 μm under cold ACSF using a vibrating microtome.

2. Incubate the slices in oxygenated ACSF for more than 1 h at room temperature before treatment.

3. For slices treatment, mount the slices in a 6-well plate modified as shown in Fig. 3, to keep each dish oxygenated (*see* **Note 11**). Dilute all treatments in ACSF in an appropriate volume to maintain slices immersed and avoid its contact with the air. This step is carried out at room temperature.

4. After treatment, transfer the slices one by one, with a soft brush or tweezers, to an Eppendorf tube containing 1 mL of the fixative and fix for 6 h. *See* **Note 12**.

5. After fixation, wash the slices with 1 mL of 0.1 M cacodylate buffer, pH 7.2 for 18 h at 4 °C.

6. Perform a secondary fixation with 1 % osmium tetroxide for 90 min and then wash with three washes, 5 min each in distilled water.

7. Stain samples with 1 % aqueous uranyl acetate for 60 min.

8. Dehydrate samples in acetone following these sequential steps: 1×50 %, 1×70 %, 2×95 %, and 3×100 % acetone, 20 min each.

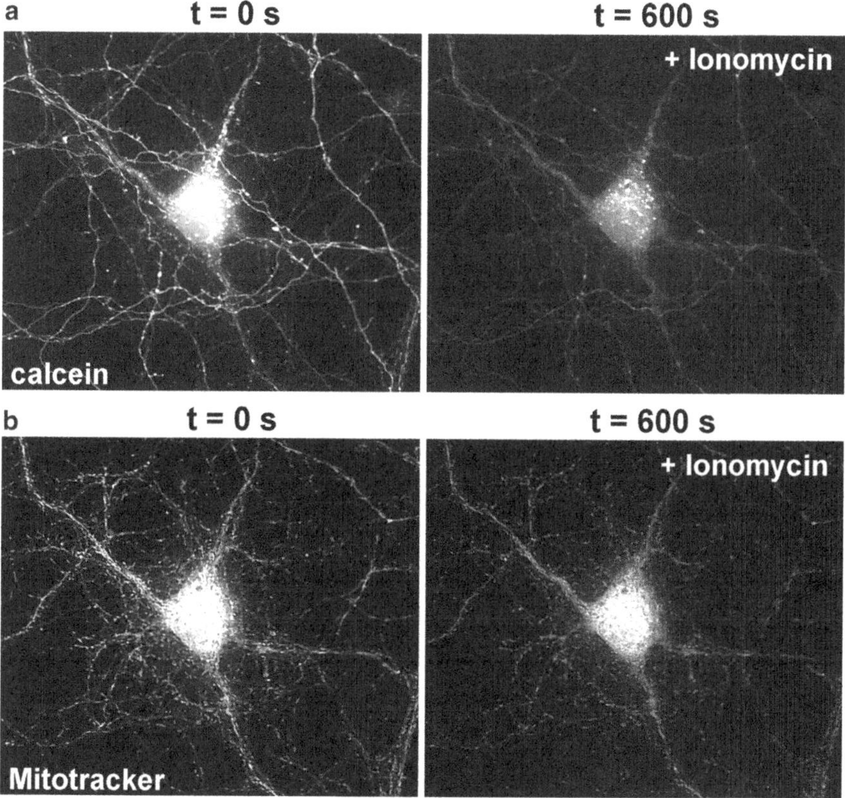

Fig. 2 Images obtained from the mPTP live cell imaging assay. Neurons were loaded with the loading Mix, containing calcein-AM, cobalt, and MitoTracker Orange. (**a**) Images show mitochondrial calcein fluorescence dissipation on neurons exposed to 0.5 μM ionomycin which indicates mitochondrial permeability transition. Images show fluorescence changes at the beginning ($t=0$) and at the end ($t=600$) of the experiment. (**b**) Images show the mitochondrial membrane potential loss in the same neuron shown in (**a**) under ionomycin stimulus

9. Place samples in 1:1 Epon:acetone overnight and embed them in 100 % Epon resin into a mold. Place mold containing samples in 60 °C oven for 24 h to polymerize.

10. Cut ultrathin sections (60–70 nm) with an ultramicrotome and collect them on 300-mesh copper electron microscopy grids (Fig. 3). Stain with 4 % uranyl acetate for 2 min and lead citrate for 5 min [17].

11. Examine the samples in a transmission electron microscope. For morphological analysis of mitochondria is recommended to capture digital images with a 16.500–20.000× magnification.

12. For mitochondrial ultrastructural analysis take 40–50 digital images per sample and evaluate membrane integrity of each mitochondrion by manually detecting structural abnormalities. Mitochondria with an overall intact structure are considered healthy, but loss of membrane continuity is considered an index of disruption. The same approach is applicable to cristae integrity analysis (Fig. 4).

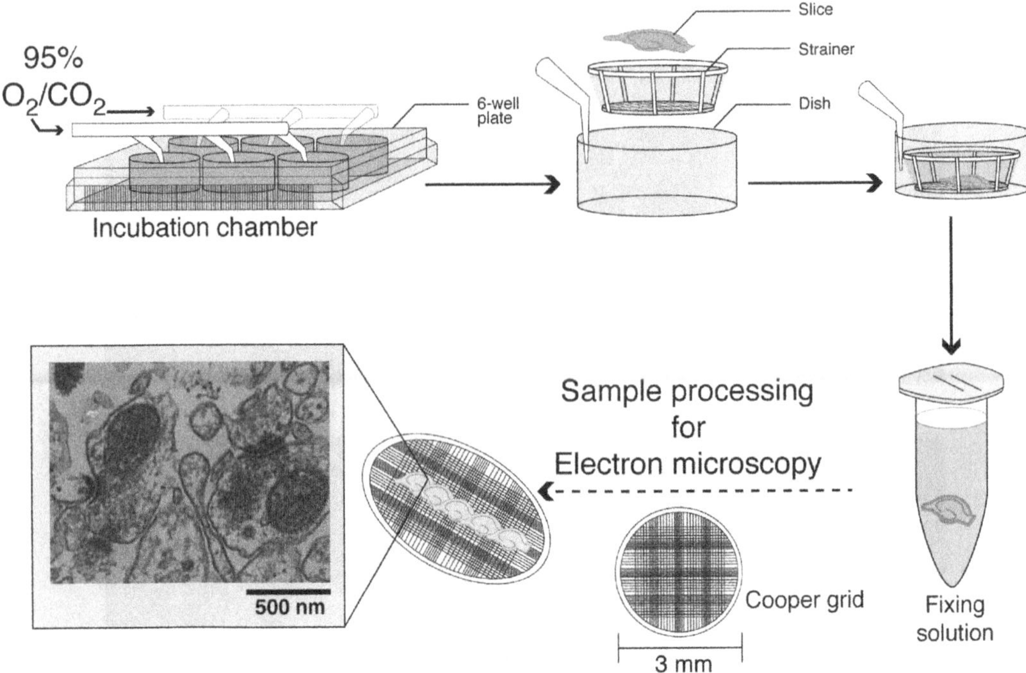

Fig. 3 Slice preparation and treatment for electron microscopy. Incubation chamber is used to treat slices separately. Each well contains a cell strainer to keep the slices protected from bubbles generated by the oxygenation system. After the incubation time, the slices are immersed in fixing solution, processed for electron microscopy and sequentially collected on copper grids for electron microscopy analysis. A representative image obtained with the microscope is shown at the end of the flowchart

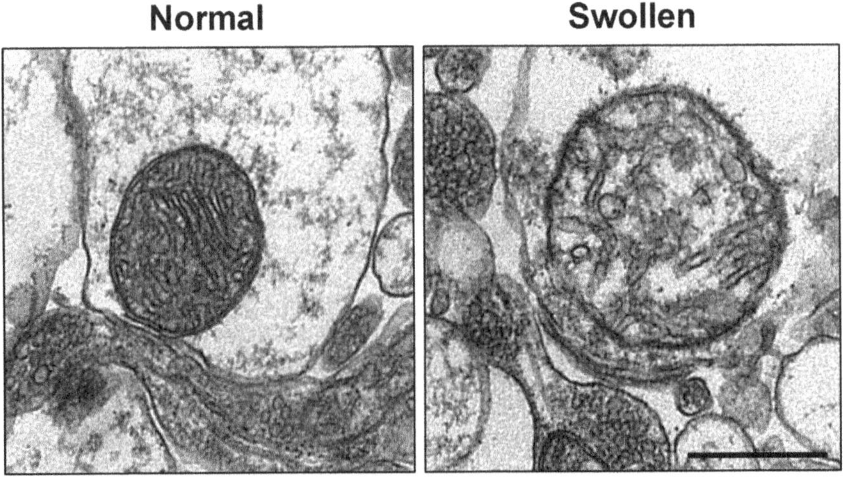

Fig. 4 Mitochondrial swelling detection by electron microscopy. Images show normal mitochondria with intact membrane and cristae (*left picture*) and a swollen mitochondrion (*right picture*) which is larger than normal and also displays membrane disruption and cristae disorganization, the common features of mitochondrial swelling by permeability transition. Bar = 500 nm

13. To evaluate mitochondrial swelling, measure morphological parameters, such as area, diameter, and perimeter [18] with ImageJ software. Set scale parameters in ImageJ: Analyze > Set scale > introduce distance in pixels and unit of length. *See* **Note 13**.

14. Set the parameters to be measured in ImageJ: Analyze > Set measurements > click in, Area, Perimeter and Feret's diameter.

15. Measure mitochondrial morphology with the polygon selection tool of ImageJ by surrounding each mitochondrion and press Ctrl + M on the keyboard to display the Results window. Plot the results in a column graph and as a scatter plot graph to analyze the whole population of mitochondria. *See* **Note 14**.

4   Notes

1. To maintain the osmolarity of the Tyrode buffer under calcium free conditions, add 3.997 g NaCl to the preparation instead 3.944 g, to change the concentration from 135 to 136.8 mM NaCl.

2. Glass coverslips of other diameters may be used depending on the chamber that is going to be mounted onto the microscope stage for the live cell assay. Some microscopes have chamber adaptors for 12, 25, or 40 mm diameter coverslip. If the microscope does not have chamber adaptors for different coverslip size, glass bottom dishes can also be used to seed cells.

3. Try to load just one cover per assay to avoid fluorescence decay during the time waiting between each experiment.

4. Other MitoTracker stains could be used to evaluate the mitochondrial membrane potential during the mPTP live cell assay, as MitoTracker Red CMXRos, but higher concentration of this probe is needed (between 200 and 1 μM). It is also important to check the bleaching of the probe depending on the microscope available.

5. Add 1 μL of 1 mM Hoechst 33342 dye to 1 mL of loading mix. It is recommended to measure Hoechst stain only at the beginning and at the end of the experiment and not along it to avoid probes bleaching.

6. If the cells were previously treated before the live imaging experiment, it is necessary to wash them to eliminate the stimulus. Also it is important to wash the cells to deplete them from B27, since it could interfere with the stimulus that is going to be used for the live cell assay.

7. Put the paraffin only at the edges of the cover to maintain an adequate cellular field to observe at the microscope. Some systems are magnetically sealed and do not need paraffin.

8. Hoechst λex/em = 350/461 nm. The ex/em peaks of calcein after hydrolysis are 494/517 nm. MitoTracker Orange λex/em = 554/576 nm.

9. For live cell imaging, the stimulus can be added manually with a micropipette or through a peristaltic pump.

10. Use 0.5 µM of the ionophore ionomycin, as a positive control of mitochondrial permeability transition. A loss of mitochondrial calcein is appreciated when this stimulus is added. As a negative control, incubate neurons for 30 min at 37 °C with 20 µM cyclosporin A (CsA) to inhibit mPTP opening before the stimulus. A control of probe bleaching is also needed. To evaluate this issue, measure the fluorescence of the probe for the duration of the experiment by simply adding vehicle.

11. An oxygenated dish can be manually made by introducing silicone tubes adapted to a white pipette tip. Fix the tip to the inside wall of the dish until the middle of the well and connect the tube to the oxygen. Be sure to keep a constant oxygen flux along the experiment, avoiding bubbles directly on the slices. Mount the slices into a cell strainer (usually used for organotypic cultures) inside each dish. Under these conditions the slices can be maintained between 6 and 8 h without affecting cell viability (*see* Fig. 3).

12. Prepare a fresh fixing solution every time you need it and only the amount that is necessary for each experiment. Do not freeze.

13. To determine distance in pixels, measure the length of the scale bar using the straight line selections tool of ImageJ and then open the set scale window and introduce the following parameters: *distance in pixels* is automatically determined by the software when the scale bar is measured with the straight line tool; the *known distance* corresponds to the length of the scale bar measured from the image; and the *unit length* also depends on the scale bar unit usually in µm. Click on "Global", and the final scale should be in pixel/µm.

14. The scatter plot graphs allow a deep analysis of the data, because they can be divided by percentiles to determine which population is affected by a determined treatment, as previously described (*see* [18]).

Acknowledgments

This work was supported by grants from the Basal Center of Excellence in Aging and Regeneration (CONICYT—PFB 12/2007) and FONDECYT N° 1120156 to NCI, and a predoctoral fellowship from CONICYT to MSA.

References

1. Kann O, Schuchmann S, Buchheim K et al (2003) Coupling of neuronal activity and mitochondrial metabolism as revealed by NAD(P)H fluorescence signals in organotypic hippocampal slice cultures of the rat. Neuroscience 119:87–100

2. Celsi F, Pizzo P, Brini M et al (2009) Mitochondria, calcium and cell death: a deadly triad in neurodegeneration. Biochim Biophys Acta 1787:335–344

3. Vos M, Lauwers E, Verstreken P (2010) Synaptic mitochondria in synaptic transmission and organization of vesicle pools in health and disease. Front Synaptic Neurosci 2:139

4. Newmeyer DD, Ferguson-Miller S (2003) Mitochondria: releasing power for life and unleashing the machineries of death. Cell 112: 481–490

5. Gunter TE, Gunter KK (2001) Uptake of calcium by mitochondria: transport and possible function. IUBMB Life 52:197–204

6. Lemasters JJ, Theruvath TP, Zhong Z et al (2009) Mitochondrial calcium and the permeability transition in cell death. Biochim Biophys Acta 1787:1395–1401

7. Rao VK, Carlson EA, Yan SS (2013) Mitochondrial permeability transition pore is a potential drug target for neurodegeneration. Biochim Biophys Acta 1842:1267–1272. doi:10.1016/j.bbadis.2013.09.003

8. Galluzzi L, Blomgren K, Kroemer G (2009) Mitochondrial membrane permeabilization in neuronal injury. Nat Rev Neurosci 10:481–494

9. Petronilli V, Penzo D, Scorrano L et al (2001) The mitochondrial permeability transition, release of cytochrome c and cell death. Correlation with the duration of pore openings in situ. J Biol Chem 276:12030–12034

10. Schinzel AC, Takeuchi O, Huang Z et al (2005) Cyclophilin D is a component of mitochondrial permeability transition and mediates neuronal cell death after focal cerebral ischemia. Proc Natl Acad Sci U S A 102:12005–12010

11. Petronilli V, Miotto G, Canton M et al (1998) Imaging the mitochondrial permeability transition pore in intact cells. Biofactors 8: 263–272

12. Gillessen T, Grasshoff C, Szinicz L (2002) Mitochondrial permeability transition can be directly monitored in living neurons. Biomed Pharmacother 56:186–193

13. Abramov AY, Duchen MR (2003) Actions of ionomycin, 4-BrA23187 and a novel electrogenic Ca2+ ionophore on mitochondria in intact cells. Cell Calcium 33:101–112

14. Sun MG, Williams J, Munoz-Pinedo C et al (2007) Correlated three-dimensional light and electron microscopy reveals transformation of mitochondria during apoptosis. Nat Cell Biol 9:1057–1065

15. Cerpa W, Godoy JA, Alfaro I et al (2008) Wnt-7a modulates the synaptic vesicle cycle and synaptic transmission in hippocampal neurons. J Biol Chem 283:5918–5927

16. Varela-Nallar L, Alfaro IE, Serrano FG et al (2010) Wingless-type family member 5A (Wnt-5a) stimulates synaptic differentiation and function of glutamatergic synapses. Proc Natl Acad Sci U S A 107:21164–21169

17. Reynolds ES (1963) The use of lead citrate at high pH as an electron-opaque stain in electron microscopy. J Cell Biol 17:208–212

18. Song DD, Shults CW, Sisk A et al (2004) Enhanced substantia nigra mitochondrial pathology in human alpha-synuclein transgenic mice after treatment with MPTP. Exp Neurol 186:158–172

Real-Time Visualization of Caspase-3 Activation by Fluorescence Resonance Energy Transfer (FRET)

Silvia Alasia, Carolina Cocito, Adalberto Merighi, and Laura Lossi

Abstract

As apoptosis occurs via a complex signaling cascade that is tightly regulated at multiple cell points, different methods exist to evaluate the activity of the proteins involved in the intracellular apoptotic pathways and the phenotype of apoptotic neurons. Detention of the activity of the enzyme caspase-3, the key executioner caspase in programmed cell death, by laser scanning confocal fluorescence microscopy and the fluorescence resonance energy transfer technology is an alternative approach to classical standard techniques, such as Western blotting, activity assays, or histological techniques, and allows working with both fixed and living cells. This technique combined with the organotypic culture approach ex vivo represents a valid tool for the study of the mechanisms of neuronal survival/death and neuroprotection.

Key words Caspase-3, Programmed cell death, Apoptosis, FRET, Organotypic cultures, Biolistic transfection

1   Introduction

Apoptosis is a form of programmed cell death (PCD) that is often considered a sort of "cell suicide" as it is activated following a gene-regulated stereotyped cascade of events. Most often apoptosis is found during normal development and tissue turnover has been recognized to be an essential process during neurogenesis. A growing body of evidence confirms that apoptosis is also responsible for the loss of neurons associated with physiological aging and neurodegeneration [1].

Though caspases are not the only cellular enzymes that participate in apoptosis, their role is essential to the completion of PCD [2–4]. In mammals, caspases are activated in a protease cascade that leads to the rapid disablement of key structural proteins and leads to cell shrinkage, blebbing, chromatin condensation, and DNA fragmentation [5]. Caspase-3 is one of the most widely studied member of the family as it plays a pivotal role in the execution of apoptosis: cleavage of caspase-3, in fact, initiates the execution

Laura Lossi and Adalberto Merighi (eds.), *Neuronal Cell Death: Methods and Protocols*, Methods in Molecular Biology, vol. 1254, DOI 10.1007/978-1-4939-2152-2_8, © Springer Science+Business Media New York 2015

or terminal pathway of apoptosis onto which converge the intrinsic and extrinsic pathways that can be triggered by different pro-apoptotic signals in neurons (and other cell types) [6, 7].

Among the most important difficulties neurobiologists have to face when studying PCD, there are the asynchrony of the process, which affects several different types of neurons at different times within the same brain area, the close relationship with proliferation during development, and the very rapid clearance of apoptotic cells from tissue [8]. In addition, as regarding neurodegenerative diseases and more specifically Alzheimer disease (AD), the evolution over a very long time span makes it quite difficult to successfully employ animal models (including transgenics) to mimic the late phases of neurodegeneration in vivo [9–11].

It therefore follows that alternative approaches to studies in vivo are highly desirable for a better comprehension of the mechanisms and functional significance of apoptosis in the normal and pathological brain.

Many of the presently available neuronal cell lines are not suitable to studies on normal brain development since their phenotype is usually different if compared to that of the cognate cells from which they originate. Brain slice models offer unique advantages over other in vitro approaches in that they (1) largely preserve the tissue architecture and maintain neuronal activity with intact functional local synaptic circuitry for variable periods of time; (2) resemble the nature and behavior of cells in vivo; (3) are easy to manipulate; (4) are adequate tools for the study of neuronal development, differentiation, function, and pathology [12]. In addition, when cultivated in an organotypic context brain slices allow studying different areas of the nervous system in their original structure, providing a big advantage over dissociated cultures in which neuronal networks are disrupted. In the case of the cerebellum, organotypic slices represent an ideal approach for the study of PCD: the cerebellar architecture is relatively simple and the cerebellar cortex contains surprisingly high numbers of granule cells (CGCs) that die during the normal course of their development mainly after activation of an apoptotic program [13].

In this chapter we describe a protocol to study caspase-3 activation in individual cerebellar neurons by the combination of laser scanning confocal fluorescence microscopy (LSCFM) and fluorescence resonance energy transfer (FRET). Thanks to the intrinsic features of LSCFM, i.e., the ability to excite small spatial volumes with submicron resolution, to provide different simultaneous readouts (intensity, spectral characteristics), to measure emissions from different channels in the same volume, and to the possibility to produce dedicated genetically encoded fluorescent probes to be monitored in the alive cells, an ever-growing range of measurement techniques can now be combined with LSCFM to study protein–protein interactions [14]. The combination of LSCFM and FRET not only offers the possibility to qualitatively

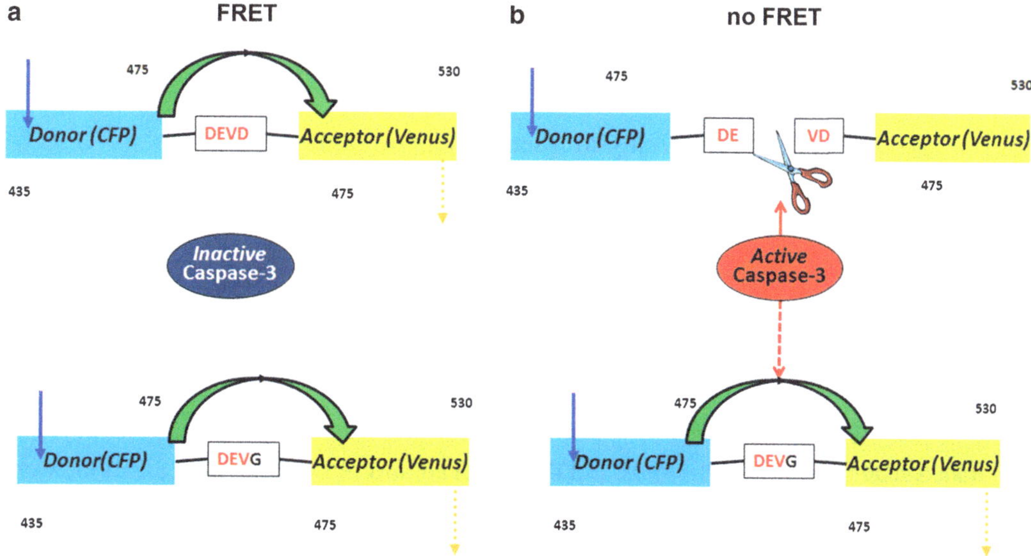

Fig. 1 Principle of detection of caspase-3 activity with the SCAT3 FRET probes. *Top*: in cells transfected pSCAT3-DEVD functional probe FRET occurs when caspase-3 is inactive (**a**). When the enzyme is in its active form, it cleaves the consensus sequence DEVD that links together the FRET pair and FRET does not occur as the donor and acceptor become separated (**b**). *Bottom*: In cells transfected pSCAT3-DEVG control probe FRET occurs independently from caspase-3 activation (**a, b**)

monitor enzyme function, but also gives quantitative information about the level of caspase-3 activation in individual cells.

The FRET probe described in this chapter (Fig. 1) is a genetically encoded FRET pair consisting of the enhanced cyan fluorescent protein (ECFP) as a donor and Venus (a mutant of yellow fluorescent protein) as an acceptor, linked by the caspase-3 recognition and cleavage sequence DEVD [15, 16].

Here, we show how to prepare cerebellar organotypic slices, to transfect them with the FRET probe, and to monitor the cleavage of the probe by caspase-3, following the calculation of the fluorescence intensity ratio of Venus to ECFP [17, 18].

This protocol can be easily adapted to the study of caspase-3 activation in tissue slices obtained from other areas of the brain or spinal cord.

2   Materials

2.1   Preparation of Cerebellar Organotypic Cultures (OCCs)

1. Postnatal mice or rats. *See* **Note 1**.

2. Sodium pentobarbital.

3. Gey's solution (Sigma Chemicals, St. Louis, MO) supplemented with glucose and antioxidants. For 500 mL: 50 % glucose 4.8 mL, ascorbic acid 0.05 g, sodium pyruvate 0.1 g.

    4. Surgical instruments for brain dissection.

    5. McIlwain tissue chopper (Brinkmann Instruments, Westbury, NY).

    6. Millicell-CM inserts (Millipore Billerica, MA).

    7. Sterile 35-mm Petri dishes.

    8. CO_2 incubator. *See* **Note 2**.

    9. Culture medium: 50 % Eagle basal medium (BME, Sigma Chemicals), 25 % horse serum (Gibco®, Life Technologies™, Carlsbad, CA), 25 % Hanks balanced salt solution (HBSS, Sigma Chemicals), 0.5 % glucose, 0.5 % 200 mM l-glutamine, 1 % antibiotic/antimycotic solution.

2.2 Transfection of OCCs

    1. Plasmids: pcDNA-SCAT3 and pcDNA-SCAT3 (DEVG) were a kind gift of Prof. Masayuki Miura [16]. *See* **Note 3**.

    2. *E. coli* competent cells.

    3. Selective LB-agar plates (ampicillin 100 μg/mL or kanamycin 50 μg/mL).

    4. LB broth.

    5. Orbital water bath shaker.

    6. Plasmid DNA maxi prep kit, e.g., GeneHelute HP Plasmid Maxiprep Kit (Sigma Chemicals) or NucleoBond Xtra Maxi PLUS (Macherey-Nagel, Düren, Germany).

    7. DNA agarose gel electrophoresis apparatus.

    8. Seashell gold carrier particles: particles can be purchased in the S1000d kit that also contains the Binding and Precipitation Buffer (Seashell Technology, LLC, La Jolla, CA).

    9. Pure ethanol.

    10. Ultrasonic homogenizer.

    11. Vortex mixer.

    12. Eppendorf microfuge.

    13. Tubing Prep Station® (Bio-Rad, Hercules, CA).

    14. Tefzel® tube (Bio-Rad).

    15. Helios Gene Gun® (Bio-Rad).

    16. Helium and nitrogen cylinders with pressure controllers.

    17. Laminar flow hood.

2.3 Imaging of OCCs and FRET Measurements

    1. Laser scanning confocal microscope. *See* **Note 4**.

    2. 4 % Paraformaldehyde in phosphate buffer (PB) 0.1 M pH 7.4.

    3. Microscope incubation chamber (Figs. 2 and 3).

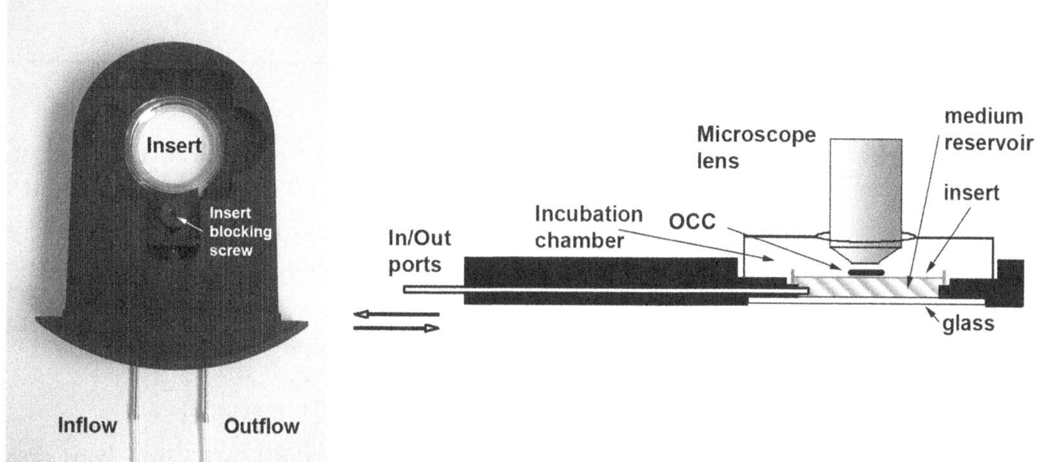

Fig. 2 Incubation chamber for conventional FM and LSCFM of OCCs. *Left*: Top view of the chamber with the Millicell-CM insert blocked into position. *Right:* Cross-sectional drawing of the chamber. The insert with its attached membrane forms the *top* of the medium reservoir that substitutes the Petri dish used in static incubation of OCCs. The *bottom* of the reservoir is made of a glass slide to permit examination with transmitted light in conventional bright field microscopy. Medium in the reservoir can be changed using the In/Out ports without moving the insert and therefore permitting long term monitoring of individual cells in slices. The chamber is inserted into a microscope stage incubator with controlled heating/O_2–CO_2 atmosphere. *FM* fluorescence microscopy, *LSCFM* laser scanning confocal fluorescence microscopy, *OCC* organotypic cerebellar culture. Reproduced with permission from [24]

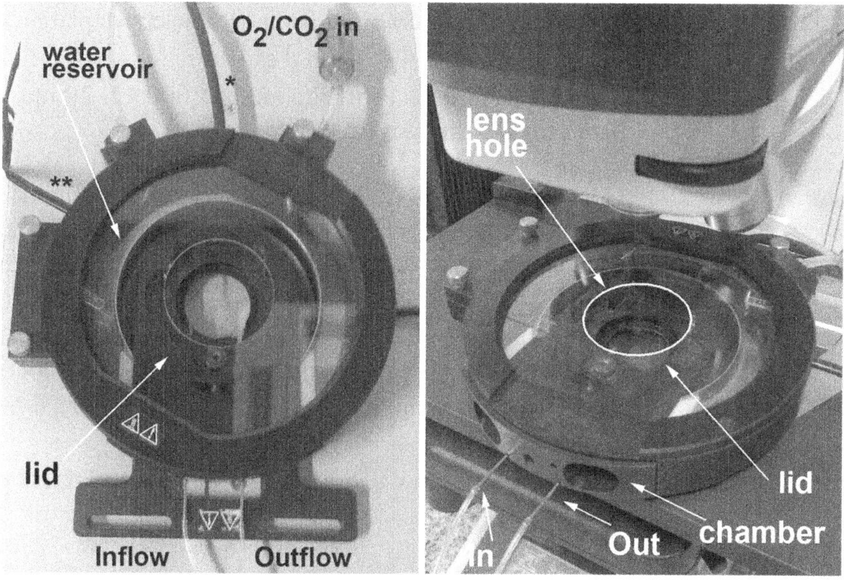

Fig. 3 Set up of the microincubator on the confocal microscope stage. *Left:* Top view of the microincubator with the incubation chamber inside. The incubation chamber is surrounded by a water reservoir into which a mixture of O_2/CO_2 is continuously bubbled through a Teflon tube (O_2/CO_2 in). The microincubator is provided with a heating resistance (*) monitored by a thermocouple (**). The top lid has a central hole for the microscope lens. *Right:* Microincubator mounted on the stage of the SP5 Laser Scanning Confocal Microscope. The perimeter of the lens hole has been highlighted by the white circle. Reproduced with permission from [24]

3 Methods

3.1 Preparation of OCCs

1. Euthanize CD1 mice at postnatal day 6–7 (P6-P7) with an overdose of sodium pentobarbital (60 mg/100 g body weight).

2. Quickly remove the brain from the skull while the head is kept submerged in ice-cooled Gey's solution and isolate the cerebellum. *See* **Note 5**.

3. Cut 350 μm-thick parasagittal slices of cerebellum with a McIlwain tissue chopper. *See* **Note 6**.

4. Plate two-three slices onto a Millicell-CM insert.

5. Place each insert has to be placed inside a 35-mm Petri dish containing 1 mL of culture medium.

6. Incubate at 34 °C in 5 % CO_2 for up to 8 days in vitro (DIV). Medium has to be changed twice a week. Allow slices to equilibrate to the in vitro conditions for at least 4-6 DIV before treatments.

3.2 Transfection of OCCs

Practise sterile technique to avoid contamination.

3.2.1 Heat-Shock Transformation to Produce Plasmid-DNA

1. Heat-shock frozen *E.coli* competent cells with 0.1–0.5 μg of the plasmid DNA for 45–50 s at 42 °C. Try different dilutions of transformed bacteria to be spread on antibiotic-containing selective LB-agar plates and incubate at 37 °C overnight.

2. Pick well isolated colonies from each selective plate to grow in LB broth liquid cultures for 12–16 h.

3. Harvest bacteria by centrifugation (4,000 × g, 20 min), to prepare cells for the lysis procedure.

3.2.2 Plasmid DNA Extraction and Concentration

1. Isolate plasmid DNA from the previously obtained recombinant *E. coli* cultures following the procedure recommended by the manufacturer of the maxi prep kit.

2. Concentrate the extracted DNA by alcohol precipitation or by the use of Nucleoboud Finalizers (contained in the NucleoBond Xtra Maxi PLUS - follow the manufacturer's protocol) at the final concentration of 1 μg/μL, ready for downstream applications.

3. Prior to use, verify size and quality of plasmid DNA by agarose gel electrophoresis.

3.2.3 Gene Gun Cartridges' Preparation

1. Connect the Tubing Prep Station® to the Nitrogen cylinder.

2. Cut a piece of the Tefzel® tube at the right length to be inserted in the Tubing Prep Station® and wash it three times with ethanol.

3. Insert the tube in the Tubing Prep Station® and dry it with a flow of nitrogen for about 15 min.

4. Add 330 µL binding buffer to 25 mg of gold particles, to a final concentration of 30 mg/mL. *See* **Note 7**.

5. Add 50 µg of plasmid DNA. *See* **Note 8**.

6. Vortex briefly.

7. Add an equal volume of precipitation buffer.

8. Vortex briefly and let stand for 3 min.

9. Spin at $10,000 \times g$ in an Eppendorf microfuge for 10 s to pellet the DNA-coated gold particles.

10. Remove the supernatant and add 500 µL of cold ethanol.

11. Vortex briefly and repeat **step 9**.

12. Remove the supernatant, add 3.5 mL cold ethanol and briefly sonicate to resuspend the gold particles (sonication minimizes the aggregation of gold particles).

13. Close the nitrogen cylinder and fill the Tefzel® tube with the gold suspension. *See* **Note 9**.

14. Let stand for 3 min.

15. Draw ethanol out from the tube using a 5 mL syringe connected to be tube.

16. Dry the tube in the Tubing Prep Station® by flushing it with nitrogen for about 5 min.

17. Close the nitrogen cylinder and cut the Tefzel® tube at the right length to enter in the barrel of the Helios Gene Gun®.

3.2.4 Transfection

1. Insert cartridges in the barrel of the Helios Gene Gun®.

2. Connect the Helios Gene Gun® with the helium cylinder, and select an operating pressure of 160 psi.

3. Place the barrel liner over the target (i.e. the Petri dish containing the slice on the Millicell-CM insert) at a distance of 1.6 mm with a spacer. Give one single shot to each culture dish.

3.3 Imaging of OCCs and FRET Measurements

1. Remove the culture medium from a Petri dish containing 2–3 slices on a Millicell-CM insert and add 1 mL of 4 % paraformaldehyde in PB. Incubate at room temperature for 1 h.

3.3.1 LSCFM on Fixed Cells

2. Wash 3×5 min in PB, at room temperature.

3. Cut the part of the membrane of the Millicell-CM inserts containing the tissue slices and place it on a microscope slide.

4. Mount using H_2O-glycerol 1:10. Do not use commercially available, fluorescence-free, mounting media (es. ProLong® antifade medium, Invitrogen™, Life Technolgies™) as they interfere with the FRET signal [19].

3.3.2 LSCFM on Live Cells

1. Fill the medium reservoir of the microscope stage incubation chamber with culture medium. *See* **Note 10**.

2. Transfer a culture on a Millicell-CM insert in the microscope stage incubator.

3. Close the chamber lid and turn on the CO_2 supply.

4. Allow culture to equilibrate and then choose a microscope field with several transfected cells using the conventional fluorescence microscopy mode of the confocal microscope.

5. Use dry objectives with high NAs to image cells with LSCMF. *See* **Note 11**.

6. Choose the better plane of focus and take a single confocal image of transfected cells with appropriate filter combinations (see below).

7. Take subsequent images at different time intervals in control and experimental cultures. Be careful to readjust focus before taking images to correct for possible mechanical drift and $X-Y$ shifting of the culture after medium changes. *See* **Note 12**.

8. Optional: At end of experiments fix cells as in Subheading 3.3.1.

3.3.3 Preliminary Evaluation of FRET Efficiency

To evaluate FRET efficiency of the probe coded by pcDNA-SCAT3 and pcDNA-SCAT3 (DEVG) plasmids in each experimental system it is necessary to perform acceptor bleaching. This involves measuring donor "de-quenching" in the presence of an acceptor and can be done by comparing donor fluorescence intensity in the same sample *before* and *after* destroying the acceptor by photobleaching. If FRET was initially present, a resultant increase in donor fluorescence will occur on photobleaching of the acceptor. The energy transfer efficiency can be quantified as

$$\textit{FRETeff} = \left(\mathbf{Dpost} - \mathbf{Dpre}\right) / \mathbf{Dpost}$$

Where **Dpost** = fluorescence intensity of the donor *after* acceptor photobleaching, **Dpre** = fluorescence intensity of the donor *before* acceptor photobleaching, FRET eff is positive when **Dpost** > **Dpre**.

1. Fix and mount cultures as described in Subheading 3.3.1.

2. Choose a region of interest (ROI) within a transfected fluorescent cell identified by conventional fluorescence microscopy.

3. Using a 63× dry objective with a 1.4 NA and setting the zoom command of the confocal to 8.0×, irradiate the ROI with the maximum dose of the 514 nm laser line (specific for the Venus fluorescent molecule) and measure, step by step, the emission intensity at 530 nm.

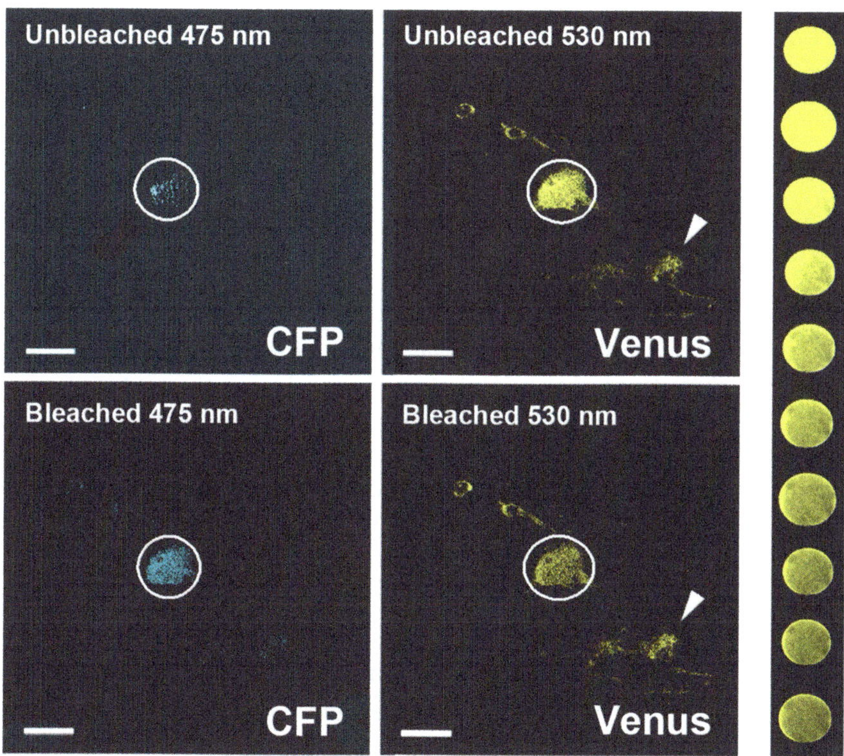

Fig. 4 Acceptor photobleaching of a SCAT3 (DEVG) transfected cerebellar granule cell in a 4 DIV OCC from a wild-type CD1 mouse. The white circle indicates the photobleached area. Ten frames of bleached area are shown in a temporal sequence (from *top* to *bottom*) in the right panel (2×). The donor (CFP) fluorescence (475 nm emission – CFP) is increased after bleaching, whereas to acceptor (Venus) fluorescence (530 nm emission—*yellow*) is reduced. The arrowhead indicates a non-photobleached area where emission intensity is unchanged. Bars = 10 μm. *DIV* days in vitro; *OCC* organotypic cerebellar culture. Reproduced with permission from [24]

4. Stop irradiation when bleaching is complete, i.e., when measuring significantly lower level of 530 nm emission compared with the value at beginning of irradiation (Fig. 4).

5. Repeat bleaching on different cells from different cultures to obtain a statistically significant sample.

6. Calculate FRET eff for each measured cell and then calculate the average value of FRET eff Positive FRET eff values indicate a well-working probe. The higher is the value, the more efficient is the probe. *See* **Note 13**.

3.3.4 Measuring Caspase-3 Activity in Fixed OCCs

1. Transfect cells with pcDNA-SCAT3.

2. Fix cells in 4 % paraformaldehyde and mount in H_2O-glycerol with a cover glass.

3. Identify fluorescent cells by conventional fluorescence microscopy.

4. Excite a field containing single cells of interest with 458 nm excitation line of the confocal microscope.

5. Measure, in each cell, the fluorescence donor emission (475 nm) and acceptor emission (530 nm). *See* **Note 12**.

6. Calculate the ratio of donor emission/acceptor emission: this value represents the level of caspase-3 activation in individual cells and can be used to compare cells and cell populations under different experimental conditions.

7. Use correct statistical methods to evaluate changes in caspase-3 activation in different type of experiments.

3.3.5 Measuring Caspase-3 Activity in Live OCCs

1. Transfect cells with pcDNA-SCAT3 and mount the culture in the microscope stage incubation chamber as described in Subheading 3.3.2.

2. Identify fluorescent cells by conventional fluorescence microscopy.

3. Excite a field containing single cells of interest with 458 nm excitation line of the confocal microscope. *See* **Note 13**.

4. Measure, in each cell, the fluorescence donor emission (475 nm) and acceptor emission (530 nm). *See* **Note 14**.

5. Calculate the ratio of donor emission/acceptor emission in single cells at different timing points: you can change the medium in the reservoir through the In/Out ports without moving the insert and thus compare the donor/acceptor ratio in different experimental environments along a time interval of several hours. *See* **Note 15**.

The protocols described here can be easily also employed on other in vitro systems, such as neuronal primary cultures. The substantial difference between the two approaches is that differently from organotypic cultures, isolated neurons are severely damaged by transfection with the Helios Gene Gun®. In this case it is possible to use, for example, lipofectin-based transfection techniques or similar approaches based on the use of membrane permeant molecules. In our hands, these techniques have a lower transfection efficiency when compared to the Helios Gene Gun®; nonetheless we have been able to obtain successful FRET measurements with the SCAT3 probe also after using this type of approach for transfection of post-mitotic neurons (Fig. 5).

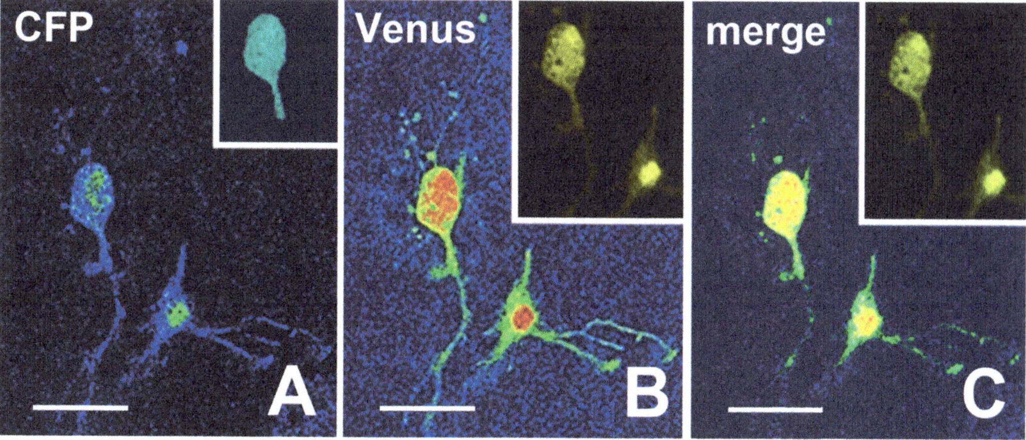

Fig. 5 Images of two transfected cerebellar granule cells in a 4 DIV OCC from a wild-type CD1 mouse. The FRET ratio has been displayed in pseudocolors using a logarithmic scale. Bars = 10 μm

4　Notes

1. As apoptotic cells are quickly removed from tissues it is important working on a system where PCD is maximal. The choice of species and age of animals depends from the type of experiments and area(s) under investigation. When working on cerebellar PCD, the peak of apoptosis occurs during the first postnatal week in mouse and rat, whereas in precocial animals (e.g., Guinea pigs) apoptosis is maximal during the last phase of embryonic development life.

2. The temperature settings of the incubator is somehow critical to the survival of cultures, as OCCs (and organotypic slices obtained from other CNS areas) survive better at temperatures below 37 °C. However, it must be taken into consideration that the neuroprotective effect of mild hypothermia on cultured neurons might somehow mask the effect of certain inductors of apoptosis [20].

3. Plasmids were produced in frozen *E. coli* competent cells (Promega, Madison, WI), by the use of LB-agar plates containing 100 μg/mL ampicillin or 50 μg/mL kanamycin according to the antibiotic resistance cassette of the plasmid, and LB broth liquid cultures. After agar plating, selected colonies were grown overnight in LB broth under continuous shaking at 37 °C.

4. We currently use a Leika SP5 confocal microscope (Leika Microsystems, Wetzlar, Germany). This instrument allows performing conventional fluorescence microscopy (FM) to directly identify transfected cells and laser scanning confocal

fluorescence microscopy (LSCFM) to detect FRET. In FM use emission filter band-pass BP470–500 nm for CFP and long-pass LP530 nm for Venus. For FRET detection, use the 458 nm excitation line from an Argon ion laser for CFP and the 514 nm line of an Argon ion laser for photobleaching of the acceptor. Take images by the use of a 40× dry objective with 0.6 NA. For photobleaching use a 63× lens (NA = 1.4). During photobleaching it is mandatory to concentrate fluorescence over a small ROI to effectively bleach the acceptor. The use of the confocal zoom is fundamental to this purpose.

5. All procedure during the preparation of cerebellar slices must be performed using ice-cooled solution. Prepare some blocks of frozen Gey's solution to be added to cool (4 °C) solution in a large Petri dish to maintain temperature a few degrees above °C during dissection. Completely remove the meninges with N.7 Dupont forceps before cutting.

6. Cutting with chopper is made easier if the cerebellum is not submerged by an excess of Gey's solution. Wipe off solution with a piece of filter paper and set section thickness to any value between 200 and 400 μm. Other cutting parameters (e.g., blade force) have to be set according to the type of chopper in use. To collect slices use a spatula with curved edges.

7. If using Seashell gold carrier particles a volume of 500 μL is added. Briefly sonicate the tube containing the gold suspension before drawing out the desired amount of gold. The original protocol from Bio-Rad uses dry gold particles that must be directly weighted from a 2 mL plastic tube. During this procedure there is the risk of a substantial loss of gold.

   For this reasons we prefer using a gold suspension. The gold particle size (1 μm) can be varied according to different experimental needs. The larger the particle the higher is the amount of DNA that can be adsorbed onto it, but the higher is the risk of cell damage and death following transfection. Very small gold particles (40 nm) have been suggested to be equally efficient in transfection with minimal cell damage [21, 22].

8. DNA is used at the final concentration of 2 μg/mg gold. If stock plasmid DNA has been produced at the suggested concentration (1 μg/μL) then add 50 μL. If the DNA stock solution is less concentrated calculate the necessary volume to add 50 μg DNA. As the total volume of the gold/DNA suspension fits to the internal volume of the Tefzel® tubing, do not add more than 150 μL of stock plasmid DNA solution. The ratio of DNA/gold can be varied according to different experimental needs.

9. This is likely the most critical step in bullet preparation. The DNA/gold suspension tends to precipitate very quickly.

It must be drawn from Eppendorf tube with a 2 mL syringe connected to the Tefzel® tube. The tube must be filled without bubbles as the stratification of the DNA/gold complex at the bottom will not be homogenous in the presence of air bubbles. This results in the production of individual bullets with a variable content of DNA/gold. Bullets with a inhomogeneous DNA/gold stratification should be discarded as they may yield inconsistent transfection or lower transfection efficiency.

10. The microscope stage incubator is provided with inflow and outflow port to permit exchange of culture medium during the live imaging experiments. It is important to pre-fill the tubing with medium, to avoid the formation of air bubbles that, once passed in the reservoir beneath the culture insert, may cause drifting in the X–Y axis, thus making it difficult to recognize individual cells in time lapse experiments.

11. The choice of the correct objective is very important for imaging studies, particularly if working with live cells. We currently use dry 20×, 40× or 63× objectives with 0.5, 0.6, and 0.7 NAs, respectively. Although higher magnification objectives give a better resolution and a lower fluorescence signal, repetitive excitation of neurons transfected with the SCAT3 FRET probe has proved to be detrimental to cell survival. Therefore, it is advisable to use a 20× objective in live imaging experiments when caspase-3 levels in individual cells have to be followed for several hours. When only a few frames need to be collected the 40× objective is the better compromise.

12. Although the on stage incubator has been designed to allow for repetitive imaging of the same microscopic field, during the course of time lapse experiments it is common to observe some lateral drift in the X–Y axes and/or changes of focus plan in the Z-axis. It is advisable to briefly re-adjust focus prior to exciting individual ROIs in the microscopic field, but this operation has to be carried out as quickly as possible to minimize laser damage to the transfected cells. It is also advisable to draw a map of the transfected cells, i.e., the ROIs in the microscopic field, to be able to identify the same neuron at different time points during the experiments.

13. Before starting to work with SCAT3 FRET probe, it can be useful to have an idea of an hypothetical "working range" of the probe in the new experimental context of use, as different parameters, such as for example the level of expression of the fluorescent reporter FRET pair, may influence probe efficiency in different types of transfected cells. By the use of the same data obtained during photobleaching preliminary experiments, calculate the average ratio between donor emission (475 nm) and acceptor emission (530 nm) in individual cells,

before and *after* bleaching. The value that the average ratio of donor emission/acceptor emission assumes *before* bleaching corresponds to the lowest theoretical possible level of caspase-3 activation in healthy cells, i.e., where there is no activation of caspase-3. On the contrary, the value of the average ratio of donor emission/acceptor emission *after* bleaching is a reliable estimation of the figure that can be obtained in cells where caspase-3 has been activated at maximal level, i.e. when cell is undergoing apoptosis. Activated caspase-3 cleaves the tetra-peptide that links together the donor and the acceptor, with a subsequent decrease in the FRET signal, and acceptor bleaching mimics this condition by quenching the fluorescence emission of Venus.

14. Measurement of fluorescence intensity can be carried out with the ad hoc software of the LSCM.

15. Choice of microscope field in live cell imaging is very important as it is advisable to imagine a relatively high number of transfected neurons in the same field to quickly reach statistical significance of the experiments. It should also be noted that it might be possible to sequentially imagine different microscopic fields from the same OCCs in different slices, and to relate them to specific areas that can be recognized in cultured slices. For example, it may be possible to separately imagine neurons in the external and internal granular layer of the forming cerebellar cortex to analyse the selective activation of caspase-3 in these two layers during the maturation of cerebellum [23].

References

1. Lossi L, Merighi A (2003) In vivo cellular and molecular mechanisms of neuronal apoptosis in the mammalian CNS. Progr Neurobiol 69:287–312

2. Yuan J (1995) Molecular control of life and death. Curr Opin Cell Biol 7:211–214

3. Yuan J, Yankner BA (2000) Apoptosis in the nervous system. Nature 407:802–809

4. Thornberry NA, Lazebnik Y (1998) Caspases: enemies within. Science 281:1312–1316

5. Cryns V, Yuan J (1998) Proteases to die for. Genes Dev 12:1551–1570

6. Blatt NB, Glick GD (2001) Signaling pathways and effector mechanisms pre-programmed cell death. Bioorg Med Chem 9:1371–1384

7. Adams JM, Cory S (2002) Apoptosomes: engines for caspase activation. Curr Opin Cell Biol 14:715–720

8. Lossi L, Mioletti S, Aimar P, Bruno R, Merighi A (2002) In vivo analysis of cell proliferation and apoptosis in the CNS. In: Merighi A, Carmignoto G (eds) Cellular and molecular methods in neuroscience research. Springer, New York, pp 235–258

9. Lee JH, Cheon YH, Woo RS et al (2012) Evidence of early involvement of apoptosis inducing factor-induced neuronal death in Alzheimer brain. Anat Cell Biol 45:26–37

10. Zhu X, Raina AK, Perry G et al (2006) Apoptosis in Alzheimer disease: a mathematical improbability. Curr Alzheimer Res 3:393–396

11. Roth KA (2001) Caspases, apoptosis, and Alzheimer disease: causation, correlation, and confusion. J Neuropathol Exp Neurol 60:829–838

12. Lossi L, Alasia S, Salio C et al (2009) Cell death and proliferation in acute slices and organotypic cultures of mammalian CNS. Prog Neurobiol 88:221–245

13. Wood KA, Dipasquale B, Youle RJ (1993) In situ labeling of granule cells for apoptosis-associated DNA fragmentation reveals different mechanisms of cell loss in developing cerebellum. Neuron 11:621–632

14. Lakowicz JR, Gryczynski II, Gryczynski Z (1999) High throughput screening with multiphoton excitation. J Biomol Screen 4: 355–362

15. Wu Y, Xing D, Luo S et al (2006) Detection of caspase-3 activation in single cells by fluorescence resonance energy transfer during photodynamic therapy induced apoptosis. Cancer Lett 235:239–247

16. Takemoto K, Nagai T, Miyawaki A et al (2003) Spatio-temporal activation of caspase revealed by indicator that is insensitive to environmental effects. J Cell Biol 160:235–243

17. Wu Y, Kim SG, Xing D et al (2007) Fluorescence resonance energy transfer analysis of bid activation in living cells during ultraviolet-induced apoptosis. Acta Biochim Biophys Sin (Shanghai) 39:37–45

18. Liu L, Wei G, Liu Z et al (2008) Two-photon excitation fluorescence resonance energy transfer with small organic molecule as energy donor for bioassay. Bioconjug Chem 19:574–579

19. Rodighiero S, Bazzini C, Ritter M et al (2008) Fixation, mounting and sealing with nail polish of cell specimens lead to incorrect FRET measurements using acceptor photobleaching. Cell Physiol Biochem 21:489–498

20. Feiner JR, Bickler PE, Estrada S et al (2005) Mild hypothermia, but not propofol, is neuroprotective in organotypic hippocampal cultures. Anesth Analg 100:215–225

21. Arsenault J, O'Brien JA (2013) Optimized heterologous transfection of viable adult organotypic brain slices using an enhanced gene gun. BMC Res Notes 6:544

22. O'Brien JA, Lummis SC (2011) Nano-biolistics: a method of biolistic transfection of cells and tissues using a gene gun with novel nanometer-sized projectiles. BMC Biotechnol 11:66

23. Lossi L, Tamagno I, Merighi A (2004) Molecular morphology of neuronal apoptosis: activation of caspase 3 during postnatal development of mouse cerebellar cortex. J Mol Histol 35:621–629

24. Merighi A, Alasia S, Gambino G, Lossi L (2012) Confocal imaging of organotypic brain slices for real time analysis of cell death. In: Méndez-Vilas A (ed) Current microscopy contributions to advances in science and technology. Formatex Research Center, Badajoz, Spain

Design and Cloning of Short Hairpin RNAs (shRNAs) into a Lentiviral Silencing Vector to Study the Function of Selected Proteins in Neuronal Apoptosis

Nadia Canu

Abstract

Double-stranded RNA-mediated interference (RNAi) is a new simple and fast research tool for shutting down genes and characterizes function of their respective proteins. Many strategies for design and delivery of siRNA to target cells are available. Here, we describe the use of lentiviral short hairpin RNA (shRNA) RNA silencing to identify the involvement of D-serine racemase (SR)- an enzyme that syntheses D-serine to modulate glutamate-*N*-methyl-D-aspartate receptor- in regulating rat cerebellar granule neurons (CGN) apoptosis. Apoptosis is induced by serum and KCl withdrawal and is detected with fluorometric caspase 3 assay.

Key words **siRNA,** ShRNA, Lentivirus, Cerebellar granule cells, D-serine racemase

1 Introduction

Gene knockout is used to study the function of specific gene, detect its protein product, and link it to physiological or pathological processes. Knockout can be deliberately made using different molecular techniques some of which, like homologous recombination, are lengthy and expensive. RNA interference (RNAi) has appeared as a novel pathway to knockdown specific mRNAs, thus preventing translation of the respective protein from occurring.

RNAi is a natural process—used in many different organisms to regulate endogenous gene expression—in which non-translated, long, double-stranded RNA (dsRNA) led to a strong, long lasting and specific silencing of selected genes [1]. Further studies revealed that small dsRNA of 21–25 bp (small interfering RNA = siRNA) derived from endonuclease Dicer-mediated processing of long dsRNAs interact with a protein complex to form the RNA-induced silencing complex (RISC) [2, 3]. This complex has nuclease activity and digests mRNA containing a base pair sequence identical to that in the siRNA. Thus, the siRNA serves

Laura Lossi and Adalberto Merighi (eds.), *Neuronal Cell Death: Methods and Protocols*, Methods in Molecular Biology, vol. 1254, DOI 10.1007/978-1-4939-2152-2_9, © Springer Science+Business Media New York 2015

as a target sequence that allows RISC to recognize specific mRNAs and to prevent their translation by cleaving them [4]. Another method to produce siRNA is based on the use of short hairpin RNAs (ShRNAs) that trigger RNAi [5, 6]. Short (60–75 bp long) DNA oligodesoxynucleotides that form hairpins are cloned into a plasmid under the control of the U6 or H1 promoter for RNA polymerase III. Transfection of such a plasmid promotes the expression of ShRNAs that induce RNAi. Non-replicating recombinant viral vectors (adeno, adeno-associated and lentiviruses) are commonly used for ShRNA expression in primary neuronal cells. Lentiviruses may be particularly suited for long-term ShRNA and expression and gene silencing in vivo since the viral DNA gets incorporated in the host genome.

Commonly used lentiviral vector systems belong to the second or third generation, ensuring safe application, as these viruses are unable to self-replicate, since the spontaneous self-assembly is prevented by distributing the least necessary number of virus elements on three and four plasmids, respectively. Here, we describe the methods used in our laboratory to silencing SR in rat CGNs as a tool for identifying the role of this enzyme during apoptosis.

2 Materials

2.1 Design, Production, and Cloning of shRNAs and Preparation of Lentiviral Vectors

1. Packaging cell line: HEK (human embryonic kidney)-293T (Invitrogen™—Life Technologies™, Gaithersburg, MD). *See* **Note 1**.

2. 15 cm plates (Becton Dickinson Labware, Franklin Lakes, NJ).

3. Dulbecco's modified Eagle's medium (DMEM, Gibco™—Life Technologies™) with 2 and 10 % fetal bovine serum (FBS). *See* **Note 2**.

4. Lipofectamine 2000 (Invitrogen™—Life Technologies™).

5. Opti-MEM 1× (Gibco™—Life Techonologies™).

6. Plasmids: pLVTHM; pCMVdR8.74; pMD2G (available from Addgene: http://www.addgene.org/) for second lentivirus generation. For third lentivirus generation, refer to Dull et al. [7].

7. 10× Tris-buffered EDTA buffer (TBE buffer): 1 M Tris, 0.9 M boric acid, 0.01 M EDTA.

8. Extraction Kit (DNA 70–10 kb): e.g., QIAquick Gel 8 (Quiagen GmbH, Hamburg, Germany).

9. Endotoxin-free plasmid maxipreps columns (Quiagen).

10. Tris-EDTA buffer (TE buffer 1×): 10 mM Tris, 1 mM EDTA, pH 8.0 with HCl.

11. Primer, 5′ forward must contain an MLu I site; Primer, 3′ reverse must contain a Cla I site.

12. Restriction endonucleases: MLu I, Cla I (New England Biolabs, Ipswich, MA).

13. T4 DNA ligase 400,000 U/mL (New England Biolabs).

14. 10× T4 DNA ligase buffer: 50 mM Tris–HCl, 10 mM MgCl$_2$, 1 mM ATP, 10 mM DTT, pH 7.5.

15. Bacterial growth strain(s): DH5α for pMD2G and pCMVdR8; HB101 for pLVTHM lentiviral vector. *See* **Note 3**.

16. LB agar ampicillin plates: Use a 2 L flask to prepare 1 L of LB broth with agar (Lennox) (Sigma Chemicals, St. Louis, MO). To 1 L of distilled water add 35 g of LB agar. Swirl to dissolve and autoclave for 15 min at 120 °C to sterilize. Cool medium to 50 °C, and add 50–100 μg/mL ampicillin. Pour into Petri dishes and allow to solidify, store at 4 °C.

17. Hank's Buffered Salt Solution (HBSS): 0.137 M NaCl, 5.4 mM KCl, 0.25 mM Na$_2$HPO$_4$, 0.1 g glucose, 0.44 mM KH$_2$PO$_4$, 1.3 mM CaCl$_2$, 1.0 mM MgSO$_4$, 4.2 mM NaHCO$_3$.

18. 20 % [w/v] sucrose in HBSS.

2.2 Primary Cerebellar Granule Neuron Culture, Induction, and Detection of Apoptosis

1. Basal medium Eagle (BME; Life Technologies™).

2. Bovine serum albumin (BSA, Sigma Chemicals).

3. Krebs-Ringer bicarbonate medium (KRB): 120 mM NaCl, 5 mM KCl, 1.22 mM KH$_2$PO$_4$, 25.5 mM, 14 mM glucose, 4.2 mM phenol red.

4. Solution A: KRB supplemented with 1.2 mM MgSO$_4$, 3 mg/mL BSA.

5. DNAse I (Sigma Chemicals).

6. Soybean trypsin inhibitor (Sigma Chemicals).

7. Trypsin type III (Sigma Chemicals).

8. L-Glutamine.

9. Gentamicin sulfate.

10. Fetal bovine serum (FBS, Gibco™).

11. CGN culture medium: BME, 10 % FBS, 25 mM KCl, 2 mM glutamine, 100 mM gentamicin sulfate.

12. 1β-Arabinofuranosylcytosine (Sigma Chemicals).

13. Caspase 3 substrate: Ac-DEVD-AMC [*N*-Acetyl-Asp-Glu-Val-Asp-AMC (7-amino-4-methyl coumarin)] (Biomol International, Plymouth Meeting, PA).

14. Caspase 3 lysis buffer A: 10 mM HEPES, pH 7.4, 42 mM KCl, 5 mMMgCl$_2$, 1mM , 1 mM PMSF, 0.5 % 3-[(3-cholamidopropyl)dimethylammonio]-1-propanesulfonic acid (CHAPS), 1 μg/mL leupeptin.

15. Caspase 3 assay buffer B: 25 mM HEPES, 1 mM EDTA, 0.1 % CHAPS, 10 % sucrose, 3 mM DTT, pH 7.5.

<table>
<tr><td>

2.3 Western Blotting and Immuno-fluorescence for D-Serine Racemase

</td><td>

1. Lysis buffer: 25 mM Tris–HCl (pH 7.6), 150 mM NaCl, 1 % NP-40, 1 % sodium deoxycholate, 0.1 % SDS.

2. 10 % SDS-PAGE (Sodium Dodecyl Sulfate—PolyAcrylamide Gel Electrophoresis)- Laemmli protocol:

(a) 10 % lower gel (resolving gel): 4.9 mL distilled H_2O, 2.5 mL 40 % acrylamide/Bis-acrylamide (29:1), 2.5 mL 1.5 M Tris, pH 8.8, 50 µL 20 % SDS, 50 µL 10 % ammonium persulfate, 10 µL TEMED (total volume = 10 mL). Mix well and quickly transfer the gel solution by using 1 mL pipette to the casting chamber between the glass plates. Once the gel has polymerized, prepare stacking gel.

(b) 3.75 % stacking gel: 2.44 mL distilled H_2O, 0.46 mL 40 % acrylamide/Bis- acrylamide (29:1), 1 mL 0.5 M Tris, pH 6.8, 40 µL 10 % SDS, 15 µL 10 % ammonium persulfate. Right before pouring the gel add 1.5 µL TEMED.

3. Agarose gel: agarose 1 % in TBE buffer 0.5×.

4. Phosphate-buffered saline 1× (PBS): 137 mM NaCl, 2.7 mM KCl, 10 mM Na_2HPO_4, 1.8 mM KH_2PO_4.

5. Normal goat serum (NGS) (Jackson ImmunoResearch, Europe Ltd., Newmarket, UK).

6. 100 % methanol.

7. Antifade mounting medium (ProLong® Gold Antifade—Life Technologies).

8. Mouse D-serine racemase antibody (BD Transduction laboratories™, San Jose, CA).

9. Affinity purified-goat D-serine racemase antibody (Santa Cruz Biotechnology, Dallas, TX).

10. Secondary TRITC-conjugated donkey anti-goat antibody (Jackson ImmunoReseach Europe Ltd.).

</td></tr>
<tr><td>

2.4 Equipment

</td><td>

1. Centrifuge (e.g., Beckman Coulter Inc, Brea, CA).

2. Tissue culture 15 cm dishes.

3. Tissue culture 6-well dishes.

4. Tissue-culture 24-well dishes for CGNs (Nunc A/s, Roskilde, Denmark).

5. Filters (0.22- or 0.45-µm).

6. Incubators preset to 37 °C (5 % CO_2).

7. Microcentrifuge.

8. PCR thermocycler.

9. SW 28 and SW 55 rotors (Beckman Coulter).

10. Sterile round-bottom polypropylene tubes 5 mL (e.g., BD Falcon, BD Biosciences, San Jose, CA).

</td></tr>
</table>

11. 50 mL tubes.

12. Centrifuge, polyallomer, 5 mL tubes (Beckman Coulter).

13. Centrifuge, polyallomer, 12 mL tubes (Beckman Coulter).

14. Microcentrifuge tubes.

15. Vortexer.

16. 96-Well plate fluorescence reader (EnVision, PerkinElmer, Wellesley, MA).

17. Spectrofluorometer (e.g., Kontron AG, Zurich, Switzerland).

18. Protein electrophoresis/Western blotting apparatus.

19. Acrylamide gel electrophoresis apparatus.

20. Fluorescence microscope.

3   Methods

3.1   Design, Production, and Cloning of shRNAs and Preparation of Lentiviral Vectors

3.1.1   Design of shRNAs

ShRNA oligonucleotide design describes the process of identifying target sequences within a gene of interest and designing the corresponding oligonucleotides to generate the ShRNA.

A number of algorithms may been utilized to predict effective siRNA sequences and design ShRNA (e.g., http://www.ambion.com/ or http://sfold.wadsworth.org/; http://eu.idtdna.com/Scitools/Applications/shRNA etc). Here are general guidelines for ShRNA design based on the work of Tuschl et al. [8] and Elabishir et al. [9, 10] (see also: http://www.mpibpc.gwdg.de/abteilungen/100/105/sirna.htmL).

1. Select a region of 19 nt within the gene to be silenced [in our case, rat D-serine racemase (NCBI accession number NM_198757)] do not opt for region near the start codon (within 50-100 bases), nor untranslated regions [9, 10]. *See* **Note 4**.

2. Sequences that have at least 3 A or T residues in positions 15–19 of the sense sequence appear to have increased knock-down activity. *See* **Note 5**.

3. Ensure the content of GC of the 19 bases oligonucleotide between 40 and 60 %, and a GC content of approximately 45 % is ideal.

4. Examine the 19 bases oligonucleotide for secondary structure and long base runs, both of which can interfere with the process of annealing.

5. Filter out, by appropriate database search, candidate targets that are present in other genes to avoid silencing of these loci. *See* **Note 6**.

6. Add the 7–9 nt hairpin loop sequence between sense and antisense strand [11–14]. One of the most effective loop sequences for H1 promoter is TTCAAGAGA [15].

a

```
446
GCGCAATCTCTTCTTCAAA

597
GCTCTCACCTATGCTGCTA

772
GCATCTTGGTCCATCCCAA

888
GGAGGAATGGTTGCTGGAA
```

b

```
5'-cgcgtGCGCAATCTCTTCTTCAAATTCAAGAGATTTGAAGAAGAGATTGCGCTTTTTTat-----3'
5'-cgatAAAAAAGCGCAATCTCTTCTTCAAATCTCTTGAATTTGAAGAAGAGATTGCGCa-----3
```

and the annealed oligos are:

```
5'-cgcgtGCGCAATCTCTTCTTCAAATTCAAGAGATTTGAAGAAGAGATTGCGCTTTTTTat-----3'
3'-----aCGCGTTAGAGAAGAAGTTTAAGTTCTCTAAACTTCTTCTCTAACGCGAAAAAAtagc-5'
```

Fig. 1 (**a**) Potential RNAi target sequences identified in the rat SR-coding region. (**b**) For one of the chosen sequences (GCGCAATCTCTTCTTCAAA) the complementary oligonucleotide pair for the hairpin siRNA and the annealed oligonucleotides are reported. The TTCAAGAGA loop sequence is highlighted in *yellow*, the stretch T is highlighted in *light blue* and the restriction sites (MLu I and Cla I) in red lowercase

7. Place a stretch of 5–6 T at the end of ShRNA to guarantee the termination of RNA polymerase III transcription.

8. Add to the end of two complementary oligonucleotides restriction sites (in our case MLu I at 5′ and Cla I at 3′) (*see* Fig. 1).

9. Include a negative control ShRNA. Usually ShRNA design online tools returns a scrambled sequence with the same nucleotide composition as your siRNA/shRNA input sequence.

10. Sense and antisense oligos must be phosphorylated and PAGE purified in order to increase cloning efficiency. When ordering, be sure to require that oligonucleotides are supplied after PAGE purification.

3.1.2 Annealing of shRNA Oligonucleotides

For expedience, annealing can be done in a thermal cycler.

1. Resuspend each PAGE-purified oligonucleotide in TE buffer to a concentration of 100 μM.

2. Mix the oligos for the sense strand and the anti strand at a 1:1 ratio. This will ultimately give 50 μM of ds oligonucleotide (assuming 100 % theoretical annealing).

3. Heat the mixture to 95 °C for 30 s. *See* **Note 7**.

4. Heat at 72 °C for 2 min.

5. Heat at 37 °C for 2 min.

6. Heat at 25 °C for 2 min.

7. Store on ice or at −20 °C until use.

<table>
<tr><td>

3.1.3 Cloning ShRNA Oligonucleotides into pLVTHM

</td><td>

1. Dilute the annealed oligonucleotides with TE buffer to obtain a concentration of 0.5 µM.

2. For each ligation, add the following reagents in a microfuge tube:
   - 2 µL digested (MLu I/Cla I) and dephosphorylated pLVTHM vector (100 ng/µL).
   - 4 µL diluted, annealed oligonucleotide (0.5 µM).
   - 2 µL 10× T4 DNA ligase buffer.
   - 0.5 µL BSA (10 mg/mL).
   - 11 µL Nuclease-free H_2O.
   - 0.5 µL T4 DNA ligase (400 U/µL).
   - For a 20 µL total volume.

3. Set up separate ligation using 2 µL of the negative scramble control ShRNA annealed oligonucleotide.

4. Set up separate ligation using 2 µL of digested (MLu I/Cla I) pLVTHM vector (50 ng/µL) without annealed oligonucleotide.

5. Incubate ligation mixture at room temperature for 3 h. *See* **Note 8**.

6. Transform immediately competent bacteria (with high transformation efficiency) and select on ampicillin plates.

7. Digest plasmid DNA from colonies with MLu I/Cla I and run on a 12 % DNA polyacrylamide gel in TBE 1× buffer gel (*See* **Note 9**). Positive clones will contain an approximately 60-bp insert compared to 17 bp for colonies without an insert.

8. Sequence the insert with human H1 primer (TCGNTATGTG TTCTGGGAAA) to check hairpin integrity.

9. Validate, by Western blot analysis or indirect immunofluorescence, the cloned ShRNA cassettes by transfecting pLVHTM-ShRNA as well as scramble vector in cell line that coexpresses the target gene. In our case, the cDNA for D-serine racemase together with ShRNA silencing cassette were transfected in an highly transfectable cell line (e.g., HEK-293). *See* **Note 10**.

</td></tr>
<tr><td>

3.1.4 Preparation of Lentiviral Vectors

</td><td>

You need to observe Biosafety-level-2 since application of this protocol leads to the production of pseudoviral particles capable of infecting mammalian cells.

1. For a 10 cm dish lentiviral preparation: plate lentivirus HEK-293-T cells at a density of approximately 100 cells/mm² in 10 mL of DMEM, supplemented with 10 % FBS, 12–24 h before transfection. Addition of antibiotic solution does not interfere with transfection.

2. To increase cell adherence, precoat twelve 15 cm dishes with 10 mL of poly-L-lysine, incubate for 30 min at room temperature

</td></tr>
</table>

under UV, and aspirate off the liquid. Grow the cells overnight at 37 °C in 5 % CO_2. *See* **Note 11**.

3. Aliquot in 200 µL of Opti-MEM the three plasmids into a sterile polypropylene tube. For a 10 cm dish, use:

    - –10 µg of lentivector pLVTHM.
    - –3.5 µg of pMD2G (Gag-Pol).
    - –6.5 µg of pCMVdR8.74.

4. In a separate tube, dilute 20 µL lipofectamine 2000 in 200 µL of Opti-MEM 1×.

5. Add diluted lipofectamine reagent drop-wise to the DNA solution while gently vortexing the DNA-containing tube and incubate for 30 min at room temperature. *See* **Note 12**.

6. Remove medium from cell plate, wash cells twice with Opti-MEM and add 5 mL of Opti-MEM without antibiotics. *See* **Note 13**.

7. Add the transfection mixture to each plate. Swirl the plates gently to distribute the complex and incubate overnight at 37 °C in a 5 % CO_2 atmosphere.

8. Approximately 6–8 h after transfection, remove media. Add 15 mL of fresh DMEM plus 2 % heat-inactivated FBS and penicillin-streptomycin to each plate and incubate overnight at 37 °C in a 5 % CO_2 atmosphere. *See* **Note 14**.

9. Collect the supernatant from the plates and centrifuge at $500 \times g$ for 10 min to remove cell debris and filter through 0.45 µm filters. *See* **Note 15**.

10. Add 15 mL of fresh medium to each plate and incubate overnight. Filtered supernatants can be stored for several days at 4 °C.

11. Collect media and filter as in **step 9**. *See* **Note 16**.

12. Pool collected supernatants from steps 9 and 11. Transfer to Beckman tubes using 25–29 mL per tube.

13. Concentrate viral particles by centrifuging in a Beckman SW 28 rotor at $65,000 \times g$ or 2 h at 20 °C.

14. Resuspend all pellets in a total of 1 mL of HBSS and wash tubes a second time with 1 mL of HBSS.

15. Increase the combined volume from 2 to 3 mL with HBSS and layer the resuspended pellets on 1.5 mL of a 20 % sucrose (in HBSS) cushion in Beckman tubes.

16. Centrifuge using a Beckman SW 55 rotor at $53,500 \times g$ for 1.5 h at 20 °C.

17. Resuspend the pellet in 100 µL of HBSS containing 1 % BSA and wash the tube with an additional 100 µL of HBSS containing 1 % BSA. Shake the resuspended viral preparation on a low-speed vortexer for 15–30 min. Centrifuge for 10 s to remove debris.

18. Aliquot the cleared viral solution and store at -80 °C. It can be stored for many months. Avoid repeated freeze-thaw cycles.

19. Titrate the viral preparations by biological titration using GFP, which is the marker contained in the lentivector. The fraction of GFP fluorescent cells can be counted by FACS (fluorescence activated cell sorting). GFP fluorescence may be also visualized under a fluorescence microscope. Usually 10–15 random fields of view are used to estimate the overall fraction of fluorescing cells in each well.

3.2 Primary Cerebellar Granule Neuron Cultures and Lentivirus Transduction

3.2.1 Primary Cerebellar Granule Neuron Cultures

Cultures enriched in CGNs are obtained from dissociated cerebella of 8-day-old Wistar rats according to Levi et al. [16]. The preparation of CGN cultures is carried out at *Day 1*, transduction at *Day 2*, induction and detection of apoptosis at *Days 6–7*.

1. Remove 4–5 cerebella from 8-day-old rats and slice them (0.4 mm thickness) with a mechanical tissue chopper.

2. Suspend in 10 mL solution A, centrifuge for 15 s at $150 \times g$.

3. Resuspend the tissue in 10 mL solution A containing 0.25 μg/mL trypsin III and incubate at 37 °C for 15 min in a shaking water bath at rate of 125 rpm.

4. Add to the suspension 10 mL solution A containing 12.8 μg DNAase I and 83 μg soybean trypsin inhibitor.

5. Centrifuge immediately for 15 s at $150 \times g$.

6. Resuspend the pellet in 2 mL of solution A containing 80 μg DNAase I, 0.52 mg soybean trypsin inhibitor and 2.7 mM $MgSO_4$. Triturate the tissue with a Pasteur pipette (25 strokes).

7. Allow the suspension to stand for 15 min, aspirate carefully the upper 1.5 mL, readjust the volume to about 2 mL ad dissociate as above. After allowing the suspension to stand for 15 min, take off the supernatant, leaving only 0.2 mL containing clumps and debris.

8. Transfer the supernatant into 3 mL Solution A containing 0.1 mM $CaCl_2$. After about 10 min decant the supernatant, allow to stand for another 10 min and resuspend the pellet in CGN culture medium.

9. Count the cells in the suspension.

10. Plate 4×10^5 CGNs per well in a NUNC 24-well plate in 800 μL CGN culture medium. Incubate the cells at 37 °C with 5 % CO_2.

11. After 24 h add 10 μM 1β-Arabinofuranosylcytosine to CGN culture medium to prevent proliferation of non-neuronal cells.

3.2.2 Transduction of Primary Cerebellar Granule Neuron Cultures

1. Transduce CGN cells with lentivirus. For each well, prepare 50 μL of virus suspension diluted in CGN culture medium (*See* **Note 17**). To transduce CGN reduce the volume of the

medium to one-third; add the recombinant lentivirus at different dilutions. Allow the virus to adsorb for 1–2 h, thus render back the medium to its original volume. Then cultivate neurons up to 6–7 days in vitro (DIV) when apoptosis will be induced (Fig. 1).

3.3 Induction and Detection of Apoptosis

3.3.1 Induction of Apoptosis

Induction of apoptosis is carried out in serum-free medium at low (5 mM) KCl [17].

1. Wash cultures twice and maintain in serum-free low (5 mM) KCl CGN culture medium for 8 h.

2. Wash and maintain control cultures in serum-free CGN culture medium for 8 h.

3.3.2 Detection of Apoptosis

Measure caspase 3 activity as follows:

1. Wash 500,000 CGNs with PBS once.

2. Add 100 μL of caspase 3 lysis buffer A to lyse cells.

3. Combine 25 L of lysate with 75 μL of caspase 3 assay buffer B containing 30 μM Ac-DEVD-AMC.

4. Incubate for 20 min at room temperature.

5. Measure fluorescence at an excitation of 380 nm and an emission of 460 nm using a spectrofluorometer (Fig. 2).

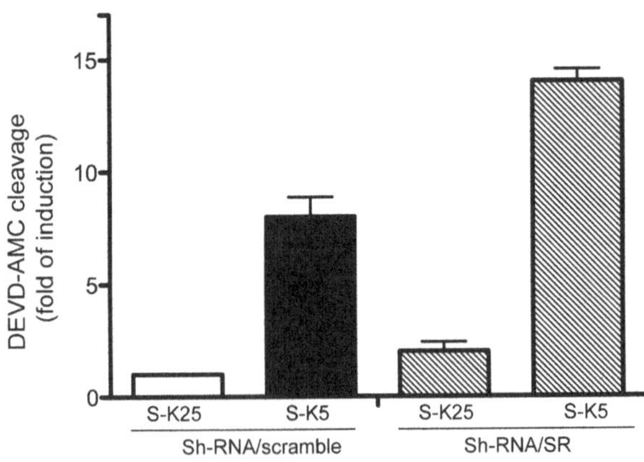

Fig. 2 In vitro CGNs (2 DIV) were transducted either with Sh-RNA/scramble and Sh-RNA/SR lentivirus at MOI 40. At 6 DIV they were induced to undergo apoptosis by serum and KCl deprivation (S-K5); control cells were maintained in serum-free medium supplemented with 25 mM KCl (S-K25). Eight hours after apoptosis induction neurons were lysed and assayed for DEVD-MCA cleavage. Fold-induction of caspase-3 activity is the mean (±SEM) of triplicate determinations from three independent experiments. Note that silencing of SR increases caspase-3 activity in CGNs undergoing apoptosis, suggesting that this enzyme has a protective role in survival of CGNs (*see* ref. 17)

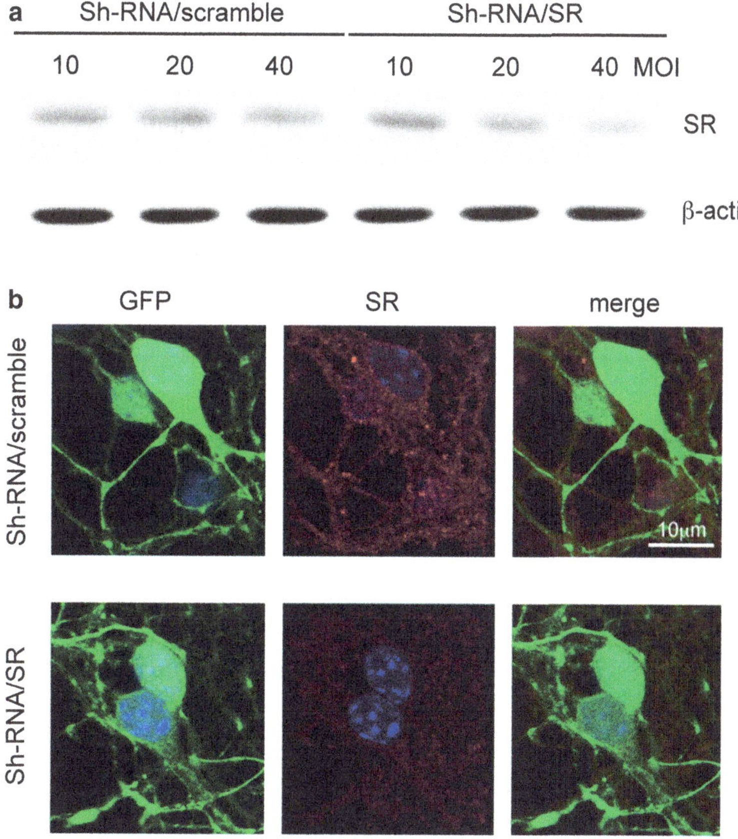

Fig. 3 (**a**) In vitro CGNs (2 DIV = day in vitro) were transducted either with Sh-RNA/scramble and Sh-RNA/SR lentivirus at MOI indicated. At 6 DIV lysates were processed for SDS-PAGE and Western blot for immunodetection of SR and β-actin as control of silencing efficiency and specificity. (**b**) Confocal microscope analysis of SR expression at 6 DIV (*red*) in Sh-RNA/scramble and Sh-RNA/SR transducted CGNs at 2DIV at MOI 40. Transducted neurons either with Sh-RNA/scramble and Sh-RNA/SR express green fluorescent protein (GFP). Nuclei are stained with Hoechst (*blue*).

3.4 Western Blotting and Immunofluorescence for D-Serine Racemase

3.4.1 Western Blotting

1. Extract total proteins by scraping cells in SDS-reducing sample buffer.

2. Boil for 5 min.

3. Perform western blot analysis with mouse anti-D-serine racemase antibody (*see* Fig. 3a).

3.4.2 Immunofluorescence

1. Fix and permeabilize CGN cultured in the chamber slide with methanol 100 % for 20 min at −20 °C.

2. Block with 4 % NGS in PBS for 1 h at room temperature.

3. Incubate slides with affinity purified-goat anti-D-serine racemase antibody diluted 1:50 in PBS overnight at 4 °C.

4. Wash three times with PBS.

5. Add secondary TRITC-conjugated donkey anti-goat antibody diluted 1:200 in PBS and incubate at room temperature for 30 min in a humid chamber.

6. Wash three times with PBS.

7. Remove excess moisture from the slide before adding anti-fade mounting medium.

8. Examine slide under fluorescence/confocal microscope (*see* Fig. 3b).

4 Notes

1. Cells should be of low-passage number and should not be used after passage 20 or if growth is slow.

2. Certain brands of FBS do not support efficient transfection and can result in low viral titers. We routinely use FBS, Qualified, Australia Origin from Gibco™.

3. There is an additional Cla I site in pLVTHM vector that is blocked by Dam methylation. The plasmid needs to be grown in a Dam+ bacterial strain such HB101 in order to use Cla I for cloning.

4. Although it is recommended to avoid to select target sequence in the untranslated regions, since regulatory protein binding to regions in and near the untranslated region might interfere with the RNAi process, in some case targets within the untranslated regions (UTR) have been also reported [18].

5. Avoid selecting target sense or antisense sequences that contain a consecutive run of three or more thymidine residues; a poly(T) tract within the sequence can potentially cause premature termination of the shRNA transcript.

6. Many online tools to design shRNA gives a link to the NCBI BLAST server to search for similarity of the suggested target against the mRNA database of the organism of interest.

7. Heating to 95 °C is essential to remove all secondary structure, disrupt the internal hairpin of each oligonucleotide and promote intermolecular annealing.

8. If you are unable to perform immediately transformation, store ligation at –20 °C.

9. See Tables 5 and 6 in Sambrook and Russell, Molecular Cloning 3rd Ed VIII, p5.42 for different acrylamide concentrations and the effective range of DNA fragment sizes separated.

10. We have transfected 200 ng of target cDNA plasmid (HA-D-serine racemase) plus 500–1,000 ng of the plasmid containing the silencing cassette per 6-well plate and harvest the cells for immunoblot analysis 48–72 h after transfection.

11. For best result and to optimize viral titer cells must be at 70–80 % confluence, equably distributed and with flat morphology before transfection.

12. The DNA-lipofectamine complex must be formed in the absence of proteins even though the complex is able to transfect cells in the presence of proteins such as 10 % FBS. Opti-MEM I is recommended for diluting both DNA and lipofectamine reagent. The ratio of 2.0 μL of lipofectamine 2000 per 1 μg of plasmid has been found to be optimal. Increasing the ratio does not further improve transfection efficiency.

13. Though the complex is able to transfect cells in the presence of proteins such as 10 % FBS, we found an improved transfection efficiency in the absence of serum.

14. To increase the lentivirus titer we have added, caffeine to a final concentration of 2–4 mM for 17–40 h post-transfection [19].

15. Do not use nitrocellulose filters, as nitrocellulose is known to bind lentivirus and reduce titers. Use 0.45 μm polyethersulfone (PES) low protein-binding filters.

16. Peak of virus production is normally achieved 24–48 h post-transfection; however collecting medium at multiple times at 36, 48, and 60 h post-transfection increases the viral yield.

17. Use several multiplicity of infection (MOI) virus stock to find the more suitable MOI to obtain silencing of you gene of interest. In addition, include a transduction with the scramble control and other appropriate positive and negative controls. Mix the virus with the medium gently by inverting the tubes several times. Do not vortex.

Acknowledgments

This work was supported by grants from Italian Ministry of Education, University and Research (PRIN 2009, 2009KP83CR-02 to NC).

References

1. Fire A, Xu S, Montgomery MK, Kostas SA, Driver SE et al (1998) Potent and specific genetic interference by double-stranded RNA in Caenorhabditis elegans. Nature 391: 806–811

2. Hannon GJ (2002) RNA interference. Nature 418:244–251

3. Dykxhoorn DM, Novina CD, Sharp PA (2003) Killing the messenger: short RNAS that silence gene expression. Nature 4:457–467

4. Hammond SM (2005) Dicing and slicing the core machinery of the RNA interference pathway. FEBS Lett 579:5822–5829

5. Yu JY, DeRuiter SL, Turner DL (2002) RNA interference by expression of short-interfering RNAs and hairpin RNAs in mammalian cells. Proc Natl Acad Sci U S A 99:6047–6052

6. Paddison PJ, Caudy AA, Bernstein E, Hannon GJ, Conklin DS (2002) Short hairpin RNAs (shRNAs) induce sequence-specific silencing in mammalian cells. Genes Devel 16:948–958

7. Dull T, Zufferey R, Kelly M, Mandel RJ, Nguyen M et al (1998) J Virol 72:8463–8471

8. Tuschl T (2006) Selection of siRNA sequences for Mammalian RNAi. CSH Protoc 2006(1). pii: pdb.prot4339. doi: 10.1101/pdb.prot4339

9. Elbashir SM, Martinez J, Patkaniowska A, Lendeckel W, Tuschl T (2001) Functional anatomy of siRNAs for mediating efficient RNAi in Drosophila melanogaster embryo lysate. EMBO J 20:6877–6888

10. Elbashir SM, Harborth J, Lendeckel W, Yalcin A, Weber K et al (2001) Duplexes of 21-nucleotide RNAs mediate RNA interference in cultured mammalian cells. Nature 411:494–498

11. Vermeulen A, Behlen L, Reynolds A, Wolfson A, Marshall W et al (2005) The contributions of dsRNA structure to Dicer specificity and efficiency. RNA 11:674–682

12. Hannon GJ, Rossi JJ (2004) Unlocking the potential of the human genome with RNA interference. Nature 431:371–378

13. Gönczy P, Schnabel H, Kaletta T, Amores AD, Hyman T et al (1999) Dissection of cell division processes in the one cell stage *Caenorhabditis elegans* embryo by mutational analysis. J Cell Biol 144:927–946

14. Jantsch-Plunger V, Glotzer M (1999) Depletion of syntaxins in the early Caenorhabditis elegans embryo reveals a role for membrane fusion events in cytokinesis. Curr Biol 9:738–745

15. Brummelkamp TR, Bernards R, Agami R (2002) A system for stable expression of short interfering RNAs in mammalian cells. Science 296:550–553

16. Levi G, Aloisi F, Ciotti MT, Gallo V (1984) Autoradiografic localization and depolarization-induced release of acidic aminoacids in differentiating cerebellar granule cultures. Brain Res 290:77–86

17. Esposito S, Pristerà A, Maresca G, Cavallaro S, Felsani A et al (2012) Contribution of serine racemase/d-serine pathway to neuronal apoptosis. Aging Cell 11:588–598

18. Anthony KG, Bai F, Krishnan MN, Fikrig E, Koski RA (2009) Effective siRNA targeting of the 3' untranslated region of the West Nile virus genome. Antiviral Res 82:166–168

19. Ellis BL, Potts PR, Porteus MH (2011) Creating higher titer lentivirus with caffeine. Hum Gene Ther 22:93–100

Chapter 10

Genomic Analysis Using Affymetrix Standard Microarray GeneChips (169 Format) in Degenerate Murine Retina

Sook Hyun Chung, Weiyong Shen, and Mark Gillies

Abstract

Microarray is one of the most useful tools for gene expression profiling. The growing development of microarray genechip technology has enabled increasingly sophisticated studies on the differences in gene transcription between diseased and non-diseased tissues and provides clues to their contribution to the disease in a single experiment. Thus, microarray is used in not only in clinical diagnostics, but also to understand pathological processes and identify leads for new treatments. Here, we present a detailed protocol for performing genomic analysis of retinal tissue with Affymetrix genechip microarray together with additional guidelines from the authors.

Key words Gene expression profiling, DNA microarray, Affymetrix mouse genechip ST 1.0 array

1 Introduction

1.1 Microarray

Apoptosis, "programmed cell death," is a common feature of neuronal degeneration. Since apoptosis is strongly associated with genomic instability and it requires RNA and protein synthesis [1, 2], genomic analysis of neuronal degeneration is likely to reveal significant changes in genes encoding proteins involved in apoptotic pathways. Microarray, one of the most powerful methods of gene expression profiling, was first developed in mid-1990s [3] to examine genetic differences between diseased and non-diseased cells. It is now widely used in various medical applications from clinical diagnosis to development of new treatments [3].

The Affymetrix mouse genechip array is one of the most advanced gene expression profiling methods. One genechip consists of 770,317 chemically synthesized probes, which cover approximately 28,853 well-annotated mouse genes, on a small quartz surface. The amount of hybridization between fragmented and labeled DNA samples and probes can be monitored at each probe cell (or feature), and is interpreted and analysed to produce biologically

Laura Lossi and Adalberto Merighi (eds.), *Neuronal Cell Death: Methods and Protocols*, Methods in Molecular Biology, vol. 1254, DOI 10.1007/978-1-4939-2152-2_10, © Springer Science+Business Media New York 2015

129

meaningful data. There are approximately a dozen commercially available microarrays. Of these, we would like to focus on Affymetrix Genechip array platform in this chapter.

1.2 Samples' Preparation

Preparation of high quality DNA samples is crucial in order to obtain quality results. Figure 1 shows a schematic diagram of sample preparation, although some of the steps can now be omitted due to the technological advances. Briefly, high quality total RNA extracted from a sample is used as a template to synthesize 1st strand cDNA which forms a double-stranded cDNA/mRNA hybrid molecule. After cleanup of mRNA from the hybrid, 2nd-strand cDNA complementary to the 1st strand cDNA (cRNA), is synthesized, which results in the formation of a DNA/cRNA heteroduplex. After degradation of cRNA from the duplex, second strand DNA is now exposed to random primers for amplification of cDNA complementary to the original RNA. Followed by fragmentation and biotin labeling, samples are injected to a genechip for hybridization to probes. After washing and staining, the genechip is scanned to obtain raw data. More detailed aspects of this protocol are discussed in Subheading 2.

2   Materials

All materials and instruments should be free of nuclease. Wipe your glassware, plasticware, bench, and instruments with RNaseZap (Sigma Chemicals, St. Louis, MO) or RNaseAway (Life Technologies™, Gaithersburg, MD). A dedicated genomic workspace is preferred. Prepare all your solutions with PCR grade water.

2.1 Instruments

1. Pellet pestle (Sigma Chemicals).
2. Microcentrifuge.
3. Heating block.
4. Nuclease free PCR tubes (0.2, 0.5, and 1.5 mL)
5. Pipettes and pipette tips (nuclease free and low retention tips are preferred) for 0.1–2 μL, 2–20 μL, 20–200 μL, 100–1,000 μL.
6. Vortex mixer.
7. Spectrophotometer, e.g., NanoDrop ND-1000 (Thermo Fisher Scientific, Wilmington, DE).
8. Bioanalyzer 2100 Agilent Technologies, Inc., Santa Clara, CA.
9. RNA 6000 nano kit (Agilent Technologies).
10. PCR system.
11. GeneChip Hybridization oven (Affymetrix, Santa Clara, CA).
12. GeneChip Fluidics station 450 (Affymetrix).
13. GeneChip Scanner 3000 7G (Affymetrix).

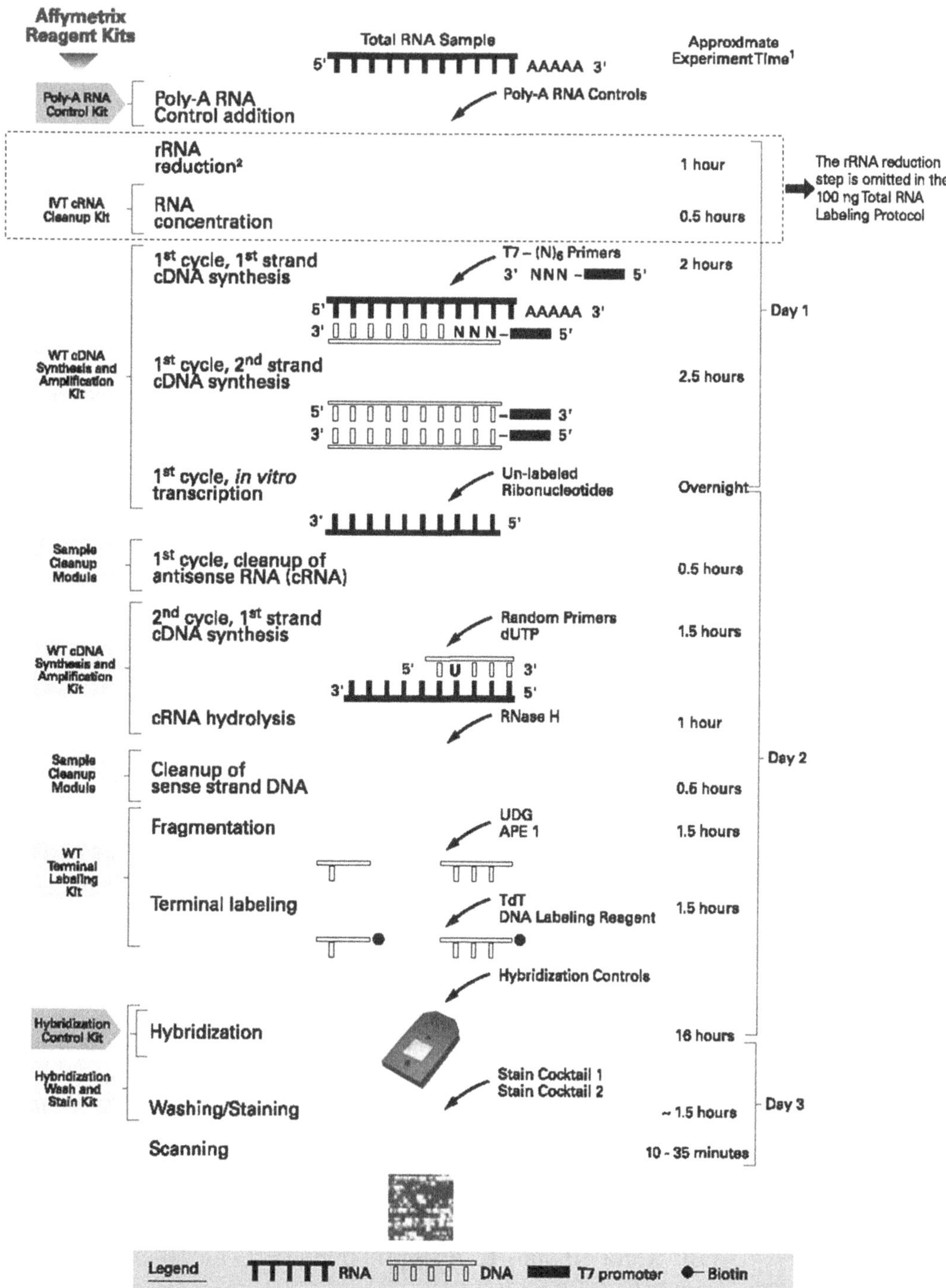

Fig. 1 A schematic diagram of whole transcript sense target labeling assay (courtesy of Affymetrix, Inc., Santa Clara, CA, USA)

2.2 Materials	1. RNeasy mini kit (Quiagen GmbH, Hamburg, Germany).
2.2.1 RNA Extraction and DNAse Treatment	2. RNase-Free DNase set (Qiagen).
	3. 14.3 M β-mercaptoethanol (β-me).
	4. Absolute ethanol.
	5. 70 % ethanol.
	6. *Optional:* RNA*later* (Qiagen).

2.2   Materials

2.2.1   RNA Extraction and DNAse Treatment

1. RNeasy mini kit (Quiagen GmbH, Hamburg, Germany).
2. RNase-Free DNase set (Qiagen).
3. 14.3 M β-mercaptoethanol (β-me).
4. Absolute ethanol.
5. 70 % ethanol.
6. *Optional:* RNA*later* (Qiagen).

2.2.2   cDNA Synthesis and Amplification

1. Applause WT-Amp ST System (NuGEN Technologies, San Carlos, CA).
2. MinElute Reaction Cleanup kit (Qiagen).

2.2.3   Fragmentation and Labeling

1. FL-Ovation™ cDNA Biotin Module V2 (NuGEN Technologies).

2.2.4   Hybridization, Washing, Staining, and Scanning

1. GeneChip hybridization, wash and stain kit (Affymetrix).
2. GeneChip Mouse Gene 1.0 ST Array (169 format) (Affymetrix).
3. Dimethyl sulfoxide (DMSO).
4. PCR-grade water.

---

3   Methods

Prepare ice bucket/cold block and program all the thermo cycles before starting.

3.1   Thermo Cycler Programs

Program 1 (Primer annealing): 65 °C for 5 min, then cool to 4 °C.

Program 2 (First strand cDNA synthesis): 4 °C for 1 min, 25 °C for 10 min, 42 °C for 10 min, 70 °C for 15 min, then cool to 4 °C.

Program 3 (Second strand cDNA synthesis): 4 °C for 1 min, 25 °C for 10 min, 50 °C for 30 min, 70 °C for 5 min, then cool to 4 °C.

Program 4 (Post second strand enhancement): 4 °C for 1 min, 37 °C for 15 min, 80 °C for 20 min, then cool to 4 °C.

Program 5 (Amplification): 4 °C for 1 min, 47 °C for 90 min, 95 °C for 5 min, then cool to 4 °C.

Program 6 (Post amplification modification 1): 4 °C for 1 min, 37 °C for 15 min, 95 °C for 5 min, then cool to 4 °C.

Program 7 (Post amplification modification 2): 4 °C for 1 min, 30 °C for 10 min, 42 °C for 60 min, 75 °C for 10 min, then cool to 4 °C.

Program 8 (Fragmentation): 37 °C for 30 min, 95 °C for 2 min, then cool to 4 °C.

Program 9 (Labeling): 37 °C for 60 min, 70 °C for 10 min, then cool to 4 °C.

Program 10 (Pre-hybridization): 99 °C for 5 min then 45 °C for 5 min.

3.2 Hybridization Oven Program	45 °C for 16–17 h at 60 rpm.

3.3 RNA Extraction

1. Excise tissue and snap freeze it in liquid nitrogen and store it at −80 °C until use or perform RNA stabilization with RNA*later*. *See* **Note 1**.

2. Add 10 µL of β-me to 1 mL of buffer RLT before start. *See* **Note 2**.

3. Add absolute ethanol in buffer RPE before use, as indicated on the bottle.

   Homogenize the tissue with 600 µL of buffer RLT. *See* **Note 3**.

4. Perform RNA extraction with on-column DNase digestion according to manufacturer's instructions.

5. Elute total RNA with 50 µL elution buffer/RNase free water. *See* **Note 4**.

6. Perform bioanalyzer according to the technical manual for quantity and RNA integrity assessment. RNA concentration higher than 40 µg/µL and RNA integrity number higher than 7 should be used in the next step.

3.4 cDNA Synthesis and Amplification

Ensure that RNA input should be between 50 µg and 200 µg.

1. Thaw A1 (First-stand primer mix), A2 (First stand buffer mix) and Nuclease free water (Applause WT-Amp ST System) from a −20 °C freezer. Briefly vortex, spin down, and place on ice.

2. Obtain A3 (First-stand enzyme mix), flick mix, briefly spin down, and place on ice.

3. Add 200 µg of total RNA sample in a volume of less than 5 µL to 0.2 mL PCR reaction tubes. *See* **Note 5**.

4. Add 2 µL of A1 to the reaction tubes. Briefly mix by pipetting up and down, and place on ice.

5. Place the reaction tubes in a thermo cycler and start program 1. *See* **Note 6**.

6. Remove the reaction tubes and place on ice.

7. Prepare first strand master mix by combining 2.5 µL of A2 and 0.5 µL of A3 for a single reaction in a 0.5 mL tube. *See* **Note 7**.

8. Add 3 µL of first strand master mix into the each reaction tube.

9. Mix by pipetting up and down few times.

10. Place the reaction tubes in a thermocycler and start program 2. *See* **Note 8**.

11. Remove the reaction tubes from the thermocycler, spin down briefly to collect condensation, and place on ice.

12. Thaw B1 (Second strand buffer mix) and place B2 (Second strand enzyme mix) and B3 (Reaction enhancement enzyme mix) on ice.

13. Briefly vortex B1, spin down, and place on ice.

14. Prepare second strand master mix by adding 9.75 μL of B1 and 0.25 μL of B2 for a single reaction (*see* **Notes 7** and **9**) in a 0.5 mL tube. Mix by pipetting up and down few times, and place on ice.

15. Add 10 μL of second strand master mix into the each reaction tube. Mix thoroughly by pipetting up and down few times, and place on ice.

16. Place the reaction tubes in a thermocycler and start program 3. *See* **Note 8**.

17. Remove the reaction tubes from the thermocycler, spin down briefly to collect condensation, and place on ice.

18. Prepare post second strand enhancement master mix by combining 3.7 μL B1 and 0.3 μL B3 for a single reaction in a 0.5 mL tube (*see* **Note 7**). Mix gently by pipetting up and down, and place on ice.

19. Add 2 μL of post-second strand enhancement master mix in the each reaction tube (*see* **Note 10**). Mix gently by pipetting up and down, and place on ice.

20. Place the reaction tubes in a thermocycler and start program 4. *See* **Note 8**.

21. Remove the reaction tubes from the thermocycler, spin down briefly to collect condensation, and place on ice.

22. Thaw C2 (SPIA amplification buffer mix) and C1 (SPIA amplification primer mix). Briefly vortex, spin down, and place on ice. Obtain C3 (SPIA amplification enzyme mix), mix thoroughly by pipetting up and down few times and place on ice.

23. Prepare SPIA amplification master mix by combining 1.25 μL C2, 1.25 μL C1, and 2.5 μL C3 for a single reaction in a 0.5 mL tube (*see* **Notes 7** and **11**). Mix thoroughly by pipetting up and down, and place on ice.

24. Add 5 μL of SPIA amplification master mix into each reaction tube. Mix thoroughly by pipetting up and down few times and place on ice.

25. Place the reaction tubes in a thermocycler and start program 5. *See* **Note 8**.

26. Remove the reaction tubes from the thermocycler, spin down briefly to collect condensation, and place on ice.

27. Thaw E1 (Post SPIA modification I primer mix). Vortex briefly, spin down, and place on ice.

28. Bring remaining post-second-strand enhancement master mix from **step 19** (*see* **Note 10**), and prepare post SPIA modification I master mix by combining 2 μL of post second strand

enhancement master mix and 4 µL E1 for a single reaction (*see* **Note 7**). Mix thoroughly by pipetting up and down, and place on ice.

29. Add 6 µL of post-SPIA modification I master mix into the each reaction tube. Mix thoroughly by pipetting up and down and place on ice.

30. Place the reaction tubes in a thermocycler and start program 6. *See* **Note 8**.

31. Remove the reaction tubes from the thermocycler, spin down briefly to collect condensation, and place on ice.

32. Thaw E2 (Post SPIA modification II buffer mix). Vortex briefly, spin down, and place on ice. Obtain E2 (post-SPIA modification II enzyme mix), mix thoroughly by pipetting up and down, and place on ice.

33. Prepare post-SPIA modification II master mix by sequentially combining 4 µL E2 and 4 µL E3 for a single reaction (*see* **Notes** 7 and **12**). Mix thoroughly by pipetting up and down, and place on ice.

34. Add 8 µL of post-SPIA modification II master mix into each reaction tube. Mix thoroughly by pipetting up and down, and place on ice.

35. Place the reaction tubes in a thermocycler and start program 7. *See* **Note 8**.

36. Remove the reaction tubes from the thermocycler, spin down briefly to collect condensation, and place on ice. You can either proceed to **step 37** or store sample at –20 °C.

37. Prepare cDNA purification kit (*see* **Note 13**) by adding absolute ethanol to Buffer PE as labeled on the bottle.

38. Add 300 µL of buffer ERC into a 1.5 mL tube. Add entire amplified cDNA to buffer ERC. Vortex gently and spin down. *See* **Note 14**.

39. Place MinElute spin column into a 2 mL collection tube and add the entire solution to the column. Centrifuge it at maximum speed for 1 min.

40. Discard flow-through and add 750 µL of buffer PE to the column. Centrifuge it at maximum speed, and discard flow-through.

41. Transfer the column into a new 1.5 mL collection tube.

42. Place 15 µL of PCR-grade water and stand the column for 1 min at room temperature. *See* **Note 15**.

43. Centrifuge the column at maximum speed to elute purified cDNA.

44. Take 1.2 µL of purified cDNA for total yield and purity assessment with Nanodrop, and store the rest of the cDNA at −20 °C.

45. Perform Nanodrop. *See* **Note 16**.

3.5 Fragmentation and Labeling

1. Thaw purified cDNA at room temperature and place on ice.

2. Pipet 2.5 µg of purified cDNA in a volume of less than 12.5 µL into a 0.2 mL PCR tube (*see* **Note 17**). Place the PCR tube on ice.

3. Thaw FL1 (Fragmentation buffer mix) from FL-Ovation™ cDNA Biotin Module V2. Mix by vortexing, briefly spin down, and place on ice.

4. Obtain FL2 (Fragmentation enzyme mix). Flick mix, spin down, and place on ice.

5. Prepare fragmentation master mix by combining 2.5 µL FL1 and 1 µL FL2 for a single reaction in a 0.5 mL tube.

6. Add 3.5 µL of the fragmentation master mix into purified cDNA reaction tubes. Mix thoroughly by pipetting up and down, and place on ice.

7. Place the reaction tubes in a thermocycler and start program 8. *See* **Note 6**.

8. Remove the reaction tubes from the thermocycler, spin down briefly to collect condensation, and place on ice.

9. Take out 1.5 µL of fragmented cDNA for bioanalyzer assessment. *See* **Note 18**.

10. Thaw FL3 (labeling buffer mix) and FL4 (labeling reagent) at room temperature. Mix by vortexing, briefly spin down, and place on ice.

11. Obtain FL5 (labeling enzyme mix). Mix FL2 by pipetting up and down few times, and place on ice.

12. Prepare labeling master mix by combining 7.5 µL FL3, 0.75 µL FL4 and 0.75 FL5 for a single reaction in a 0.5 mL tube. Mix thoroughly by pipetting up and down, and place on ice.

13. Add 9 µL of the labeling master mix into the fragmented cDNA reaction tubes. Mix thoroughly by pipetting up and down, and place on ice.

14. Place the reaction tubes in a thermocycler and start program 9. *See* **Note 6**.

15. Remove the reaction tubes from the thermocycler, and spin down briefly to collect condensation. Fragmented and labeled cDNA can be store at −20 °C until hybridization.

3.6 Hybridization

1. Heat 20× Eukaryotic controls in a heat block at 65 °C for 5 min.

2. Thaw fragmented and labeled cDNA and place on ice.

3. Prepare hybridization master mix by combining 1.7 µL of control oligo BS (3 nM), 5 µL of 20× Eukaryotic hybridization controls, 50 µL of 2× hybridization mix, 7 µL of DMSO, 9.3 µL of nuclease-free water for a single reaction (*see* **Note 7**). Mix by vortexing and briefly spin down.

4. Add 73 µL of hybridization master mix and 27 µL of fragmented and labeled cDNA in a 0.2 mL tube.

5. Place the reaction tubes in a thermocycler and start program 10. *See* **Note 6**.

6. Remove the reaction tubes from the thermocycler and centrifuge for 1 min at maximum speed.

7. Remove genechip arrays from fridge and equilibrate at room temperature. Label the genechip clearly with sample ID.

8. Inject 80 µL of hybridization mix with cDNA into a genechip array through lower septa on the back of the chip. *See* **Note 19**.

9. Apply tough spot onto both septas to avoid leak while hybridization.

10. Place the genechips in hybridization oven (*see* **Note 20**), and start hybridization at 45 °C for 17 h at 60 rpm.

11. Turn on the computer which has fluidics software program. *See* **Note 21**.

12. Open gene chip operating system (GCOS), select run-fluidics, and choose priming protocol.

13. Add washing buffers A and B to the corresponding inlet position.

14. Choose "all module" in GCOS and select RUN.

15. While priming, scan genechip array barcode and input all experimental data including sample IDs into GCOS.

16. Aliquot 600 µL of stain cocktail 1 and 2, and 800 µL of array holding buffer into 1.5 mL tubes (*see* **Note 22**). Mix by vortexing, briefly spin down (*see* **Note 23**), and place all tubes on fluidics station in position 1, 2, and 3, respectively.

17. Remove the genechip arrays from the hybridization oven and remove tough spots from the septas. Retain hybridization mix and cDNA through lower septa. Store hybridization mix and cDNA at −80 °C. *See* **Note 24**.

18. Place arrays in fluidics station with array window facing out, and run fluidics.

19. When fluidics script is finished, remove the array and inspect for bubbles. If bubbles are present, manually add 100 μL of array holding buffer through the lower septa. *See* **Note 25**.

20. Apply new tough spots to the septas to avoid leak of array holding buffer. *See* **Note 26**.

21. Place the array in scanner (*see* **Note 27**) and click Run Scanner in GCOS.

22. Obtain the scanned data and start bioinformatics.

4 Notes

1. If you perform RNA stabilization with RNA*later*, ensure that the tissue is fully immersed into RNA*later* and the thickness of tissue should be less than 0.5 cm for efficient penetration of RNA*later*.

2. Use fume hood when handling β-me.

3. Wipe the tip of the pestle with RNaseZap before use. Resuspend pellets with pestle for efficient homogenization.

4. Aliquot RNA extracts and store at –80 °C to avoid freeze-thaw cycle.

5. Adjust the volume with nuclease-free water up to 5 μL if necessary.

6. Preheat the thermocycler lid before start.

7. Prepare the master mix by 10 % more than the number of reactions you need. If you have eight samples, for example, prepare volume of master mix for 8.8 samples.

8. Cool down the thermocycler before start.

9. Do not return B1 (Second strand buffer mix) to a freezer since the post second strand enhancement protocol requires B1 (**step 18**).

10. Leave the remaining post-second-strand enhancement master mix on ice, since SPIA modification reaction requires it (**step 28**).

11. Ensure that C3 is added at the last moment before adding the master mix into each reaction tube.

12. E3 enzyme mix is quite viscous. Ensure to slowly pipette out E3 and mix it thoroughly with the master mix by pipetting up and down.

13. Store MinElute spin column at 4 °C to avoid enzyme degradation.

14. Check the color of the mixture. Yellow indicates pH ≤7.5. If the color is orange or violet, add 10 μL of 3 M sodium acetate and mix gently. The color of the mixture will turn to yellow.

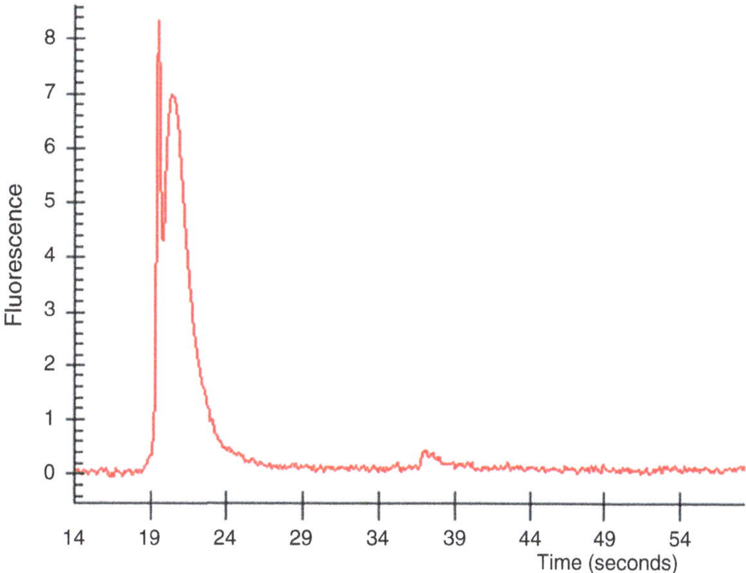

Fig. 2 A typical bioanalyzer profile of fragmented single stranded DNA (courtesy of Affymetrix, Inc., Santa Clara, CA, USA)

15. Ensure that the water is dispensed directly onto the column membrane for complete elution of cDNA.

16. 260/280 ratio should be ≥1.8.

17. Adjust the volume with nuclease-free water up to 12.5 µL if necessary.

18. Bioanalyzer assessment should be performed for size analysis. Ensure that the range in peak size of fragmented cDNA is approximately between 22 and 45 nucleotides. *See* Fig. 2.

19. Insert a 200 µL pipette tip into the upper septa for venting of the air from the genechip chamber, and inject hybridization mix with cDNA sample through the lower septa. *See* Fig. 3. Check probe array on glass and ensure there is no bubble in the probe array. If bubbles are present, pipette up and down several times to remove them.

20. Ensure to balance the hybridization even when placing genechips.

21. Do not stop hybridization oven while preparing fluidics.

22. Stain cocktail 1 is light sensitive. Ensure to wrap the tube containing cocktail 1 with aluminum foil.

23. Ensure that no bubbles are in the tubes. Presence of bubbles may affect staining.

24. Hybridization mix and cDNA sample can be reused if required.

25. Ensure that no bubbles are present in the array. Bubbles can affect scanning of the data.

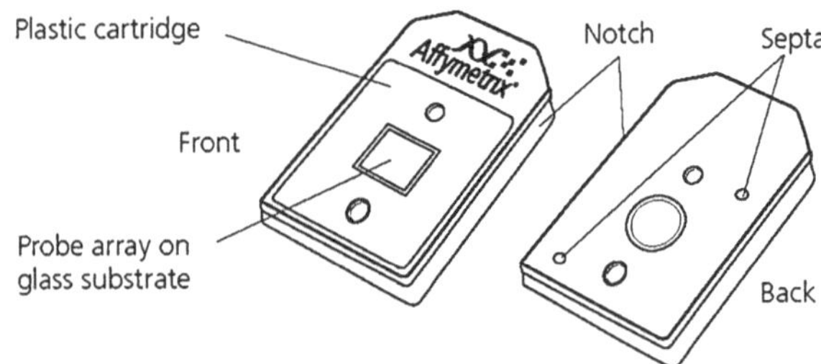

Fig. 3 A diagram of GeneChip probe array (courtesy of Affymetrix, Inc., Santa Clara, CA, USA)

26. If septas leak, the scanner can be damaged. Wipe excess liquid from the array with kimwipe.

27. Do not manually open or close the scanner door.

Acknowledgements

The authors would like to thank Martha Manion for Affymetrix copyright permission and Robert Solomon for technical support.

References

1. Hutchins JB, Barger SW (1998) Why neurons die: cell death in the nervous system. Anat Rec 253:79–90

2. Zhivotovsky B, Kroemer G (2004) Apoptosis and genomic instability. Nat Rev Mol Cell Biol 5:752–762

3. Hoopes L (2008) Genetic diagnosis: DNA microarrays and cancer. Nat Educ 1:3

Chapter 11

Genomic Analysis of Transcriptional Changes Underlying Neuronal Apoptosis

Sebastiano Cavallaro

Abstract

Neuronal apoptosis represents an intrinsic «suicide» program, by which a neuron orchestrates its own destruction. Although engagement of apoptosis requires transcription and protein synthesis, the complete spectrum of genes involved in distinct temporal domains remained unknown until the advent of genomics. In the last 10 years, the genome sequences and the development of high-throughput genomic technologies, such as DNA microarrays, have offered the unprecedented experimental opportunities to explore the transcriptional mechanisms underlying apoptosis from a new systems-level perspective. This review goes over this genomic approach and illustrates the use of microarray methodology to dissecting the multigenic program underlying neuronal apoptosis.

Key words Apoptosis, DNA microarray, Gene, Mechanisms, Pathway, Transcription, Transcriptome

1 Introduction

Neuronal apoptosis represents an intrinsic «suicide» program, by which a neuron orchestrates its own destruction. It is characterized by specific morphological and biochemical events, including fragmentation of nuclear DNA, breakdown of the cellular cytoskeleton, and the bulging out of the plasma membrane (blebbing), which may lead to the detachment of the so-called apoptotic bodies [1, 2]. During normal nervous system development, physiologically appropriate neuronal loss contributes to a sculpting process that removes approximately one-half of all neurons born during neurogenesis [3]. Neuronal loss subsequent to this developmental window is physiologically inappropriate for most systems and can contribute to neurological deficits, e.g., neurodegenerative diseases such as Alzheimer's and Parkinson disease [1, 4, 5]. Elucidating the molecular mechanisms underlying neuronal apoptosis hence may contribute to our understanding of basic developmental biology and to human neuropathology.

Laura Lossi and Adalberto Merighi (eds.), *Neuronal Cell Death: Methods and Protocols*, Methods in Molecular Biology, vol. 1254, DOI 10.1007/978-1-4939-2152-2_11, © Springer Science+Business Media New York 2015

Although an extensive number of studies have implicated individual genes or genetic pathways during apoptosis, the complete spectrum of genes involved in distinct temporal domains remained mostly unknown until the advent of genomics. In the last 10 years, the genome sequences and the development of high-throughput genomic technologies, such as DNA microarrays, have offered novel experimental opportunities to explore the transcriptional mechanisms underlying apoptosis in different paradigms [6–16]. This chapter illustrates how a genomic approach based on expression DNA microarrays can be used to dissecting the multigenic program underlying apoptosis of neurons. This approach has been initially used in primary cultures of rat cerebellar granule neurons (CGNs), a model of election for the study of neuronal apoptosis [7]. In this in vitro paradigm, rat CGNs undergo rapid apoptotic cell death within 24 h after removal of serum and lowering of extracellular potassium from 25 to 5 mM [17]. Engagement of apoptosis requires transcription and protein synthesis and the process becomes irreversible during the first 6 h following induction. Before this "commitment point" CGNs can be rescued by the activation of specific signal transduction pathways or by the treatment with specific neurotrophic factors, such as insulin-like growth factor-1 (IGF-1) [17, 18] and pituitary adenylyl cyclase-activating polypeptide 38 (PACAP) [19]. The first hours following the induction of apoptosis, therefore, represent a time frame that is suitable for studying the transcriptional program controlling neuronal apoptosis and survival. In addition to rat CGNs, a number of other in vitro and in vivo experimental models can be used to explore the transcriptional mechanisms underlying neuronal apoptosis. For space limitations this chapter will not detail these experimental models, but focus only on the DNA microarray methodology used to analyze gene expression on a genomic scale.

2   Materials

2.1   General and Safety Notes

Follow good laboratory practices and follow all waste disposal regulations when disposing waste materials. Wear appropriate personal protective equipment. Calibrate pipettes at least yearly to avoid over-drawing. Prepare all solutions using molecular biology-grade, RNase- and DNase-free water and use analytical grade reagents. Prepare and store all reagents at room temperature (unless indicated otherwise). To prevent contamination of reagents by nucleases, always wear powder-free laboratory gloves and eye/face protection, and use dedicated solutions and pipettors with sterile nuclease-free aerosol-resistant tips.

2.2   Warnings

Cyanine dye reagents are potential carcinogens. Avoid inhalation, swallowing, or contact with skin. LiCl (a component of the Agilent

2× Hybridization Buffer) is toxic and a potential teratogen, it is harmful if inhaled, swallowed, or contacts skin. Lithium dodecyl sulfate (LDS) (a component of the Agilent 2× Hybridization Buffer) is harmful by inhalation and irritating to eyes, respiratory system, and skin. Triton (a component of the Agilent 2× Hybridization Buffer and wash buffers) is harmful if swallowed or contacts the eye.

2.3 Equipment and Reagents

1. Microarray scanner (Agilent Technologies, Santa Clara, CA). *See* **Note 1**.

2. Hybridization chamber (Agilent Technologies).

3. Hybridization chamber gasket slides (Agilent Technologies, 4 microarray/slide).

4. Hybridization oven (Agilent Technologies).

5. Hybridization oven rotator for Agilent microarray hybridization chambers (Agilent Technologies).

6. Nuclease-free 1.5 mL microfuge tubes (e.g., Life Technologies, Grand Island, NY).

7. Magnetic stir bars and plate (2×).

8. Microcentrifuge (e.g., Eppendorf AG, Hamburg, Germany).

9. Slide-staining dish, with slide rack (3×).

10. Vacuum concentrator (e.g., Thermo Scientific™).

11. Circulating water baths or heat blocks set to 37, 40, 60, 65, 70, and 80 °C.

12. Forceps.

13. Ice bucket.

14. Vortex mixer.

15. Timer.

16. Nitrogen purge box for slide storage.

17. Low Input Quick Amp Labeling Kit, One-Color (Agilent Technologies).

18. RNA Spike-In Kit, One-Color (Agilent Technologies).

19. Gene Expression Hybridization Kit (Agilent Technologies).

20. Gene Expression Wash Buffer Kit (Agilent Technologies).

21. Ethanol (DNase/RNase free).

22. RNeasy Mini Kit (Qiagen, Germantown, MD).

23. Kit for total RNA extraction from cells or tissues (e.g., TRIzol Reagent Life Technologies™, Grand Island, NY). *See* **Note 2**.

24. Spectrophotometer (e.g., Thermo Scientific™, NanoDrop products, Wilmington, DE). *See* **Note 3**.

25. Agilent 2100 bioanalyzer with the RNA 6000 Nano LabChip kit (optional). *See* **Note 4**.

3   Methods

The workflow for sample preparation and microarray processing is shown in Fig. 1.

3.1   Template Preparation

The time required for this procedure is about 0.5 h. Agilent One color Spike-In mix contains ten in vitro synthesized, polyadenylated transcripts in predetermined ratios. These RNAs are transcript constructed by the cloning of a unique 55-mer sequence into the human adenovirus type 6 E1A 13S gene. By specifically hybridizing to complementary control probes on Agilent's microarrays, they allow to monitor microarray workflow for linearity, sensitivity, and accuracy. Fixed amounts of Spike-In RNA are mixed to sample RNA before the labeling procedure. Table 1 provides the dilutions of Spike Mix for a range of total RNA input amounts.

For inputs not shown in Table 1, the amount of Spike Mix is proportional to the amount of RNA input.

To prepare the Spike Mix dilution appropriate for 50 ng of total RNA starting sample:

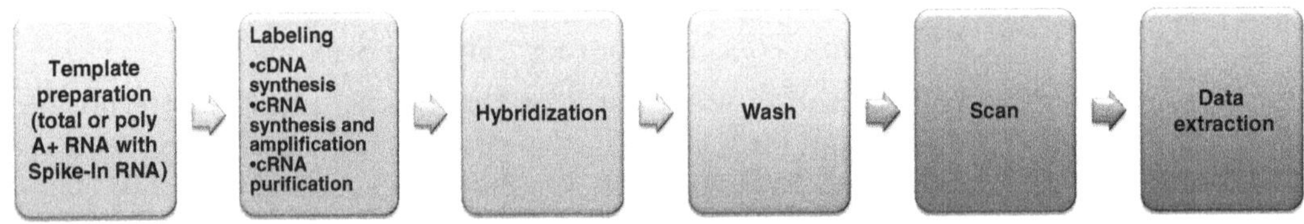

Fig. 1 Workflow for sample preparation and array processing

Table 1
Dilutions of Spike Mix for one color (cyanine 3)-labeling

Starting amount of RNA		Serial dilution of Spike-In mix				Volume of diluted Spike Mix used in each
Total RNA (ng)	PolyA + RNA (ng)	First	Second	Third	Fourth	labeling reaction (μL)
10		1:20	1:25	1:20	1:10	2
25		1:20	1:25	1:20	1:4	2
50		1:20	1:25	1:20	1:2	2
100		1:20	1:25	1:20		2
200		1:20	1:25	1:20		2
	5	1:20	1:25	1:20		2

1. Label a new sterile 1.5-mL microcentrifuge tube "Spike Mix First Dilution."

2. Mix the thawed Spike Mix vigorously on a vortex mixer.

3. Heat at 37 °C in a circulating water bath for 5 min.

4. Mix the Spike Mix tube vigorously again on a vortex mixer.

5. Spin briefly in a centrifuge to drive contents to the bottom of the tube.

6. Put 2 µL of Spike Mix stock into the First Dilution tube.

7. Add 38 µL of Dilution Buffer provided in the Spike-In kit (1:20).

8. Mix thoroughly on a vortex mixer and spin down quickly in a microcentrifuge to collect all of the liquid at the bottom of the tube. This tube contains the First Dilution. It can be stored for up to 2 months in a freezer at –80 °C.

9. Label a new sterile 1.5-mL microcentrifuge tube "Spike Mix Second Dilution."

10. Put 2 µL of First Dilution into the Second Dilution tube.

11. Add 48 µL of Dilution Buffer (1:25).

12. Mix thoroughly on a vortex mixer and spin down quickly in a microcentrifuge to collect all of the liquid at the bottom of the tube. This tube contains the Second Dilution.

13. Label a new sterile 1.5-mL microcentrifuge tube "Spike Mix Third Dilution."

14. Put 2 µL of Second Dilution into the Third Dilution tube. Add 38 µL of Dilution Buffer (1:20).

15. Mix thoroughly on a vortex mixer and spin down quickly in a microcentrifuge to collect all the liquid at the bottom of the tube. This tube contains the Third Dilution.

16. Label a new sterile 1.5-mL microcentrifuge tube "Spike Mix Fourth Dilution.

17. Into the Fourth Dilution tube, add 10 µL of Third Dilution to 30 µL of Dilution Buffer for the Fourth Dilution (1:4).

18. Mix thoroughly on a vortex mixer and spin down quickly in a microcentrifuge to collect all of the liquid at the bottom of the tube. This tube contains the Fourth Dilution (now at a 40,000-fold final dilution).

19. Add 2 µL of Fourth Dilution to 50 ng of total RNA sample as listed in Table 1 and continue with cyanine 3 labeling described below.

3.2 Labeling Reaction

The time required for this procedure is about 5.5 h.

3.2.1 cDNA Synthesis

1. Add 10–200 ng of total RNA to a 1.5-mL microcentrifuge tube in a final volume of 1.5 μL.

2. Add 2 μL of diluted Spike Mix. The tube now contains a total volume of 3.5 μL.

3. Add 1.8 μL of diluted T7 Primer mix from the Low Input Quick Amp Labeling Kit (0.8 μL T7 Primer with 1 μL of nuclease-free water) to the tube. The tube now contains a total volume of 5.3 μL.

4. Denature the primer and the template by incubating the tube at 65 °C in a circulating water bath for 10 min.

5. Place the reactions on ice and incubate for 5 min.

6. Prepare a cDNA by adding 2 μL of 5× First-Strand Buffer, 1 μL of 0.1 M , 0.5 μL of 10 mM dNTP Mix, and 1.2 μL of Affinity Script RNase Block Mix. Add this cDNA Master Mix (4.7 μL) to each sample tube and mix by pipetting up and down. Each tube now contains a total volume of 10 μL.

7. Incubate samples at 40 °C in a circulating water bath for 2 h.

8. Move samples to a 70 °C circulating water bath and incubate for 15 min to inactivate the AffinityScript enzyme.

9. Move samples to ice. Incubate for 5 min.

10. Spin sample briefly in a microcentrifuge. If you do not immediately continue to the next step, store the samples at –80 °C.

3.2.2 cRNA Synthesis and Amplification

1. Add 6 μL of a Transcription Master Mix (containing 0.75 μL of nuclease-free water, 3.2 μL of 5× Transcription Buffer, 0.6 μL of 0.1 M DTT, 1 μL of NTP Mix, 0.21 μL of T7 RNA Polymerase Blend, 0.24 μL of Cyanine 3-CTP) to each sample tube. Gently mix by pipetting. Each tube now contains a total volume of 16 μL.

2. Incubate samples in a circulating water bath at 40 °C for 2 h. If you do not immediately continue to the next step, store the samples in a freezer at –80 °C.

3.2.3 cRNA Purification

The time required for this procedure is about 0.5 h. Use the Qiagen RNeasy Mini Kit to purify the amplified cRNA samples as follows:

1. Add 84 μL of nuclease-free water to your cRNA sample. The tube now contains a total volume of 100 μL.

2. Add 350 μL of Buffer RLT and mix well by pipetting.

3. Add 250 μL of ethanol and mix thoroughly by pipetting.

4. Transfer the 700 μL of the cRNA sample to an RNeasy Mini Spin Column in a 2 mL collection tube. Spin the sample in a centrifuge at 4 °C for 30 s at 13,000×g. Discard the flow-through and collection tube.

5. Transfer the RNeasy column to a new 2 mL collection tube and add 500 μL of Buffer RPE (containing ethanol) to the column. Spin the sample in a centrifuge at 4 °C for 30 s at 13,000×*g*. Discard the flow-through. Reuse the collection tube.

6. Add another 500 μL of Buffer RPE to the column. Centrifuge the sample at 4 °C for 60 s at 13,000×*g*. Discard the flow-through and the collection tube.

7. Elute the purified cRNA sample by transferring the RNeasy column to a new 1.5 mL collection tube. Add 30 μL RNase-free water directly onto the RNeasy filter membrane. Wait 60 s, then centrifuge at 4 °C for 30 s at 13,000×*g*.

8. Maintain the cRNA sample-containing flow-through on ice and discard the RNeasy column.

9. Use the NanoDrop ND-1000 UV–Vis spectrophotometer to quantify the cRNA. Blank the instrument by pipetting 1 μL of nuclease-free water. Then pipette 1 μL of the cRNA sample onto the instrument sample loading area and calculate Cyanine 3 dye concentration (pmol/μL), RNA absorbance ratio (A_{260}/A_{280}), and cRNA concentration (ng/μL). Use the concentration of cRNA (ng/μL) to determine the μg cRNA yield as follows: (concentration of cRNA)×30 μL (elution volume)/1,000 = μg of cRNA.

Use the concentrations of cRNA (ng/μL) and cyanine 3 (pmol/μL) to determine the specific activity as follows: concentration of Cy3 / Cy3 per μg cRNA×1,000 = pmol Cy3 per μg cRNA. The recommended cRNA yield (μg) and specific activity (pmol Cy3 per μg cRNA) for hybridization with a 4× format microarray (*see* below) are 1.65 and 6, respectively.

3.3 **Hybridization**

The time required for this procedure is about 17 h. Use the Agilent Gene Expression Wash Buffer Kit.

1. Prepare the 10× Blocking Agent by adding 500 μL of nuclease-free water to the vial containing lyophilized 10× Gene Expression Blocking Agent supplied with the Gene Expression Hybridization Kit. Gently mix on a vortex mixer and briefly spin the tube for 5–10 s.

2. Prepare the hybridization samples by adding 1.65 μg of -labeled, linearly amplified cRNA, 11 μL of 10× Gene Expression Blocking Agent, Nuclease-free water (up to 52.8 μL) and 2.2 μL of 25× Fragmentation Buffer. The total volume is 55 μL.

3. Incubate at 60 °C for exactly 30 min to fragment the cRNA.

4. Immediately cool on ice for 1 min.

5. Add 55 µL of 2× Hi-RPM Hybridization Buffer to stop the fragmentation reaction.

6. Mix well by careful pipetting up and down. Do not introduce bubbles to the mix.

7. Spin for 1 min at $13,000 \times g$ in a microcentrifuge to drive the sample off the walls and lid and to aid in bubble reduction.

8. Place sample on ice and load onto the array as soon as possible.

9. Prepare the hybridization assembly by loading a clean gasket slide into the Agilent SureHyb chamber base with the label facing up and aligned with the rectangular section of the chamber base.

10. Slowly dispense 100 µL of the hybridization sample onto the gasket well in a "drag and dispense" manner.

11. Slowly put the Agilent expression 4×44 microarray slide "active side" down, parallel to the SureHyb gasket slide, so that the "Agilent"-labeled barcode is facing down and the numeric barcode is facing up. Make sure that the sandwich-pair is properly aligned.

12. Place the SureHyb chamber cover onto the sandwiched slides and slide the clamp assembly onto both pieces. Firmly hand-tighten the clamp onto the chamber. Vertically rotate the assembled chamber to wet the gasket and assess the mobility of the bubbles. If necessary, tap the assembly on a hard surface to move stationary bubbles. Place assembled slide chamber in a hybridization oven set to 65 °C for 17 h. Set your hybridization rotator to rotate at 10 rpm.

3.4   Wash

The time required for this procedure is about 0.5 h. Use the Agilent Gene Expression Wash Buffer Kit. *See* **Note 5**.

1. Add 2 mL of 10 % Triton X-102 to Gene Expression Wash Buffer 1 and Gene Expression Wash Buffer 2.

2. Dispense 1,000 mL of Gene Expression Wash Buffer 2 directly into a sterile bottle and pre-warmed it in a 37 °C water bath.

3. Wash all dishes, racks, and stir bars with Milli-Q water. Run copious amounts of Milli-Q water through the staining dish. Empty out the water collected in the dish.

4. Completely fill a slide-staining dish #1 with Gene Expression Wash Buffer 1 at room temperature.

5. Place a slide rack into slide-staining dish #2. Add a magnetic stir bar. Fill slide-staining dish #2 with enough Gene Expression Wash Buffer 1 to cover the slide rack. Place this dish on a magnetic stir plate.

6. Place an empty dish #3 on the stir plate and add a magnetic stir bar. Do not add the pre-warmed (37 °C) Gene Expression Wash Buffer 2 until the first wash step has begun.

7. Remove the hybridization chamber from incubator and record time. Place the hybridization chamber assembly on a flat surface and loosen the thumbscrew, slide off the clamp assembly, and remove the chamber cover. Remove the array-gasket sandwich from the chamber base by grabbing the slides from their ends. Keep the microarray slide numeric barcode facing up as you quickly transfer the sandwich to slide-staining dish #1. Without letting go off the slides, submerge the array-gasket sandwich into slide-staining dish #1 containing Gene Expression Wash Buffer 1.

8. With the sandwich completely submerged in Gene Expression Wash Buffer 1, separate the slides and let the gasket slide drop to the bottom of the staining dish. Grasp the top corner of the microarray slide and then put it into the slide rack in the slide-staining dish #2 that contains Gene Expression Wash Buffer 1. Transfer the slide quickly so avoid premature drying of the slides. Touch only the barcode portion of the microarray slide or its edges. Stir using setting 4 for 1 min.

9. During this wash step, remove Gene Expression Wash Buffer 2 from the 37 °C water bath and pour into the slide-staining dish #3.

10. Transfer slide rack to slide-staining dish #3 containing Gene Expression Wash Buffer 2 at elevated temperature. Stir using setting 4 for 1 min.

11. Slowly remove the slide rack minimizing droplets on the slides. It should take 5–10 s to remove the slide rack. If liquid remains on the bottom edge of the slide, dab it on a cleaning tissue.

12. Put the slides in a slide holder and scan it as soon as possible to minimize the impact of environmental oxidants on signal intensities. If necessary, store slides in orange slide boxes in a nitrogen purge box, in the dark.

3.5 Scan

1. After selecting the appropriate settings (Profile AgilentHD_GX_1color for Agilent C Scanner) scan the slide. *See* **Note 6**.

3.6 Data Extraction

Data extraction is the process by which probe feature information is extracted from microarray scan data, allowing obtaining gene expression values for each of the RNA measured. In this process a grid template containing the general information about image acquisition and analysis (such as the diameter, the x- and y-coordinates of each feature, which is associated to a unique identifier), is overlay to the .tif image produced by the scanner. After the extraction is completed successfully, it is possible to obtain a quality control report as well as all the metric evaluation of the expression analysis. This information will then be analyzed by further software (such as GeneSpring) to identify differentially regulated genes and pathways.

4 Notes

1. The Agilent microarray scanner is the most appropriate for the microarray type and platform described in this chapter. Other scanners can be used, such as the GenePix Personal 4100A microarray scanner and the GenePix Pro 6.0 acquisition and data-extraction software (Molecular Devices, Sunnyvale, CA). Before use, please verify the scanner properties for compatibility: scan region (61×21.6 mm), scan resolution (5 mm), dye channel, and PMT settings. The method described here is suitable for the Agilent 4×44 format expression microarrays. A long list of microarrays of this format are available for different organisms. Other Agilent microarrays, containing a larger number of features could be used with minor changes in the amount of cRNA hybridized or scanner settings.

2. Use pure high quality, intact template total or poly A + RNA. RNA that is not pure, as measured by A_{260}/A_{230} ratio, can lead to poor results and must be purified.

3. RNA quality should be determined before labeling by the use of a spectrophotometer (A_{260}/A_{280} ratio >1.8).

4. The assessment of total RNA quality can be performed by the Agilent 2100 bioanalyzer with the RNA 6000 Nano LabChip kit. This instrument provides a RNA Integrity Number (RIN) that represents a quantitative value for RNA integrity and facilitates the standardization of quality interpretation. RNA with a minimum RIN threshold of nine should be used. Sample required per reaction is 200 ng of Poly A(+) RNA, or 50–2,000 ng of total RNA.

5. The microarray wash procedure for the Agilent one-color platform must be done in environments where ozone levels are 50 ppb or less.

6. Additional information and updates on the methodology described here can be accessed on the Agilent web site at www.agilent.com.

References

1. Arends MJ, Wyllie AH (1991) Apoptosis: mechanisms and roles in pathology. Int Rev Exp Pathol 32:223–254

2. Jellinger KA (2006) Challenges in neuronal apoptosis. Curr Alzheimer Res 3:377–391

3. Oppenheim RW (1991) Cell death during development of the nervous system. Annu Rev Neurosci 14:453–501

4. Pettmann B, Henderson CE (1998) Neuronal cell death. Neuron 20:633–647

5. Mattson MP (2006) Neuronal life-and-death signaling, apoptosis, and neurodegenerative disorders. Antioxid Redox Signal 8:1997–2006

6. Cavallaro S, Calissano P (2006) A genomic approach to investigate neuronal apoptosis. Curr Alzheimer Res 3:285–296

7. Cavallaro S, D'Agata V, Alessi E et al (2004) Gene expression profiles of apoptotic neurons. Genomics 84:485–496

8. Cavallaro S (2007) Neuronal apoptosis revealed by genomic analysis: integrating gene expression profiles with functional information. Neuroinformatics 5:115–126

9. Paratore S, Parenti R, Torrisi A et al (2006) Genomic profiling of cortical neurons following exposure to beta-amyloid. Genomics 88:468–479

10. Kristiansen M, Menghi F, Hughes R et al (2011) Global analysis of gene expression in NGF-deprived sympathetic neurons identifies molecular pathways associated with cell death. BMC Genomics 12:551

11. Ohkawara T, Nagase H, Koh CS et al (2011) The amyloid precursor protein intracellular domain alters gene expression and induces neuron-specific apoptosis. Gene 475:1–9

12. Desagher S, Severac D, Lipkin A et al (2005) Genes regulated in neurons undergoing transcription-dependent apoptosis belong to signaling pathways rather than the apoptotic machinery. J Biol Chem 280:5693–5702

13. Koh CH, Peng ZF, Ou K et al (2007) Neuronal apoptosis mediated by inhibition of intracellular cholesterol transport: microarray and proteomics analyses in cultured murine cortical neurons. J Cell Physiol 211:63–87

14. Kondo M, Shibata T, Kumagai T et al (2002) 15-Deoxy-delta(12,14)-prostaglandin J(2): the endogenous electrophile that induces neuronal apoptosis. Proc Natl Acad Sci U S A 99:7367–7372

15. Ng JM, Chen MJ, Leung JY et al (2012) Transcriptional insights on the regenerative mechanics of axotomized neurons in vitro. J Cell Mol Med 16:789–811

16. Choy MS, Chen MJ, Manikandan J et al (2011) Up-regulation of endoplasmic reticulum stress-related genes during the early phase of treatment of cultured cortical neurons by the proteasomal inhibitor lactacystin. J Cell Physiol 226:494–510

17. D'Mello SR, Galli C, Ciotti T et al (1993) Induction of apoptosis in cerebellar granule neurons by low potassium: inhibition of death by insulin-like growth factor I and cAMP. Proc Natl Acad Sci U S A 90:10989–10993

18. Galli C, Meucci O, Scorziello A et al (1995) Apoptosis in cerebellar granule cells is blocked by high KCl, forskolin, and IGF-1 through distinct mechanisms of action: the involvement of intracellular calcium and RNA synthesis. J Neurosci 15:1172–1179

19. Cavallaro S, Copani A, D'Agata V et al (1996) Pituitary adenylate cyclase activating polypeptide prevents apoptosis in cultured cerebellar granule neurons. Mol Pharmacol 50:60–66

Chapter 12

High-Throughput Cell Death Assays

Matthew E. Pamenter and Gabriel G. Haddad

Abstract

High-throughput screens (HTS) are powerful tools that permit the rapid evaluation of thousands of samples in a cost-effective manner and minimize sample and reagent consumption. Such assays have recently begun to be utilized to evaluate cell death modalities and also the cytoprotective efficacy of compounds against a wide variety of stresses. Here we describe the design, preparation, and undertaking of HTS-appropriate assays that utilize simple and cost-effective fluorophore- and luminescence-based functional readouts of cell viability. These assays permit the examination of 96–384 compounds in a single multiwell plate with highly robust statistical significance at a fraction of the financial and work cost of traditional approaches.

Key words Cell line, Fluorophore, Ischemia, Luciferase, Microplate, Z'-factor

1 Introduction

Cell death assays are widely used both in the examination of cell death modalities induced by a given insult and also in the search for cytoprotective candidate compounds against a wide variety of clinically relevant stresses. Most commonly used cell death assays are not suitable for large-scale screening approaches and this has led to the development of a specialized field of study focused on the design of assays optimized for the rapid analysis of large numbers of candidate compounds [1, 2]. Such high-throughput screens (HTS) are assays designed to rapidly evaluate very large numbers of compounds for a biological interaction of interest and recently they have begun to inform the study of cell death modalities and the efficacy of cytoprotective compounds against a wide variety of stressors [3]. HTS rely on simple, cost-effective cell viability assays that are easily "scaled up" to evaluate cells in multiwell plates and permit the examination of compounds on a scale that vastly exceeds traditional single-sample approaches at a far more cost-effective price point. HTS rely on outstanding statistical separation between

Laura Lossi and Adalberto Merighi (eds.), *Neuronal Cell Death: Methods and Protocols*, Methods in Molecular Biology, vol. 1254, DOI 10.1007/978-1-4939-2152-2_12, © Springer Science+Business Media New York 2015

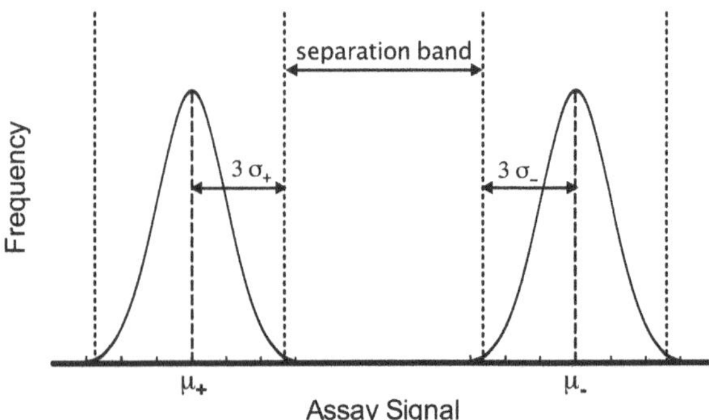

Fig. 1 Illustration of data variation and separation between positive and negative controls in a typical HTS assay. σ_+ and σ_- are the standard deviations of positive and negative controls, and μ_+ and μ_- are the mean values of positive and negative controls. Modified from [4]

positive and negative controls such that the means of these two groups are typically separated by >12 standard deviations. At this degree of separation, a single replicate is sufficient to infer statistical significance, permitting researchers to assay upward of 100,000 compounds in a few weeks. The robustness of an HTS is determined by a Z'-factor score, which is an evaluation of the separation of the means between positive and negative controls and the tightness of the standard deviations of the means (Fig. 1) [4]. Typically the positive controls are the insult-treated cells, and the negative controls are cells treated in serum-free culture media (i.e., healthy, growth-arrested cells).

HTS are typically designed to incorporate microplate-based assays that can be read on a standard laboratory spectrophotometer or luminometer. Such assays are usually developed initially in 96- or 384-well plates but can be scaled up to automated systems that support up to 9,600-well plates and utilize volumes of solution so minute that sonic impulses are required to accurately dispense nano-volumes into the wells [5]. Each sample well should contain an approximately equal number of cells that behave in a homogenous fashion, and therefore immortalized cell lines are typically used for HTS because such cells are strongly adherent to the growth matrix (i.e., the bottom of the wells), hardy, and divide rapidly [6]. These properties allow for cultures to be quickly scaled up, producing sufficient cells to screen a large number of compounds in a short period of time. Lastly, HTS of cell viability usually rely upon simple cell death assays to act as "readouts" of cell state. Typically fluorophore- or luciferase-based assays are utilized for such screens [7, 8], as these assays require only simple reagent addition to the wells (cf. as opposed to many molecular assays of cell viability that require multiple rinsing or incubation steps before reading).

HTS offer remarkable advantages over traditional single-replicate cell viability assays such as flow cytometry, fluorescence microscopy, and immunohistochemistry. In particular, HTS are far cheaper and simpler than such traditional cell viability assays, and thus permit far more compounds to be evaluated in a short period of time while minimizing reagent consumption. For example, evaluating the efficacy of a 1,000 candidate compounds at inhibiting the expression of the commonly used apoptotic markers using traditional microscopy approaches would require months of effort and considerable materials and manpower costs to treat and evaluate samples, while evaluation of 100,000 compounds would take an entire research career using traditional methods. Conversely, the same large number of compounds can be evaluated using an HTS on multiwell plates in days to weeks by a single researcher. HTS are limited somewhat in that they are generally restricted to evaluations in cell lines, which are considerably removed from more physiologically relevant in vivo experiments and results from HTS must therefore eventually be scaled down and validated in more complex systems. However, HTS approaches can be used in smaller replicates to rapidly assay primary cell cultures and may eventually be adaptable to evaluate isolated tissues. In this chapter we describe the basic steps of preparing an HTS, including (1) cell selection and cell culture preparation, (2) scaling up of cell cultures and setting up multiwell microplates, (3) assay selection and experimentation, and (4) data evaluation.

2 Materials

2.1 Cell Culture Media and Treatment Media Components

Prepare all solutions using ultrapure water and analytical grade reagents. The choice of cell culture media will vary between cell lines and a neuronal cell line culture media is utilized as an example for our protocol. Prepare all reagents at room temperature and store at 4 °C unless otherwise indicated.

1. Complete cell culture media: Dulbecco's modified Eagle medium (DMEM), 10 % bovine calf serum (BCS), 100 U/mL penicillin/streptomycin. Filter using standard cell culture filters.

2. Serum-free cell culture media: DMEM, 100 U/mL penicillin/streptomycin. Filter using standard cell culture filters.

3. Artificial cerebral spinal fluid (ACSF): 129 mM NaCl, 5 mM KCl, 1.3 mM $CaCl_2$, 1.5 mM $MgCl_2$, 21 mM $NaHCO_3$, 10 mM glucose, 315 mOsM, pH 7.4.

4. Ischemic solution (IS): 64 mM K^+, 51 mM Na^+, 77.5 mM Cl^-, 0.13 mM Ca^{2+}, 1.5 mM Mg^{2+}, 3 mM glucose, 0.1 mM glutamate, 315 mOsM, pH 6.5, 1.5 % O_2, 15 % CO_2, balance N_2 [9, 10].

2.2 Other Materials and Hardware Components

1. Standard sterile cell culture facilities (laminar flow fume hood, incubator, media-filtration apparatus, low-temperature centrifuge, etc.).

2. Standard cell culture flasks.

3. Sterile 96- or 384-well microplates. Solid white or black with clear bottoms (for fluorescence-based assays) or solid bottoms (for luminescence-based assays).

4. 8-Channel 100 µL (for 384-well plates) and 1,000 µL (for 96-well plates) multi-pipetters.

5. Plate spectrophotometer with fluorescence or luminescence reading capabilities (as appropriate for the chosen cell viability assay). For fluorescence-based assays ensure that the plate reader scans in the required wavelength specified for the chosen fluorescent probe. For repeated measures in real time, plate readers that maintain physiological temperatures are required.

6. 0.05 % Trypsin-EDTA solution (commercially available).

7. Propidium iodide (PI).

8. ATP luciferase assay: PerkinElmer ATPlite Luminescence Assay System kits (PerkinElmer, Waltham, MA).

9. 4 % paraformaldehyde (optional).

10. Orbital plate shaker.

11. Light microscope.

12. Vortex.

3 Methods

3.1 Preparation of Culture Media and Cell Cultures

For the purposes of demonstration we have chosen the HT22 murine hippocampal neuronal cell line, which is grown in DMEM and split with 0.05 % trypsin. These cells divide extremely rapidly and are an excellent subject for HTS assays (Fig. 2). Feeding rate, cell growth, desired confluence, etc. will vary depending on the cell line used and researchers should refer to the suppliers' instructions for information specific to their chosen cell type.

1. Prepare complete cell culture media (normal growth media) and filter sterilize. Sterile-filtered media may be stored at 4 °C for several weeks.

2. Thaw an aliquot of frozen cell suspension. *See* **Note 1**.

3. Suspend the contents of the cryovial in an appropriate volume of pre-warmed culture media (approximately 5, 10, or 20 mL of media for 25, 75, or 150 cm² flasks, respectively) and add this suspension to the appropriate cell culture flask. *See* **Note 2**.

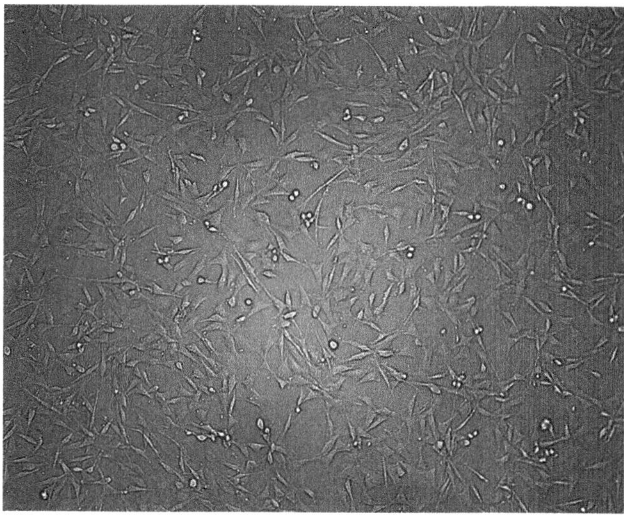

Fig. 2 Representative image of ideal cell density of HT22 neurons in a single well of a 96-well plate prior to experimentation

4. Place the flask in a 5 % CO_2 incubator at 37 °C and allow cells to grow and divide.

5. Feed cells 2–3 times per week and split cells when they reach approximately 70–80 % confluence.

3.2 Seeding Cells onto Microplates

1. On the day before the experiment, detach cells from the growth matrix. Use 0.05 % trypsin with EDTA for 10 min to split HT22 neurons. The appropriate lysis agent and concentration will vary between cell types. *See* **Note 3**.

2. Resuspend cells in ~5 volumes of complete cell culture media in a 15 mL Falcon tube. *See* **Note 4**.

3. Inspect the cell suspension under a light microscope to ensure that cells are isolated and not clumped.

4. Centrifuge the resulting cell suspension at $125 \times g$ for 5–7 min at room temperature.

5. Aspirate the supernatant and resuspend the cell pellet in complete cell culture media diluted to the desired seeding density. *See* **Note 5**.

6. Add cell suspension to each well of the microplate (use ~30–40 μL of the cell suspension for 384-well plates and 200–300 μL of the cell suspension for 96-well microplates) (*see* **Notes 6 and 7**). Be sure to leave several blank wells and also several wells with media only (no cells) to serve as on-plate controls.

7. Replace the lids of the microplates and incubate the cells and cover slips overnight in the normal tissue culture incubator.

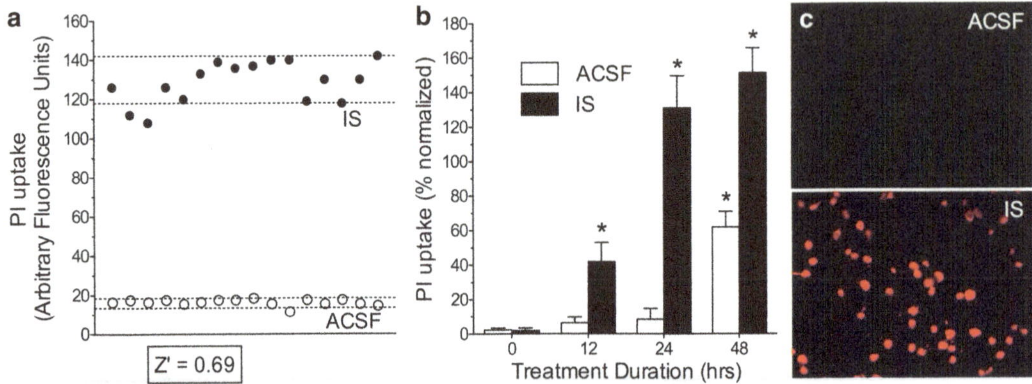

Fig. 3 (**a**) Representative PI exclusion measurements from HT22 neurons treated with ACSF (cell death negative control) or IS (cell death positive control) for 24 h in a 96-well microplate. *Dashed lines* are coefficients of variance for each treatment. (**b**) Effect of experimental duration on PI exclusion. (**c**) Sample images of PI uptake (red fluorescence) into HT22 neurons treated with ACSF (*top panel*) or IS (*bottom panel*) for 24 h. Data are mean ± SD from 20 wells for each treatment condition on a single plate. *Asterisks* indicate significant difference from control at $t=0$ min ($P<0.001$)

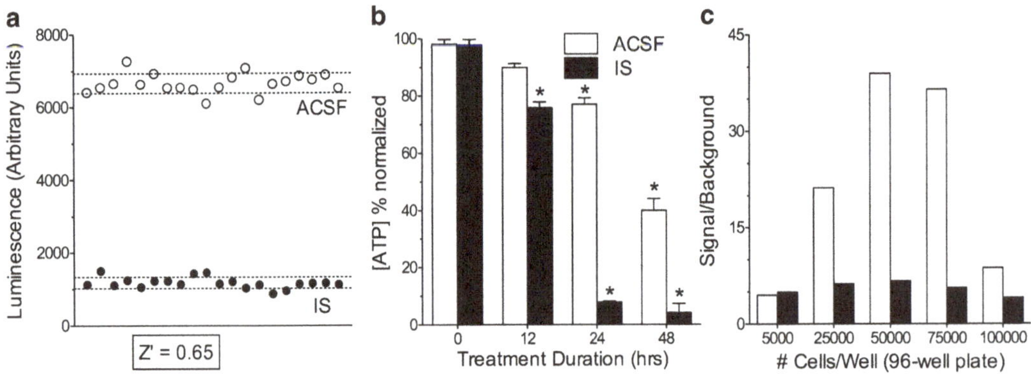

Fig. 4 (**a**) Representative total (ATP) measurements from HT22 neurons treated with ACSF (cell death negative control) or IS (cell death positive control) for 24 h in a 96-well microplate. *Dashed lines* are coefficients of variance for each treatment. (**b**) Effect of experimental duration on [ATP]. (**c**) Effect of initial cell seeding density on signal-to-background noise ratio of measurements of [ATP]. Data are mean ± SD from 20 wells for each treatment condition on a single plate. *Asterisks* indicate significant difference from control at $t=0$ min ($P<0.001$)

3.3 Treating Cells for Experimentation

For the purposes of demonstration we have chosen to treat HT22 neurons with either ACSF (cell death negative control) or an in vitro mimic of the ischemic mammalian penumbra (IS—cell death positive control) [11]. We assessed cell viability using both a fluorescence-based assay (PI exclusion; Fig. 3) and a luminescence-based assay (ATP luciferase; Fig. 4). PI is a bulky molecule that is too large to penetrate the plasma membrane of healthy cells but which readily permeate dying cells with damaged plasma membranes. Upon penetrating a cell, PI binds to nuclear DNA and fluoresces brightly in the red wavelength.

ATP luciferase measures the total ATP content of a cell and requires lysis of cells to free cellular ATP before measurement. If multiple time points are to be examined then care should be taken in determining the appropriate number of plates to seed with cells. Fluorescence-based assays can be read multiple times during an experiment; however, luminescence-based assays can only be read once. Therefore for these later assays, separate plates will be required for each time point, including controls (i.e., $t = 0$ min).

1. On the day of experimentation, view the microplates under a microscope to ensure that cells are evenly distributed in the wells and appear healthy (Fig. 2). *See* **Note 8**.

2. For fluorescence-based assays only: Prepare serum-free treatment media (preferably pre-equilibrated to 37 °C and 5 % CO_2 overnight) with the appropriate concentration of the fluorescent probe. For PI assays use 5 µg PI per mL media. Protect from light.

3. Remove the culture media and rinse cells to remove all residual serum. *See* **Note 9**. Inspect wells to ensure that cells have not detached during the rinsing process. Note any wells that have been compromised.

4. Add an equal volume of the treatment media. *See* **Notes 10** and **11**.

5. Optional: For fluorescence-based assays: read pretreatment fluorescence on the plate spectrophotometer to determine baseline fluorescence if desired (*see* Subheading 3.4).

6. Optional: For luminescence-based assays: read control plates on the plate spectrophotometer to determine baseline luminescence if desired (*see* Subheading 3.4).

7. Return the microplates to the incubator and incubate samples for the desired experimental period.

3.4 Data Acquisition and Analysis

Fluorescence-based assays of live cells should be performed at physiological temperatures. Luminescence-based assays may be performed at room temperature.

1. For fluorescence-based assays: Following treatment, read fluorescence at the appropriate wavelength. For our PI assay these wavelengths are excitation/emission = 485/630 nm (Fig. 3). *See* **Note 12**.

2. For luciferase-based assays: Add cell lysis reagent to each well and shake plate on an orbital plate shaker at 700 rpm for 5 min. The volume of lysis reagent added will vary depending on the specific assay kit chosen.

3. Add luciferase reagent to each well and shake plate on an orbital shaker at 700 rpm for 5 min. Protect plates from light once this reagent has been added.

4. Let plate sit in the dark for 15–30 min.

5. Read luminescence on a plate reader (Fig. 4).

6. Determine Z' score according to the following equation: $Z' = 1 - |(3\sigma_+ + 3\sigma_-)/(\mu_+ - \mu_-)|$, where σ_+ and σ_- are the standard deviations of positive and negative controls, and μ_+ and μ_- are the mean values of positive and negative controls. A Z'-factor of >0.5 is considered to be an excellent and robust HTS [4].

4 Notes

1. Cells are typically stored at −80 to −160 °C and can be significantly damaged if warmed up too rapidly. To minimize temperature shock, thaw an aliquot of frozen cell suspension by rapidly rubbing the cryovial between gloved hands.

2. Cultured cells are often a relatively fragile preparation and cell death can result from the handling of cells prior to experimentation and examination, which can confound results. To help ensure that cells are healthy during the initial experimental setup, equilibrate media in a cell culture incubator in vented-top sealed cell culture flasks overnight. This will allow the media to reach the temperature and pH of the incubator (typically 37 °C and 5 % CO_2) without compromising the sterility of the solution, and thereby reduce the stress to cells during setup.

3. Use 1–2 mL trypsin for 75 and 150 cm^2 cell culture flasks. Care should be taken to make sure that trypsin is evenly distributed over all cells. Cells treated with trypsin should be kept in the incubator at 37 °C for the duration of the treatment to speed enzymatic cleavage of adherent proteins.

4. The serum in the media inactivates the trypsin and prevents unwanted cell lysis. Add ~4–5 volumes of complete cell culture media directly to each cell culture flask for each volume of trypsin. Repeatedly sluice this media around the bottom of the flask to detach all cells. Repeated pipetting of the media may be used to detach cells that are more adherent. Removal of cell clumps is important to ensure homogenous cell distribution in the multiwell plates and holding the tip of a 5 or 10 mL pipette against the wall of the flask and repeatedly pipetting the cell suspension can remove cell clumps. The sheer force of the culture media being forced out between the pipette tip held flush to the flask wall will facilitate clump breakage.

5. The desired seeding density will vary depending on the growth rate and desired confluency of the cells at the time of experimentation. Typically, it is ideal to seed cells at a density such that following 24 h (overnight growth) they will be at the desired density. If cells are seeded too thinly the resulting signal will be weak. Thin seeding can also result in uneven growth

between wells. If cells are seeded too densely then they may become overgrown and more easily detach in "sheets" from the bottoms of the wells. Detached cells may be washed away during rinse stages, leading to minimal signal from a given well and experimental error. It is necessary to determine the ideal seeding density for each cell type used. A good starting point is to seed ~50,000 cells per well in 96-well plates, and ~10,000 cells per well in 384-well plates.

6. Adherent cells will begin to clump rapidly in a cell suspension and care should be taken to resuspend the cells regularly by gentle pipette mixing while they are being dispensed into microplates. With 384-well plates in particular, this step can be time consuming and so 8-well multi-pipette dispensers should be used where possible to facilitate even cell distribution between wells. If available, automated liquid handling and cell dispensing machines are preferable as they permit the rapid seeding of numerous plates and reduce human errors in pipetting. However, liquid-handling systems typically require a large dead volume of cell suspensions, making them less ideal for experiments where the volume of available cells is limited.

7. It is important to consider the design of the microplate in setting up your assay. In general it is best to minimize exposure to light sources and light transmission between wells. Therefore black or white microplates are desirable over clear microplates. Fluorescence-based assays require plates with bottoms that permit light transmission (i.e., clear bottom plates), while luminescence-based assays require solid-bottom plates to reduce light contamination. If using a luminescence-based assay with solid-bottom plates then cells should also be seeded into additional clear-bottomed plates at the same seeding density to permit visual examination of cell density and health prior to experimentation. These "test" plates should be rinsed and treated in the same fashion as the solid-bottom experimental plates in order to control for treatment permutations and ensure that cell density is maintained throughout washing and experimentation.

8. (Optional) It may be desirable to halt the growth of cells at some point prior to experimentation. If so, 2–16 h prior to experimentation, replace normal growth media with serum-free media to halt cell division and phase-lock cells.

9. Effective removal and rinsing of culture medium in a consistent fashion between wells are critical to maximizing the signal-to-noise ratio and minimizing the coefficient of variance of the results. Serum can protect cells from a given stress and may also interact with fluorophores or luciferase probes to quench the signal and interfere with measurements. Numerous options are available to researchers to rinse cells. The most time consuming of these is to carefully remove the growth media

and rinse the cells by manual pipetting. This is a reasonable approach if working with up to a few dozen wells in a single plate but is overly time consuming for larger sample sizes. A better manual option that works well particularly in 384-well plates is to hold the plate upside down and whip the plate downwards before abruptly stopping this motion. This "whiplash-like" action will effectively remove most cell culture media from all wells of the plate and adherent cells will not detach. Multiwell pipetters can then be used to pipette a rinse media (usually pre-warmed serum-free culture media) into the wells and rinses can be repeated using the same approach. If available, cells seeded into multiwell microplates can be more gently washed with 4–5 volumes of rinse media using a plate washer such as the TECAN PW96/384 Washer (TECAN, San Jose, CA).

10. Assay consistency is critical to developing a robust HTS and researchers should be aware of the time bottlenecks of their procedure. Often the biggest bottlenecks occur at the readout stage as most plate spectrophotometers will take several minutes to read a single plate. Thus the setup of the initial assay should be staggered by 15–30 min between each plate to account for this and ensure that samples are treated for the same duration.

11. Luminescence-based assays require the addition of cell lysis reagents and luciferase reagents following treatment. Plan ahead to ensure that adequate space remains in the wells of each plate to permit these additions, and such that vortexing the plate to mix the wells does not result in liquid spillover, and contamination of adjacent wells.

12. (Optional) Fluorophore-treated cells can also be fixed by addition of 4 % paraformaldehyde before measurement. This approach permits experimental plates to be preserved for future examination as desired.

Acknowledgements

This work was supported by NIH grants 5P01HD032573 to GGH and an NSERC postdoctoral fellowship to MEP. The authors have no conflicts of interest to declare.

References

1. Pope AJ, Hertzberg R (1999) HTS 2010: a retrospective look at screening in the first decade of the new millennium. J Biomol Screen 4:231–234

2. Fox S, Wang H, Sopchak L, Khoury R (2001) High throughput screening: early successes indicate a promising future. J Biomol Screen 6:137–140

3. Varma H, Lo DC, Stockwell BR (2008) High throughput screening for neurodegeneration and complex disease phenotypes. Comb Chem High Throughput Screen 11:238–248

4. Zhang JH, Chung TD, Oldenburg KR (1999) A simple statistical parameter for use in evaluation and validation of high throughput screening assays. J Biomol Screen 4:67–73

5. Oldenburg KR, Zhang J, Chen T et al (1998) Assay miniaturization for ultra-high throughput screening of combinatorial and discrete compound libraries: a 9600-well (0.2 microliter) assay system. J Biomol Screen 3:55–62

6. Pagliaro L, Praestegaard M (2001) Transfected cell lines as tools for high throughput screening: a call for standards. J Biomol Screen 6:133–136

7. Lavery P, Brown MJ, Pope AJ (2001) Simple absorbance-based assays for ultra-high throughput screening. J Biomol Screen 6:3–9

8. Maffia AM 3rd, Kariv II, Oldenburg KR (1999) Miniaturization of a mammalian cell-based assay: luciferase reporter gene readout in a 3 microliter 1536-well plate. J Biomol Screen 4:137–142

9. Yao H, Shu Y, Wang J et al (2007) Factors influencing cell fate in the infarct rim. J Neurochem 100:1224–1233

10. Yao H, Sun X, Gu X et al (2007) Cell death in an ischemic infarct rim model. J Neurochem 103:1644–1653

11. Pamenter ME, Perkins GA, McGinness AK et al (2012) Autophagy and apoptosis are differentially induced in neurons and astrocytes treated with an in vitro mimic of the ischemic penumbra. PLoS One 7:e51469

Part II

Cell Death in Neuropathology and Experimental Neuropathology

Chapter 13

Staining of Dead Neurons by the Golgi Method in Autopsy Material

Stavros J. Baloyannis

Abstract

Golgi silver impregnation techniques remain ideal methods for the visualization of the neurons as a whole in formalin fixed brains and paraffin sections, enabling to obtain insight into the morphological and morphometric characters of the dendritic arbor, and the estimation of the morphology of the spines and the spinal density, since they delineate the profile of nerve cells with unique clarity and precision. In addition, the Golgi technique enables the study of the topographic relationships between neurons and neuronal circuits in normal conditions, and the following of the spatiotemporal morphological alterations occurring during degenerative processes. The Golgi technique has undergone many modifications in order to be enhanced and to obtain the optimal and maximal visualization of neurons and neuronal processes, the minimal precipitations, the abbreviation of the time required for the procedure, enabling the accurate study and description of specific structures of the brain. In the visualization of the sequential stages of the neuronal degeneration and death, the Golgi method plays a prominent role in the visualization of degenerating axons and dendrites, synaptic "boutons," and axonal terminals and organelles of the cell body. In addition, new versions of the techniques increases the capacity of precise observation of the neurofibrillary degeneration, the proliferation of astrocytes, the activation of the microglia, and the morphology of capillaries in autopsy material of debilitating diseases of the central nervous system.

Key words Golgi method, Neuronal degeneration, Dendritic pathology, Dendritic spines, Neuronal death

1 Introduction

The morphological study of neurons in health and disease requires special histological techniques, which might visualize the neurons in a three dimensional (3D) arrangement, including the dendritic arbor, the majority of the dendritic branches, the spines, the axon and the axonal collaterals, the neuronal networks, and the dendritic and axonal bands and tracks. In addition, the histological techniques, which are applied in neurosciences, must enable the precise study of the topographic relationships between neurons

Laura Lossi and Adalberto Merighi (eds.), *Neuronal Cell Death: Methods and Protocols*, Methods in Molecular Biology, vol. 1254, DOI 10.1007/978-1-4939-2152-2_13, © Springer Science+Business Media New York 2015

167

and neuronal circuits and the morphometric estimation of the dendritic spines, a fact which is of essential value in the study of ageing and neuronal degeneration.

It is well known that nerve cells have a unique morphology with many dendritic branches studded with spines and axons, which vary in length from few micrometer to many centimeter. The visualization of the nerve cells as a whole, including the cell body and the many complex processes, is essential in the study of neuronal circuits and the analysis of mechanism of neural information processing in normal conditions as well as in neurodegenerative disorders. For the neuroscientist, it is important to study and follow the spatiotemporal morphological alterations which occur during the degenerative processes, affecting neuronal synapses, dendrites, axons, and axonal collaterals and inducing death of cell bodies resulting eventually in the degradation of the neuronal networks sometimes in distant parts of the brain.

The silver impregnation technique, which is very important in elucidating nervous system anatomy, was introduced by Golgi in 1873, more than 100 years ago as "reazione nera" (black reaction) [1–3]. It is based on the immersion of the nerve tissue in a solution of potassium chromate and potassium dichromate, followed by impregnation with silver nitrate, resulting in the formation of silver chromate. This technique remains an ideal method for the visualization of the neurons, enabling to obtain insight into the morphological and morphometric characters of the dendritic arbor, the estimation of the morphology of the spines and the spinal density in formalin fixed brains and paraffin sections, although the mechanism of staining is not interpreted in detail [4–6].

The based technique was extensively applied and evaluated precisely by Santiago Ramon y Cajal, in the detailed description of the histological organization of the brain and in the validation of the under his authorship "neuronal doctrine" on the histological structure of the nervous system [7, 8]. The importance of the method was established in 1906 when Camillo Golgi and Santiago Ramon y Cajal shared the Nobel Prize, which was awarded for the description of the histological organization of the central nervous system (CNS), based mostly on silver techniques [9, 10].

For many years, in the hands of the neuroscientists [11], the Golgi technique has been considered to be instrumental for the precise description of neurons and neuronal networks, glia and blood vessels, since it delineates the profile of nerve cells, glial cells and blood capillaries, with unique clarity and precision. Gradually, the Golgi technique has undergone many modifications in order to be enhanced [11] and obtain the maximal visualization of the neurons and the neuronal processes, the minimal precipitates and the abbreviation of the time required for the procedure [12–14]. However, staining results are frequently selective and hardly provisional [15].

Eventually, the term Golgi technique has started to be referred to the several staining methods which are based on the preparation of neural tissue with potassium dichromate, followed by exposure to heavy metals [16]. Among them, the most frequently applied in clinical neuropathology and morphological investigation are the modified Golgi stain, the rapid Golgi technique, the modified Golgi–Cox method, the Golgi–Braitenberg method, and others [16], which sometimes are called Neo Golgi methods [17], or gold-toning methods whenever silver is replaced by gold [18]. These new versions have contributed greatly in optimizing the quality of impregnation, making staining faster, easier, brighter, richer, and less patchy. In addition, neurons stained with these new methods are evenly dispersed and uniformly distributed by removing of excess staining solution [19], making the visualization of the morphological details more efficient [19, 20] and the accurate study and description of special structures of the brain more precise [21–24].

For many years the association of Golgi techniques with electron microscopy, has enlarged our views on neuronal development, migration [24] maturation, synaptic plasticity and neuronal alterations in senility and degenerating processes [25–27]. Particularly, Golgi techniques and electron microscopy assist as a harmonious and valuable combination for the visualization the concrete morphology of neurons and particularly the microstructures such as dendritic spines and growth cones [28], which represent the smallest, most sensitive, and vulnerable structural unit of the neuron, rapidly involved in many diseases and degenerative processes of the brain [28–30].

In the endeavor for visualization of the sequential stages of the neuronal degeneration and death, silver impregnation techniques have played a prominent role for many years [31–33] enabling the study of degenerating axons [34], synaptic boutons [35], axonal terminals [36], and organelles of the soma of the nerve cells [37]. The new versions of the techniques contributed substantially in accumulation of valuable data from detailed morphological analysis of autopsy material in debilitating diseases such as Alzheimer's disease [37], revealing "Dark" neurons, neurofibrillary degeneration, neuritic plaques, microglial cells, and capillaries [38, 39]. They may be helpful in assessing and staging neurotoxicity [40] and in understanding the role of astrocytic proliferation in degenerating conditions [41, 42]. The neuroscientist applying these techniques may insert further in the pathways of neuronal disintegration [43], apoptosis and death [44], which may occur under various conditions and etiologies [45, 46].

In this chapter one of the new Golgi methods is described.

2   Materials

All solutions must be fresh-prepared.

1. 10 % formalin solution. *See* **Note 1**.

2. Potassium dichromate solution: Dissolve 7 g potassium dichromate in 300 mL of cold distilled water.

3. 1 % silver nitrate (AgNO3): Dissolve 1 g silver nitrate in 100 mL distilled water.

4. 10 % methylene blue.

5. Clear and dark glass jars. *See* **Note 2**.

6. Flat glass.

7. Ethanol.

8. Paraffin.

9. Sliding microtome.

10. Glass Petri dishes.

11. Xylene.

12. Flat brush.

13. Large glass cover slips.

14. Mounting medium: Entellan® new, Merck-Millipore, Darmstadt, Germany.

15. Transmitted light microscope.

16. Tilting microscope stage (*optional*).

3   Methods

The degeneration of spines appears as an early degenerative phenomenon, with substantial consequences in cognition and psychological homeostasis of the patients.

We apply the semi-rapid Golgi technique in an attempt to study the morphological alterations of dendrites and dendritic spines in dementias, including vascular dementia, Parkinson's disease associated with dementia, frontal dementia, frontotemporal dementia, and Alzheimer's disease at any stage.

3.1   Semi-rapid Golgi Method

1. Excise the brain from the skull. *See* **Note 3**.

2. Immediately immerse the brain in a freshly prepared 10 % formalin solution. The volume of the fixative solution should be calculated so to be ten times of the brain's volume.

3. Suspend sample inside a glass jar for 30 days at room temperature in darkness (*see* **Note 4**). It is worth to renew the formalin solution every 10 days. The silver impregnation might be better if the time of fixation would be prolonged. *See* **Note 5**.

4. Hand-cut the brain in thick (how much) coronal sections and immediately immerse slabs in freshly prepared potassium dichromate solution.

5. Keep in potassium dichromate for 10 days in darkness, at room temperature.

6. Immerse specimen in freshly prepared 1 % silver nitrate solution in a dark glass jar and in darkness for 10 days, at a temperature of 16 °C. We enhance usually the solution of the silver nitrate with few grains of copper. *See* **Note 6**.

7. Place specimens on a flat glass and remove the deposits of silver by brushing them gently for 2 min.

8. Dehydrate rapidly in absolute alcohol for 2 min.

9. Embed in melting paraffin.

10. With a rotating microtome, alternatively cut sections of 100 and 25 μm.

11. Collect sections in a glass Petri dish containing absolute alcohol.

12. Post-stain some sections with methylene blue for a clear visualization of all of the neurons, as a variation of Golgi-Nissl method.

13. Carefully transfer sections with a flat brush to another glass dish containing xylene for clearing.

14. As soon as sections become translucent, mount them with Entellan® between two large cover slips.

15. Left to dry at least for 3 days and observe under a transmitted light microscope.

16. Use a titling microscope stage to have a 3D visualization of neurons and dendritic arbors.

3.2 Morphology and Quantitation

Several morphological parameters can be evaluated for a characterization of neuronal alterations in postmortem samples from patients reported to suffer from neurodegenerative diseases.

According to the shape, the size, and the morphology of the dendritic arborization it is possible to classify neurons into (a) large pyramidal, (b) large polyhedral, (c) small triangular, and (d) stellate (Table 1 and Figs. 1 and 2).

The volume of the cell body of pyramidal or triangular neurons is estimated by measuring the height and the diameter, the volume of round neurons by measurement of the diameter and of ellipsoid ones by measurement of the two diameters, according to the relevant mathematical type (Table 2).

The main parameters that can be calculated for dendritic arborisation are reported analytically in Table 2. For a precise calculation of the dendritic branches the Sholl's method is useful [47].

Table 1

Classification of neuronal shapes and main features of dendritic arborization

Neuron types	Shape (soma)	Dendritic arborization	Apical dendrite	Basal dendrites
Large pyramidal	Triangular/prismatic	Asymmetric	Well developed	Short
Large polyhedral	Polyhedral/prismatic			
Small triangular	Triangular/prismatic			
Stellate	Global, ellipsoid, or fusiform	Symmetric	N/A	N/A

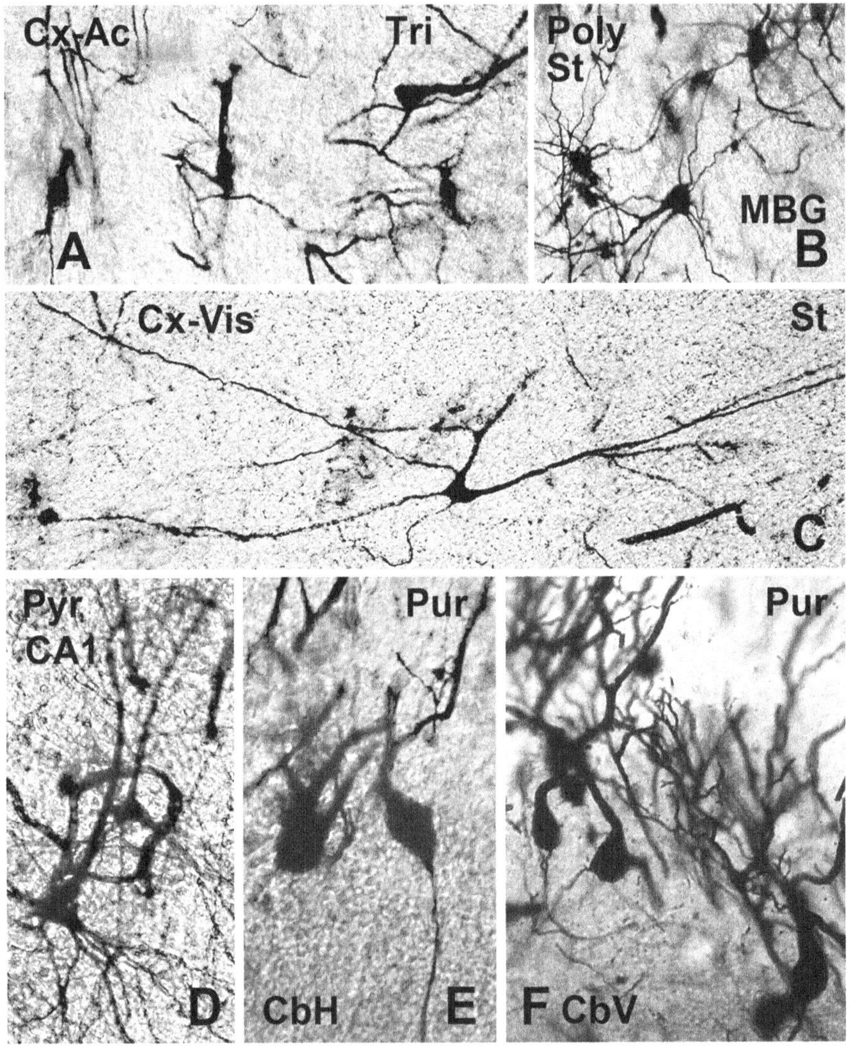

Fig. 1 Alterations of Golgi-stained neurons in Alzheimer's disease. (**a**) Layer III triangular (*Tri*) neurons of the acoustic cortex (*Cx-Ac*) display a prominent loss of dendritic branches. (**b**) Large polyhedral (*Poly*) and global stellate (*Ste*) neurons of the medial geniculate body (*MGB*) show a marked poverty of secondary and tertiary dendritic branches and depletion of spines. (**c**) Stellate neuron (*Ste*) of the visual cortex (*Cx-Vis*) reveals a dramatic decrease of spine density. (**b**) Pyramidal (*Pyr*) neurons of the CA1 area of the left hippocampus display a marked loss of secondary dendritic branches. (**e, f**) Purkinje cells (*Pur*) of the nodule of the cerebellar vermis (*CbV*) and left hemisphere (*CbH*) show a tremendous loss of dendritic branches and spines. (**a–d**) Semi-rapid Golgi method; (**e, f**) Golgi-Nissl method. Orignal magnifications: (**a**) 600×; (**b–e**) 1,200×

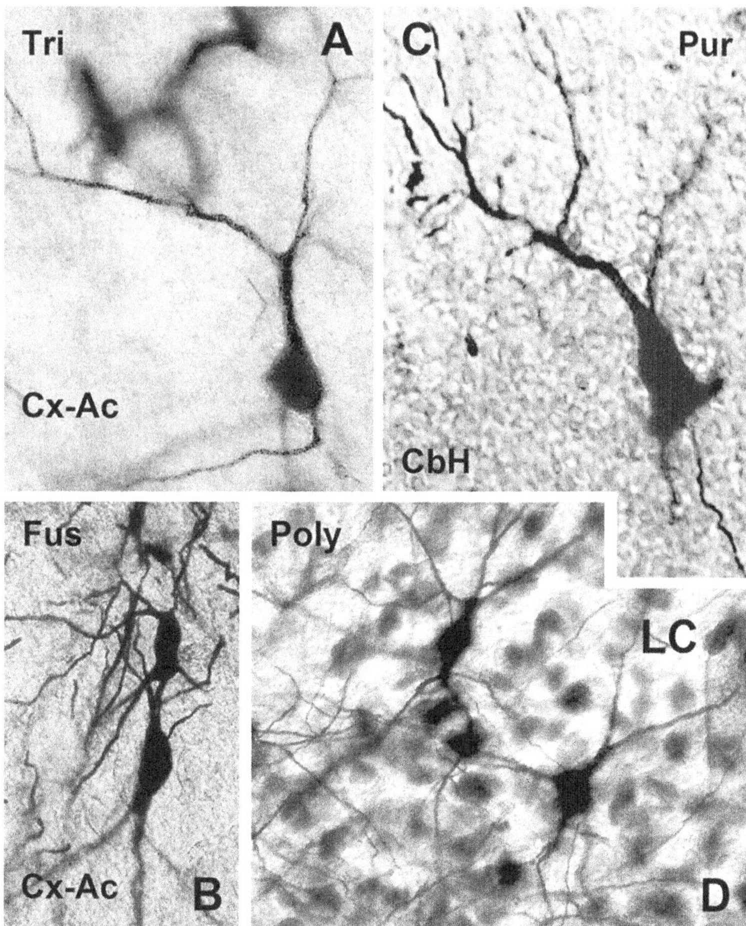

Fig. 2 Alterations of Golgi-stained neurons in vascular dementias (**a–c**) and Parkinson's disease (**d**). (**a**) Triangular neuron (Tri) of the layer III of the acoustic cortex (*Cx-Ac*) in a case of vascular dementia. The secondary dendrites bifurcate at right angles. The dendritic branches are depleted of spines. (**b**) Fusiform neurons of the layer II of the acoustic cortex (*Cx-Ac*) in a case of frontal dementia are characterized by loss of tertiary dendritic branches and spines. (**c**) Purkinje cells (*Pur*) of the superior surface of the left cerebellar hemisphere of a case of vascular dementia. A marked abbreviation of the dendritic arbor is seen due to loss of the majority of tertiary dendritic branches. (**d**) Polyhedral neurons (*Poly*) of the locus coeruleus (*LC*) in a case of Parkinson-plus disease (Parkinson's disease associated with dementia) are characterized by the substantial loss of dendritic branches and the depletion of spines. (**a–d**) Semi rapid Golgi method; (**c–d**) Golgi-Nissl method. Original magnifications 1,200×

Thus, concentric cycles centred on the cell bodies are drawn, at intervals of 15 μm, and all dendritic branches and segments are counted on each one of the cycles. Alternatively, it is possible to use automated reconstruction systems.

Table 2
Morphometric features to be considered in neuronal classification according to cell compartment and level of analysis

Cell compartment	Morphometric features	
	I Level	II Level
Soma	Type/shape Size	
Dendritic arbor	Shape	Presence of basal/apical dendrites Number of primary branches Number of secondary branches Number of tertiary branches Angles between primary/secondary branches Type of bifurcations (right or acute angles)
	Size	Lenght of dendritic segments Diameters of dendritic segments Length of terminal dendritic branches Ratio between the numbers of primary, secondary, and tertiary branches
Spines	Shape Size Volume Linear density/dendritic segment Ratio between numbers in secondary branches/ tertiary branches Number in terminal dendritic branches Linear density in terminal dendritic branches	

Golgi staining combined with morphological/morphometric alterations of neurons has provided to be very useful in diagnosing some common neuropatholgical conditions.

Some exemplificative findings are briefly discussed below and reported in Figs. 1 and 2.

3.2.1   Telencephalon   In the acoustic cortex of a substantial number of patients suffering from vascular dementia Golgi staining reveals a tremendous (80–90 %) decrease in the number of Cajal–Retzius cells in the molecular or plexiform layer in comparison with normal controls [48]. Decrease in spine density is noticed in the majority of the neurons being more pronounced in the triangular and round neurons of layers II, III, and V (Fig. 2a). Similar alterations have been described in late [49–51] and early cases [27] of Alzheimer's disease (Fig. 1a), as well as in frontal dementia [52] (Fig. 2b), in frontotemporal dementia, particularly in Pick disease and in primary progressive aphasia [53]. Also the study of the superior, middle, and inferior temporal gyri in Alzheimer' s disease revealed a control-matched significant decrease in the number of Cajal–Retzius cells, particularly in the anterior part of the gyri [27], whereas in the visual cortex (Fig. 1c) an extensive dendritic pathology with a dramatic decrease of spine density was demonstrated [54]. Neuronal loss and dendritic alterations were also evident in hippocampus (Fig. 1d) with reduced density of thorny excrescences on the apical and basilar dendritic tree of CA3 pyramidal neurons [55].

3.2.2 Diencephalon, Encephalic Trunk, and Cerebellum

Neuronal loss in Alzheimer's disease was also revealed in the medial geniculate bodies (Fig. 1b) and inferior colliculi, where the large polyhedral or round neurons showed a marked poverty of secondary and tertiary dendritic branches [56]. In the cerebellum, which, as a rule, demonstrates a minimal Alzheimer's pathology, the semi-rapid Golgi method demonstrated a tremendous loss of dendritic branches and spines associated with an impressive decrease in spine density in the majority of Purkinje cells of the hemispheres (Fig. 1e), the nodule of the vermis (Fig. 1f), and the flocculus [26, 57, 58]. In vascular dementia [59], on the other hand, a marked loss of climbing fibers is noticed in the molecular layer of the cerebellar cortex associated with loss of the majority tertiary dendritic branches of Purkinje cells (Fig. 2c).

In cases of Parkinson's disease associated with dementia [60], a substantial loss of dendritic branches, dendritic spines and axonal retrograde collaterals was seen in the locus coeruleus (Fig. 2d), in comparison with parkinsonian patients without dementia and normal controls.

4 Conclusion

Golgi staining methods are still very valuable tools for the neuropathologist when applied together with a careful morphometric analysis of neuronal soma and dendritic features.

It must be emphasized that semi-rapid Golgi method is instrumental in revealing dendritic pathology and neuronal death under *various* etiologies and pathological conditions [61].

It must be stressed out that still the diagnosis of the major neuropathological conditions relies on histopathological examination of autopsy samples. When a correct statistical approach is followed, these techniques not only remain of primary diagnostic importance, but have proved to be useful in underlining some mechanisms at the basis of the neuronal alterations/loss, such as the important role of the vascular factor in the pathogenesis of Alzheimer's disease

[62], or the mechanisms involved in the post-lesion remodeling of neuronal networks, in the recovering brain on the basis of neuronal plasticity [63] and therapeutic interventions [64].

5   Notes

1. The Golgi silver impregnation technique may also be used in experimental animals following an in vivo perfusion. In rats the carotid artery is cannulated under pentobarbital anesthesia (5 mg/100 g weight injected). The animals are subsequently sacrificed by perfusion with a fixative composed of 1 % paraformaldehyde (PFA) and 2.5 % glutaraldehyde in cacodylate buffer adjusted at pH 7.35. The brains are quickly removed and immediately immersed in the same fixative for 3 h. Then samples may be taken for electron microscopy. The remained brain is immersed in 10 % of fresh prepared formalin for 1 week at room temperature and then in freshly prepared potassium dichromate solution for 1 week followed by immersion in freshly prepared 1 % silver nitrate solution in a dark glass jar and in darkness for 10 days, at a temperature of 16 °C according to semi-rapid Golgi method.

2. The glass jars are cylinders of dark jar, having a diameter of 50 cm and height of 80 cm. They are placed in a dark place. Dark plastic cylinders may also be used.

3. It is important to remove the brain as soon as possible, especially if Golgi technique is combined with electron microscopy. Practically it might be possible to remove the brain within 4–8 h after death. The time of fixation is very important. After the excision from the skull, the brain is placed in the fixing solution immediately. A prolonged fixation in formalin, more than 30 days, may give better results.

4. In order to suspend the brain in the jar we pass a plastic string under the basilar artery (between the middle of the pons and the basilar artery) and we fix the ends of the string on the walls of the jar. The brain is suspended in the middle of the jar and it does not touch the walls. Therefore, the brain retains its normal shape and doesn't undergo any deformity or distortion.

5. The time of fixation should be prolonged if the volume of the brain is high. In general, a long fixation in formalin results to better visualization of neurons than a short one.

6. The addition of 0.3 g of grains of copper in 100 mL of 1 % solution of silver nitrate optimizes the quality of the impregnation, increasing substantially the visualization of the neurons, the dendritic arbor and the dendritic spines.

References

1. Golgi C (1873) Sulla struttura della sostanza grigia del cervello. Gazz Med Ital Lombarda 33:244–246

2. Golgi C (1989) On the structure of nerve cells, 1898. J Microsc 155:3–7

3. Mazzarello P, Buchtel H, Badiani A (1999) The hidden structure: a scientific biography of Camillo Golgi. Oxford University Press, Oxford

4. Peters A (1955) Experiments on the mechanism of the silver staining. Part I. Impregnation. J Microsc Sci 96:84–102

5. Peters A (1955) Experiments on the mechanism of the silver staining. Part II. Development. J Microsc Sci 96:103–115

6. Peters A (1955) Experiments on the mechanism of the silver staining. Part III. Quantitative studies. J Microsc Sci 96:301–316

7. Ramon y Cajal S (1954) Neuron theory or reticular theory? Objective evidence of the anatomical unity of the nerve cells. Consejo Superior de Investigaciones Científicas, Madrid

8. Ramon y Cajal S (1909) Histologie du Système Nerveux de l'Homme et des Vertèbres. Instituto Ramon y Cajal, Madrid, pp 774–838

9. López-Piñero JM (1993) Cajal y la estructura histológica del sistema nervioso. Invest Cienc 197:6–13

10. Corral-Corral I, Corral-Corral C, Corral-Castañedo A (1998) Cajal's views on the Nobel Prize for physiology and medicine (October 1904). J Hist Neurosci 7:43–49

11. Ramon-Moliner E (1970) The Golgi–Cox technique. In: Nauta WJH, Ebbesson SOE (eds) Contemporary research methods in neuroanatomy. Springer, New York, pp 32–55

12. Van der Loos H (1956) Une combinaison de deux vieilles méthodes histologiques pour le système nerveux central. Mschr Phychiatr Neurol 132:330–334

13. Morest DK, Morest RR (1966) Perfusion-fixation of the brain with chromeosmium solutions for the rapid Golgi method. Am J Anat 118:811–831

14. Zhang H, Weng SJ, Hutsler JJ (2003) Does microwaving enhance the Golgi methods? A quantitative analysis of disparate staining patterns in the cerebral cortex. J Neurosci Meth 124:145–155

15. Pasternak JF, Woolsey TA (1975) On the "selectivity" of the Golgi–Cox method. J Comp Neurol 160:307–312

16. Scheibel ME, Scheibel AB (1978) The methods of Golgi. In: Robertson RT (ed) Neuroanatomical research techniques. Academic, New York, pp 90–114

17. Blackstad TW, Osen KK, Mugnaini E (1984) Pyramidal neurones of the dorsal cochlear nucleus: a Golgi and computer reconstruction study in cat. Neuroscience 13:827–854

18. Rosoklijaa GB, Petrushevskie VM, Stankovc A et al (2014) Reliable and durable Golgi staining of brain tissue from human autopsies and experimental animals. J Neurosci Meth 230:20–29

19. Levine ND, Rademacher DJ, Collier TJ et al (2013) Advances in thin tissue Golgi–Cox impregnation: fast, reliable methods for multi-assay analyses in rodent and nonhuman primate brain. J Neurosci Meth 213:214–227

20. Somogyi P, Hodgson AJ, Smitha AD (1979) An approach to tracing neuron networks in the cerebral cortex and basal ganglia. Combination of Golgi staining, retrograde transport of horseradish peroxidase and anterograde degeneration of synaptic boutons in the same material. Neuroscience 4:1805–1852

21. Ryugo DK, Fekete DM (1982) Morphology of primary axosomatic endings in the anteroventral cochlear nucleus of the cat: a study of the end bulbs of Held. J Comp Neurol 210:239–257

22. Ryugo DK, Parks TN (2003) Primary innervation of the avian and mammalian cochlear nucleus. Brain Res Bull 60:435–456

23. Friedland DR, Los JG, Ryugoa DK (2006) A modified Golgi staining protocol for use in the human brain stem and cerebellum. J Neurosci Meth 150:90–95

24. Felten DL, Cummings JP (1979) The raphe nuclei of the rabbit stem. J Comp Neurol 187:199–243

25. Rakic P (1971) Neuron-glia relationship during granule cell migration in developing cerebellar cortex. A Golgi and electron microscopic study in Macacus rhesus. J Comp Neurol 141:283–312

26. Baloyannis SJ, Manolidis SL, Manolidis LS (2000) Synaptic alterations in the vestibulocerebellar system in Alzheimer's disease – a Golgi and electron microscope study. Acta Otolaryngol 120:247–250

27. Baloyannis SJ (2005) Morphological and morphometric alterations of Cajal–Retzius cells in early cases of Alzheimer's disease: a Golgi and electron microscope study. Int J Neurosci 115:965–980

28. Fiala JC, Spacek J, Harris KM (2002) Dendritic spine pathology: cause or consequence of neurological disorders? Brain Res Rev 39:29–54

29. Koyamaa Y, Nishidab T, Tohyamac M (2013) Establishment of an optimised protocol for a

Golgi–electron microscopy method based on a Golgi–Cox staining procedure with a commercial kit. J Neurosci Meth 218:103–109

30. Penzes P, Cahill ME, Jones KA et al (2011) Dendritic spine pathology in neuropsychiatric disorders. Nat Neurosci 14:285–293

31. Bielschowsky M (1904) Silberimpregnation der neurofibrillen. J Psychol Neurol 3:169–188

32. Glees P (1946) Terminal degeneration within the central nervous system as studied by a new silver method. J Neuropathol Exp Neurol 5: 54–59

33. Goodpasture C, Bloom SE (1975) Visualization of nucleolar organizer regions in mammalian chromosomes using silver staining. Chromosoma 53:37–50

34. Nauta WJH, Gygax PA (1951) Silver impregnation of degenerating axon terminals in the central nervous system. (1) Technique (2) Chemical notes. Stain Technol 26:5–11

35. Fink RP, Heimer L (1967) Two methods for selective silver impregnation of degenerating axons and their synaptic endings in the central nervous system. Brain Res 4:369–374

36. de Olmos JS, Ingram WR (1972) An improved cupric-silver method for impregnation of axonal and terminal degeneration. Brain Res 33:523–529

37. Gallyas F, Wolff JR, Bottcher H et al (1980) A reliable and sensitive method to locate terminal degeneration and lysosomes in the CNS. Stain Technol 55:299–306

38. Gallyas G, Hsu M, Buzsaki G (1993) Four modified silver methods for thick sections of formaldehyde-fixed mammalian central nervous tissue: 'Dark' neurons, perikarya of all neurons, microglial cells and capillaries. J Neurosci Meth 50:159–164

39. Switzer RC, Campbell SK, Murdock TM (1993) A histologic method for staining Alzheimer pathology. US Patent 5192688

40. de Beltramino CA, Olmos JS, Gallyas F et al (1993) Silver staining as a tool for neurotoxic assessment. In: Erinoff L (ed) Assessing neurotoxicity of drugs of abuse. Abuse monograph 136. National Institute of Drug, Rockville, MD, pp 101–126

41. Khurgel M, Switzer RC, Teskey GC et al (1995) Activation of astrocytes during epileptogenesis in the absence of neuronal degeneration. Neurobiol Dis 2:23–35

42. O'Callaghan JP, Jensen KF (1992) Enhanced expression of glial fibrillary acidic protein and the cupric silver degeneration reaction can be used as sensitive and early indicators of neurotoxicity. Neurotoxicology 13:113–122

43. Leonard C (1981) Silver degeneration methods. In: Johnson JE Jr (ed) Current trends in morphological techniques. CRC Press, Boca Raton, FL, pp 93–140

44. Switzer RC, Wheat DL, Turner JC et al (1999) Patterns of programmed cell death in rat brain nuclei during postnatal days 1–10 as revealed with a silver degeneration stain. Soc Neurosci Abstr 25:1776

45. Black JE, Kodish IM, Grossman AW et al (2004) Pathology of layer V pyramidal neurons in the prefrontal cortex of patients with schizophrenia. Am J Psychiatry 161:742–744

46. Cook SC, Wellman CL (2004) Chronic stress alters dendritic morphology in rat medial prefrontal cortex. J Neurobiol 60:236–248

47. Sholl DA (1953) Dendritic organization in the neurons of the visual and motor cortices of the cat. J Anat 87:387–406

48. Baloyannis SJ (2005) The acoustic cortex in vascular dementia: a Golgi and electron microscope study. J Neurol Sci 229–230:51–55

49. Baloyannis SJ, Manolidis SL, Manolidis LS (1992) The acoustic cortex in Alzheimer's disease. Acta Otolaryngol S494:1–13

50. Baloyannis SJ, Costa V, Mauroudis I et al (2007) Dendritic and spinal pathology in the acoustic cortex in Alzheimer's disease: morphological and morphometric estimation by Golgi technique and electron microscopy. Acta Otolaryngol 127:351–354

51. Baloyannis SJ (2009) Dendritic pathology in Alzheimer's disease. J Neurol Sci 283: 153–157

52. Baloyannis SJ, Manolidis SL, Manolidis LS (2001) The acoustic cortex in frontal dementia. Acta Otolaryngol 121:289–292

53. Baloyannis SJ, Mauroudis I, Manolides SL et al (2011) The acoustic cortex in frontotemporal dementia: a Golgi and electron microscope study. Acta Otolaryngol 131:359–361

54. Mavroudis IA, Fotiou DF, Manani MG et al (2011) Dendritic pathology and spinal loss in the visual cortex in Alzheimer's disease: a Golgi study in pathology. Int J Neurosci 121:347–354

55. Baloyannis SJ, Mauroudis I, Manolides SL et al (2009) Synaptic alterations in the medial geniculate bodies and the inferior colliculi in Alzheimer's disease: a Golgi and electron microscope study. Acta Otolaryngol 129: 416–418

56. Tsamis IK, Mytilinaios GD, Njau NS et al (2010) Properties of CA3 dendritic excrescences in Alzheimer's disease. Curr Alzheimer Res 7:84–90

57. Mavroudis IA, Fotiou DF, Adipepe LF et al (2010) Morphological changes of the human Purkinje cells and deposition of neuritic plaques and neurofibrillary tangles on the cerebellar cortex of Alzheimer's disease. Am J Alzheimers Dis Other Demen 25: 585–591

58. Mavroudis IA, Manani MG, Petrides F et al (2013) Dendritic and spinal pathology of the Purkinje cells from the human cerebellar vermis in Alzheimer's disease. Psychiatr Danub 25:221–226

59. Baloyannis SJ (2007) Pathological alterations of the climbing fibres of the cerebellum in vascular dementia: a Golgi and electron microscope study. J Neurol Sci 257:56–61

60. Baloyannis SJ, Costa V, Baloyannis IS (2006) Morphological alterations of the synapses in the locus coeruleus in Parkinson's disease. J Neurol Sci 248:35–41

61. Baloyannis SJ (2013) Recent progress of the Golgi technique and electron microscopy to examine dendritic pathology in Alzheimer's disease. Future Neurol 8:239–242

62. Baloyannis SJ, Baloyannis IS (2012) The vascular factor in Alzheimer's disease: a study in Golgi technique and electron microscopy. J Neurol Sci 322:117–121

63. Brown CE, Wong C, Murphy TH (2008) Rapid morphologic plasticity of peri-infarct dendritic spines after focal ischemic stroke. Stroke 39:1286–1291

64. Li S, Kang L, Zhang C et al (2013) Effects of dihydrotestosterone on synaptic plasticity of hippocampus in male SAMP8 mice. Exp Gerontol 48:778–785

Chapter 14

Image Analysis Algorithms for Immunohistochemical Assessment of Cell Death

Stan Krajewski, Jeffrey Wang, Tashmia Khan, Jonathan Liu, Chia-Hung Sze, and Maryla Krajewska

Abstract

Light microscopy allows for the inexpensive and fast detection of neuronal/glial cell demise and estimation of infarct and traumatic lesion volumes; the direct correlates of cell death. Quantitative assessment of brain tissue damage following stroke, traumatic brain injury (TBI) or neurodegenerative diseases, and recovery after therapeutic intervention has been facilitated by recent developments in computer-assisted image analysis technologies that enable more objective and accurate morphometric quantification of cell injury in whole brain sections. In this chapter, the proposed workflow describes what tasks need to be fulfilled to visualize and gauge cell death characterization by histological stains and immunohistochemical markers.

Key words Artifact-free brain fixation/processing into cryo- or paraffin blocks, Digital pathology, TBI neuropathology and hemorrhage scores, IHC, Algorithm-based morphometry of death events

1 Introduction

Cell death is an essential biological process that plays an important physiological and pathological role within an organism. Part of the ongoing efforts is concentrated on understanding cell death processes, as well as identifying and measuring useful cell death biomarkers for each type of cell death. Although a clear-cut distinction between different modes of cell demise is disputable [1, 2], the Nomenclature Committee on Cell Death (NCCD) recommended, among others, mechanism-based morphological criteria to define death modalities such as apoptosis, mitotic catastrophe, necrosis, autophagy, anoikis, and paraptosis [3].

To assess the extent of brain tissue loss and the effectiveness of therapeutic strategies, we need to be able to visualize and gauge cell death in the brain following stroke, traumatic brain injury (TBI) or neurodegenerative diseases. Light microscopy allows for the inexpensive and fast detection of neuronal/glial cell demise and estimation of infarct and traumatic lesion volumes.

Laura Lossi and Adalberto Merighi (eds.), *Neuronal Cell Death: Methods and Protocols*, Methods in Molecular Biology, vol. 1254, DOI 10.1007/978-1-4939-2152-2_14, © Springer Science+Business Media New York 2015

Quick fixation of the brain tissue is necessary to prevent degradation or processing of tissue antigens by protease to avoid pathological changes in response to premortal hypoxia. Transcardial perfusion fixation is recommended to avoid artifacts introduced during removal of the unfixed brain from the skull, and handling-induced surface artifacts produced during necropsy (i.e. "dark neurons"). Brain perfusion with post-fixation ensures the best possible preservation of the brain for histological and immunohistochemical stainings, reducing variation in the reproducibility of immunohistochemistry (IHC).

IHC techniques permit visualization of multiple proteins and other antigens on consecutive sections in specific cells/histological regions, blood–brain barrier (BBB) deterioration, providing an advantage over liquid tissue assays [4]. Morphological criteria identified with Masson's trichrome stain assist in visualizing degenerating neurons [5] (Fig. 1a, b). It is important, however, to note that histological methods can identify cell death only at a single time point.

Recent developments in computer-assisted image analysis technologies [4–6] enable more objective and accurate assessment of histological stains and immunohistochemical markers, facilitating morphometric quantification of cell injury on the entire brain sections [7] (Fig. 1a, b). Our [5] and others' [4, 8] data provided evidence for a strong correlation between the computer-aided assessment and pathologist's scoring.

In mouse experimental studies, we characterized neuronal damage in vivo using seizure-induced brain injury caused by kainic acid (KA) [7] and in two models of acute brain injury: stroke caused by middle cerebral artery occlusion (MCAO) [9] and traumatic brain injury caused by controlled cortical impact (CCI) [7, 10].

We used rabbit polyclonal antibodies to cleaved caspase-3, phospho-c-Jun (Ser73), phospho-SAPK/JNK (Thr183/Tyr185), mouse monoclonal antibody to growth arrest- and DNA damage-inducible gene 153 (GADD 153) [also known as C/EBP homologous protein (CHOP)] and TUNEL (Terminal Deoxyribonucleotidyl Transferase-mediated dUTP Nick End Labeling) assay to characterize apoptotic cell death [5, 7, 9]. The c-Jun N-terminal protein kinase (JNK) signaling pathway modulates the activity of several Bcl-2 family proteins, promoting apoptosis [10]. Transcription factor CHOP, an indicator of endoplasmic reticulum (ER) stress, is known to control expression of various apoptosis genes (*reviewed in* refs. 11, 12).

It is noteworthy that activation of caspase-3 has been documented also under non-apoptotic conditions, i.e. in adaptive responses to synaptic activities in the nervous system (*reviewed in* ref. 13). Therefore, bright-field visualization of cleaved caspase-3 expression in the context of histologically interpretable brain tissue sections is highly informative and superior to dark-field illumination used in immunofluorescence (IMF).

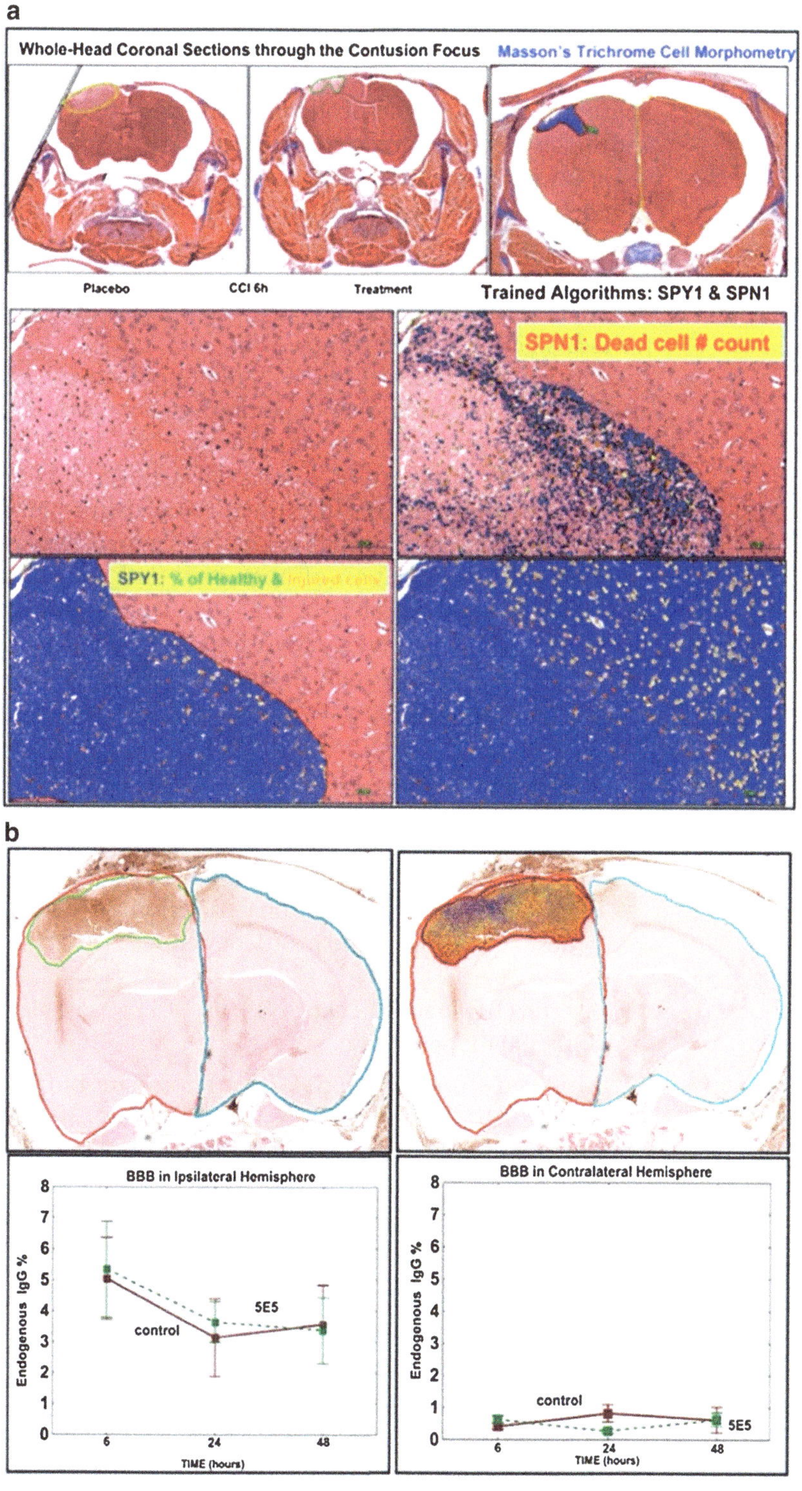

Fig. 1 Examples of TBI-CCI experiments in genetically impaired animals with neuronal conditional deletion of death-regulating gene, evaluated by digital pathology tools. (**a**) Whole-head Masson's trichrome stained sections after tuning procedure of algorithms for the count of cell death/degeneration ratio and annotation of the impact area for volume analysis. (**b**) BBB impairment in IgG-IHC staining. The results of algorithm work are shown on the right mirror image with the mark-up annotation of the pathological area and on the diagrams below comparing the areas of ipsi- and contralateral hemisphere for the % of IgG signal

To assess neuronal preservation, a monoclonal antibody to neuron-specific nuclear protein NeuN was applied. In addition, antibodies detecting microtubule-associated proteins MAP2 and phospho-tau (Thr231) were utilized to characterize neuronal degeneration [7]. Loss of immunoreactivity due to early processing of MAP2 following TBI or ischemia could serve as a territorial marker to separate the lesion focus from the penumbra and unaffected areas of the brain thus providing more precise parameters for assessment of lesion volume (Fig. 2a, b).

Partial restoration of MAP2 immunoreactivity at later time points provides additional quantitative information about the pathophysiological recovery of some neurons, whereas the loss of MAP2 protein is indicative of neuronal cell death. TBI often leads to diffuse axonal injury. JNK protein was shown to be activated in damaged axons, and inhibition of JNK activity was reported to diminish the accumulation of both total and phosphorylated tau in injured axons (*reviewed in* ref. 14). Digital pathology image analysis and morphometry algorithms are potent tools to assess these dynamics in genetically engineered mouse models and novel neuroprotective drug tests.

2 Materials

2.1 Specimens, Buffers, Histology or IHC Reagents and Detection Systems

2.1.1 Specimens

1. Brains removed from wild-type (WT) or genetically engineered mice (here with pan-neuronal deletion of death regulating genes) according to the protocol below.

2.1.2 Anesthesia and Perfusion Buffers

1. Avertin (BD Franklin Lakes, NY): use at 125–250 mg/kg—about 350–500 μL per mouse.

2. Modified PBS (mPBS; pH 7.6): Phosphate-buffer saline containing 120 mM NaCl, 11.5 mM NaH_2PO_4, 30 mM K_2HPO_4, prepared fresh each week. To prepare 2 L: weigh 14.026 g NaCl, 2.76 g NaH2PO4, and 10.88 g K_2HPO_4 and dissolve in about 1 L of ddH_2O. Adjust final pH to 7.6 by adding 1 *N* HCl or NaOH, accordingly. Make up to 2 L with ddH_2O.

3. 78.7 mM activated sodium orthovanadate, zVAD-fmk-Caspase inhibitor1 (Chemicon International, Temecula, CA) stock solution. *See* **Note 1**.

4. 0.22 M Z-Asp-2,6-dichlorobenzoyloxymethylketone/Caspase-1 Inhibitor V (Z-Asp) stock solution.

5. Perfusion buffer with protease and phosphatase inhibitor cocktail (*see* **Note 1**): 5 mM NaF, 10 mM glycerol 2-phosphate disodium salt hydrate (BGP), 1 mM activated sodium orthovanadate in water. Add 0.05 μL/mL zVAD-fmk-Caspase inhibitor1 stock solution, 1 μL/mL Z-Asp stock solution, 1 μL/mL PMSF. Use 3–5 mL/mouse.

6. Z-Fix (Anatech Inc., Battle Creek, MI). *See* **Note 2**.

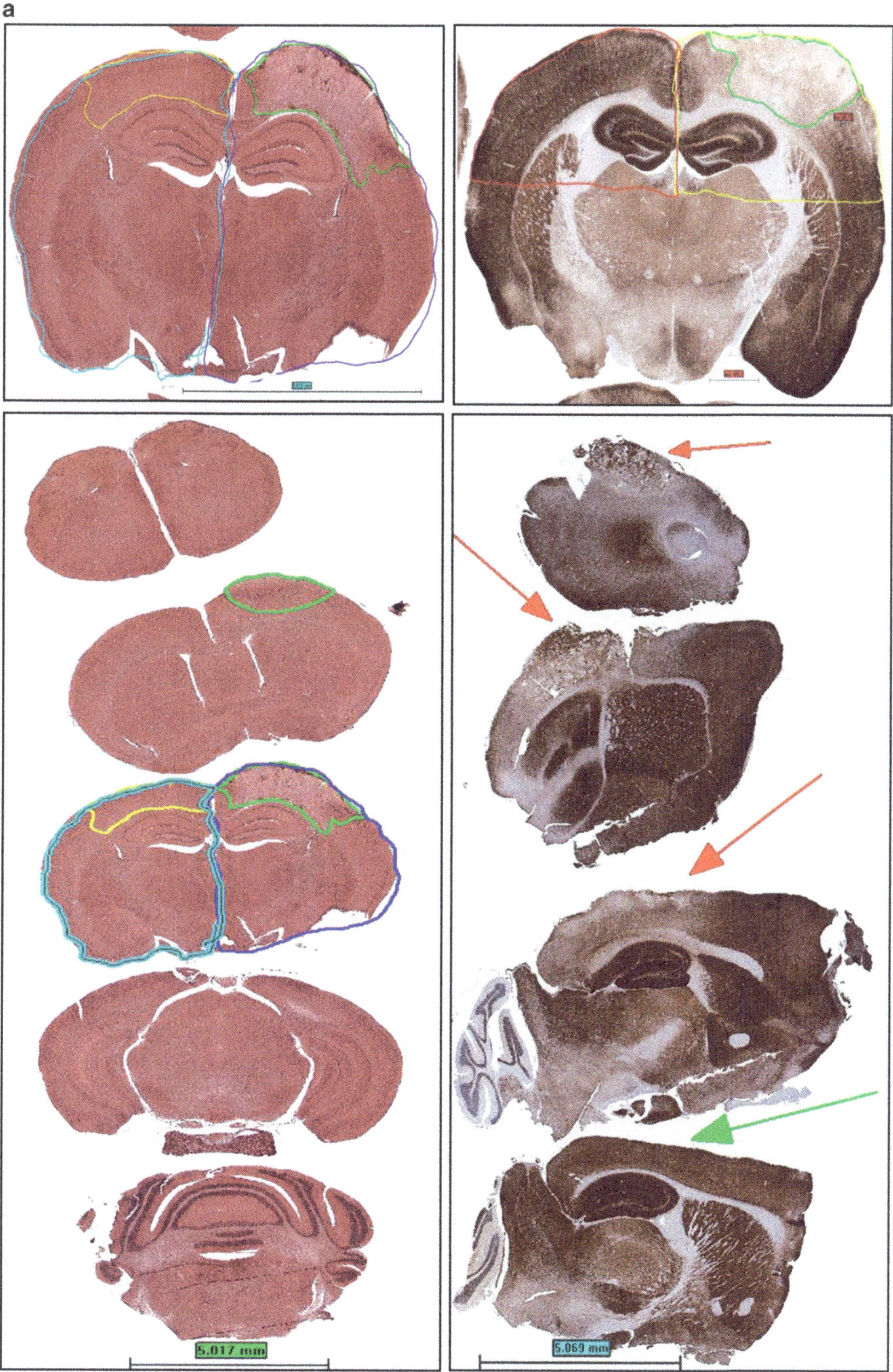

Fig. 2 Staining methods and brain slice orientations for impact volume measurements. (**a**) Coronal or sagittal sectioning orientation could be chosen and either Masson's trichrome histochemical (*left*) or MAP2-HRP-IHC stains (*right*; DAB/*brown color*) could be easily annotated and evaluated. (**b**) Examples of MAP2-Alexa-488 IMF staining in Z-Fix, sucrose-cryoprotected mouse brain after 1 h (*left images*; cross sections) and 24 h (*right image* with sagittal plane of brain sections) post TBI-CCI. Note similar loss of MAP2-IMF signal in the cortex and hippocampus as on the panel **a** with ZFPE-HRP-DAB IHC sections (impact focus pointed by *red arrows* versus control, intact hemisphere indicated by *green arrow*)

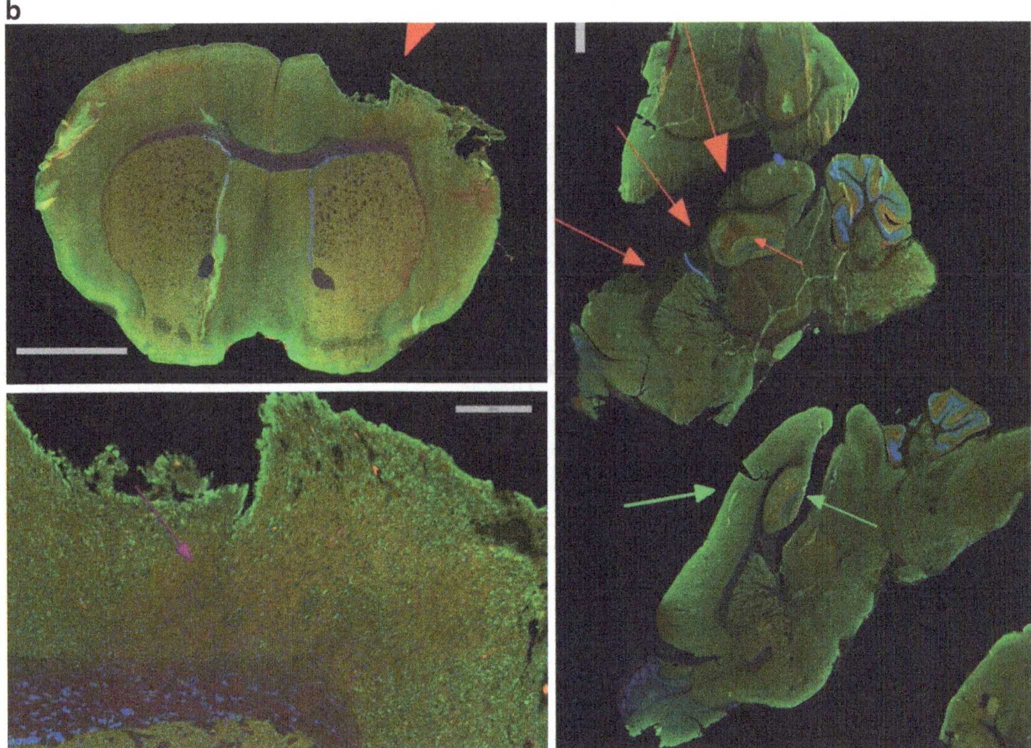

Fig. 2 (continued)

**2.2 Immunohisto-
chemical Reagents
and Detection Kits**

1. Primary antibodies: rabbit polyclonal antibodies to cleaved caspase-3 (Cell Signaling Technology, Danvers, MA and Imgenex, San Diego, CA), phospho-c-Jun (Ser73) (Cell Signaling Technology), phospho-SAPK/JNK (Thr183/Tyr185) (Cell Signaling Technology); mouse monoclonal antibodies to GADD 153 (CHOP) (Santa Cruz Biotechnology, Santa Cruz, CA), neuron-specific nuclear protein NeuN (clone A60 Millipore Chemicon, Ramona, CA, USA), MAP2 (clone HM-2; Sigma-Aldrich, St. Louis, MO), phospho-tau (Thr231; Thermo Fisher Scientific, Lafayette, CO, and anti-tau (phosho T181; Abcam, Cambridge, UK)).

2. Antigen retrieval buffers, antibody diluent and protein block: Dako Target Retrieval Solutions pH 6 and 9 (Dako, Carpenteria, CA); Antibody Diluent w/Background Reducing Components (Dako) and Serum-Free Protein Block (Dako).

3. TUNEL Apoptosis detection kit (e.g. ApopTag Peroxidase in situ Apoptosis Detection Kit Millipore, Chemicon).

4. Anti-mouse or -rabbit DakoEnVision + System- HRP Labeled Polymers (Dako).

5. 3,3′-diaminobenzidine (DAB) + Liquid HRP Substrate Chromogen System (Dako).

6. Hematoxylin–Eosin stain.

7. Masson's trichrome stain (American Master*Tech Scientific Inc., Lodi, CA).

8. Mounting media (non-aqueous for paraffin sections; fluorescence-free aqueous medium for cryosections).

2.3 Special Instrumentation and Equipment

1. 50 mL tubes.

2. Automated tissue processor of any type for paraffin histology. *See* **Note 3**.

3. Microtome.

4. Microtome blades.

5. Cryostat.

6. Cryostat blades.

7. 30 % sucrose-PBS buffer.

8. 10 % thimerosal stock solution.

9. Silanized (APES or TES or positively charged) slides.

10. Slide holders.

11. Automated slide stainer (Pittsburgh, PA; *optional*).

12. Vacuum oven for histology.

13. Pressure cooker for histology.

14. Aperio ePathology—Leica Biosystems (http://www.leicabiosystems.com/pathology-imaging/epathology/). *See* **Note 4**.

15. Surgery annex provided with gas anesthesia equipment.

16. Stereotaxic instrument: e.g. Benchmark™ Angle Two Stereotaxic Instrument or Benchmark™ Stereotaxic Impactor (myNeuroLab Inc., St. Louis, MO; http://www.laboratoryequipment.com/).

3 Methods

3.1 Artifact-Free Preparation of Brains for IMF on Fixed-Frozen Cryosections (FFC) or Immunohistochemical Applications Using Tissue Sections from Z-Fix-Fixed Paraffin-Embedded (ZFPE) Blocks (See Note 5)

This protocol can be applied to any study involving rodents' brains to explore neuroanatomy, focal or global ischemia, novel drug treatment and drug toxicity, brain injury, and brain tumor models. The technical information provided here is related to mouse studies and these recommendations have to be modified when larger animals are being used. The protocol is based on the principle that the 2–5-day delay in brain removal following the perfusion permits successful preservation of the brain and prevents peri-mortal artifacts resulting from manual handling of the unfixed brain surface during immediate removal after an experiment, thus allowing for proper assessment of neuronal cell death.

1. Anesthetize mouse with Avertin and perfuse transcardially first with mPBS, pH 7.4 (3–5 mL/mouse), followed by Z-Fix. In studies focused on brain proteases (caspases, calpains, or

metaloproteinases) and phosphatases, start perfusion with buffer containing additionally protease and phosphatase inhibitor cocktail.

2. After perfusion decapitate the animals, remove the skin and muscles on the side of the skull, and fix the entire head using excess volume of Z-Fix in a 50 mL tube. To allow faster fixative penetration, make a small incision at the base of skull, by inserting one scissor tip pointing down to the post-decapitation opening of the cervical spinal cord canal and the second one into the external ear canal in the petrous bone contralateral to the side of the brain injury. The incision should not damage the brain tissue. During the first 24 h leave the entire head immersed in Z-Fix at room temperature; continue the fixation in the cold room for the next 3–5 days.

3. On day 2 or 3 remove the bones, and postfix the whole brain for the next 2–3 days (total fixation time 5 days). Store in PBS in the cold room. You may collect specimens from the entire experiment, storing the brains or brain slices assembled in the embedding cassettes in PBS or 30 % sucrose-PBS for about 1 month before tissue processing paraflin and embedding or preering.

4. After fixation, rinse the brains in 3 changes of mPBS (10–15 min each) to eliminate the fixative, and then perform routine cassetting of the tissues.

5. Cut each brain in 1.5–2.0 mm thick coronal or sagittal slices fitting in one cassette to be frozen or processed/embedded into a single cryo- or paraffin block. Thus, all slices from the entire brain will be sectioned, placed and stained on the same slide ensuring identical staining conditions. The cassetting can be done at any convenient time within 3–4 weeks after fixative removal, and the samples can be maintained in mPBS at 4 °C for long-term storage.

6. To obtain fixed-frozen cryosections (FFC), cryoprotect the brain slices placed in embedding cassettes before you freeze the brains. For this purpose, transfer all cassettes to 30 % sucrose-mPBS buffer changing it daily for 3–4 days at 4 °C. Add 1 µL/mL buffer of thimerosal stock solution to the last change of 30 % sucrose buffer. Freeze using conventional freezing techniques for cryosectioning.

3.2 IMF on FFCs or HRP-IHC on ZFPE Sections

Our brain preparation and processing protocol can be applied successfully to both IMF and HRP-IHC methods, as demonstrated in Fig. 2a, b. *See* **Note 6**.

A detailed description of the routine protocols that were used for IMF or HRP-IHC immunolabeling of specimens is beyond the purpose of this chapter. We will briefly describe below an exemplificative protocol for immunostaining of ZFPE sections. Readers are invited to look at our previous publications [5, 7, 9, 15] for additional information.

Some sections can be stained for reference with hematoxylin–eosin or Masson's trichrome stain.

1. Deparaffinize and hydrate ZFPE sections according to routine protocols.

2. Perform antigen retrieval with Dako Target Retrieval Solutions pH 6 either or 9 according to the manufacturer's recommendations. *See* **Note 7**.

3. Wash in mPBS (3 × 5 min).

4. Incubate in Antibody Serum-Free Protein Block for 45–60 min.

5. Dilute primary antibodies at optimal titer with Diluent w/ Background Reducing Components. *See* **Note 8**.

6. Incubate for 1.5 h at room temperature, if using low dilution and highly concentrated primary antibodies (most of monoclonal antibodies are in this category). For high-affinity antibodies (predominantly polyclonal) use higher dilutions and incubate overnight either at room temperature or 4 °C.

7. Wash in mPBS (3 × 5 min).

8. Incubate with goat anti-mouse or goat anti-rabbit polymer-based EnVision-HRP-enzyme conjugate for 45 min, not exceeding 1 h.

9. Wash in mPBS (3 × 5 min).

10. Develop the HRP reaction with DAB+ Liquid HRP Substrate Chromogen System following the manufacturer's recommendations.

11. Alternative perform TUNEL staining with ApopTag Peroxidase in situ Apoptosis Detection Kit following the manufacturer's recommendations.

12. Wash in mPBS (3 × 5 min).

13. Counterstain nuclei with hematoxylin or nuclear red (*optional*).

14. Dehydrate and mount.

3.3 Quantitative Analysis

In our studies, we undertook computer-assisted quantification of specific histological and immunohistochemical parameters that characterize processes associated with cell death. We demonstrated the utility of image analysis algorithms for color deconvolution, colocalization, and nuclear morphometry to characterize cell death events in the brain tissue specimens subjected to immunostaining for cytoplasmic, nuclear and stromal biomarkers, and detection of fragmented DNA or showing features of neurodegeneration in Masson's trichrome staining (Fig. 1a) [5, 7]. Recently, our particular interest

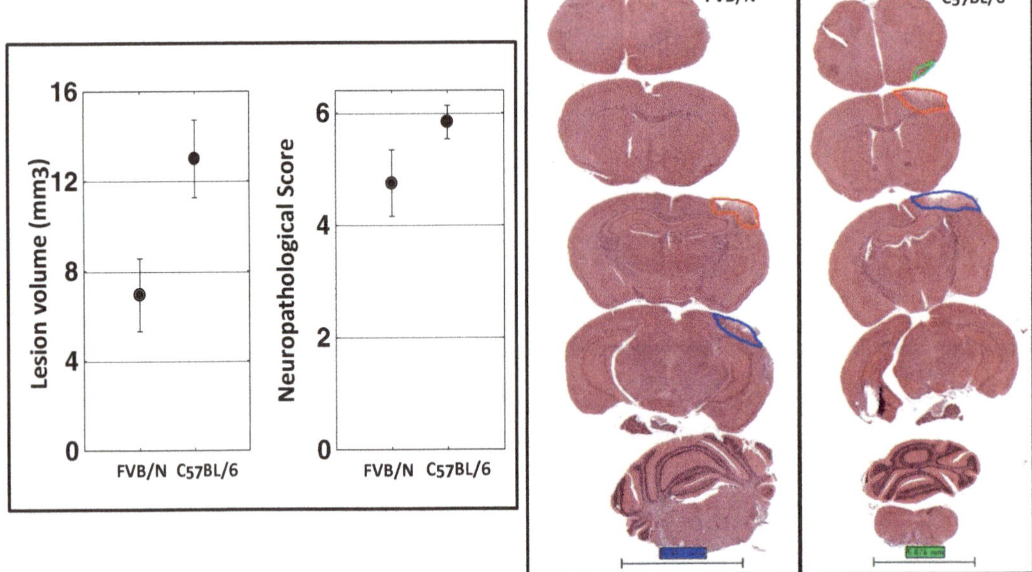

Fig. 3 Mouse strain susceptibility to brain trauma: FVB/N versus C57Bl/6 WT mouse lines. The comparative investigation of the lesion size (volume) and Neuropathological score (Npt score) in two commonly used WT-mouse strains for genetical manipulation, revealed that C57Bl/6 mice are more susceptible to brain trauma than FVB/N mice and responded with larger pathological focus and elevated Npt score to the same impact velocity (3 m/s)

focused on the numerical assessment of the brain tissue damage in the standardized brain injury model of CCI (Figs. 1, 2, and 3). To support the computerized measurements of lesion volume and the assessment of the degree of neuronal death, in addition we have developed a novel neuropathological (Fig. 3) and hemorrhage scores. In association with the lesion volume, BBB damage and the counts of cell death (Figs. 1b, 2a, b, 3), these scoring systems proved to be helpful in interpretation of the results of variety of behavioral tests and permit more precise stratification of the TBI experimental groups.

Once your digital slide images are captured, before applying morphometric algorithms to register your experiment specific variables, the sequence of adjustments should be performed, which are described below.

1. Scan slides at an absolute magnification of 200× or 400× [resolution of 0.50 and 0.25 µm/pixel, respectively] using the Aperio ScanScope CS, XT or AT systems, typically scanning from 5 to up to 120 slides per session, respectively. Based on our experience, blurring on a small part of an image does not affect algorithm performance.

2. Calibrate the background illumination levels using a prescan procedure.

3. Evaluate the acquired whole-tissue digital images for image quality. Images, which were not captured due to a failure of the "tissue finding" function to detect weak immunostaining, need to be rescanned using the "faint" mode.

4. Label the acquired images, place them in dedicated project folders, and store in designated local servers. You can view and analyze the slides remotely using desktop personal computers employing the web-based ImageScope viewer.

5. Apply the Spectrum Analysis algorithm package and Image Scope analysis software (Aperio Technologies, Inc.) to quantify IHC and histochemical stainings. These algorithms make use of a color deconvolution method [16] to separate stains.

6. Using the software, calibrate individually each stain and record the average red, green, and blue (RGB) optical density (OD) vectors for each applied chromogen and counterstain. When you use DAB as a chromogen, utilize the following vectors for the RGB channels to apply a color deconvolution algorithm: for color 1 hematoxylin, 0.65, 0.70, and 0.29; for color 3 DAB, 0.27, 0.57, and 0.78 [16].

7. Determine weak, moderate, and strong positive thresholds that conform to intensity ranges on a scale of 0–255 (black to white, respectively). Set other algorithm parameters to achieve concordance with manual scoring on a number of high-power fields, including intensity thresholds for positivity and parameters that control cell segmentation using the nuclear algorithm. Reducing the curvature threshold separates coalescing nuclei. We recommend retraining algorithm for different biomarkers or antibodies, and staining conditions.

8. Using a pen tool, annotate regions of interest (ROIs) for quantitative analysis of the applied biomarkers. The algorithms calculate the area of positive staining, the average positive intensity (API and OD (1), the percentage of weak (1+), medium (2+), and strong (3+) positive staining) as well as the number of 1+, 2+, 3+, and all positive cells in the annotated ROI. You can calculate density of immunopositive cells as a ratio (n/mm^2) of positive cell number (n) to the annotated area (mm^2).

9. Use a colocalization algorithm based on the deconvolution method to separate the stains and classify each pixel according to the number of stains present. For colocalization, specify a detection threshold for each stain, and the algorithm reports the percentage of area for which each stain combination is detected: 1, 2, 3, 1+2, 1+3, 2+3, 1+2+3, none (up to 3 stains are supported). The algorithm also provides an eight-color mark-up image for visualization of the colocalized states.

10. Save the results in annotation format. To provide a visual representation of the numerical results, you can display digital images of slides side by side with pseudocolor mark-up images.

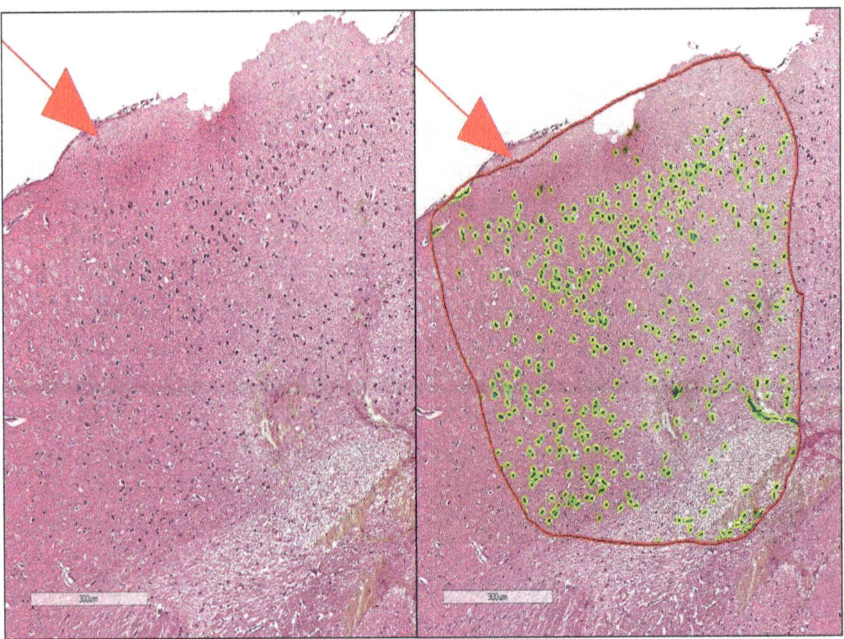

Fig. 4 Exemplificative analysis of the total cell, multifunctional Next Generation Algorithm vC2. Mirror images from the impact focus and penumbra with annotation of the dead/degenerating neurons by multifunctional, next generation Cell/Cytoplasm v2 algorithm (Aperio Technology, Inc.), performed on the Masson trichrome stained brain section. Note that none of the penumbra intact neurons is annotated by this algorithm, which tuning depict solely all pathologically changed and degenerating neurons in the lesioned cortex after TBI

11. Algorithms of the next generation are more complex and combining in one the functions of multiple, single, basic algorithms (membrane, cytoplasmic, nuclear), thus permitting simultaneous registration of signals from all cell compartments. The novel, whole Cell/Cytoplasm algorithm vC2 from Aperio Technology Inc. is particularly powerful in detecting and counting dying neurons in Masson's trichrome stain (Fig. 4).

3.4 Assessment of Brain Lesion Volume

1. Cut the brains into five equally spaced 2 mm coronal slabs with three cross sections placed on the rostral and caudal edges and in the middle of the cortical lesion.

2. Stain the sections using Masson's trichrome stain.

3. Digitalize the sections using the Aperio scanning system.

4. Using the pen tool, annotate virtual slides by encircling the lesion edges (penumbra) between intact and pathologically changed brain tissue; the lesion area (mm²) is reported in the annotation window.

5. Determine the rostral–caudal dimensions of the foci through the use of a stereotaxic atlas for the mouse brain [18], which

can be viewed electronically side by side with the coronal section images. Distance (mm) between coronal coordinates for bregma +1.54 (rostral position) and –3.88 (caudal position) determine the distribution of the lesion volume.

6. Calculate lesion volume by summing each lesion area multiplied by the distance between each coronal slice. The proposed calculation is based on Cavalieri's method [19] modified for the purpose of volumetric analysis performed on digital slides.

3.5 Neuropatho-logical Scoring of Brain Injury

1. View digital images of Masson's trichrome-stained brain specimens to assess brain injury using the following four-tier neuro-pathological scoring system:

   (a) Grade 0—no observable injury

   (b) Grade 1—restricted cortical impact area with small petechial bleedings and variable degree of acute ischemic neuronal degeneration but no cortical tissue loss

   (c) Grade 2—necrotic lesion core and variable tissue loss of the injured cortex with penumbra demonstrating neuronal degeneration limited to cortex only

   (d) Grade 3—sporadic larger hematomas and ischemic, necrotic, spongiotic changes crossing the corpus callosum and involving ipsilateral hippocampal sectors CA1–CA3

   (e) Grade 4—confluent impact core encompassing large cortical area, entire hippocampus and deeper thalamic structures in the upper quadrant of the ipsilateral hemisphere.

2. Score each brain coronal section that contains a pathological focus. Calculate the total score by adding up scores for all sections recorded for each animal.

4 Notes

1. To prepare the zVAD-fmk stock solution dissolve 1.84 g of sodium orthovanadate in 45 mL purified water in a small beaker with a stir bar. Adjust the pH to 10 using either 1 N NaOH or 1 N HCl, with stirring. The starting pH of the sodium orthovanadate may vary with lots of the chemical. At pH 10, solution will be yellow.

   Boil solution until it turns colorless (approximately 10 min). All of the crystals should dissolve. Cool to room temperature and then readjust the pH to 10 and repeat boiling/cooling until solution remains colorless and pH stabilizes at 10. Adjust the final volume to 50 mL with purified water.

   Store the activated sodium orthovanadate in 1 mL aliquots and freeze at –20 °C.

The protease and phosphatase inhibitor cocktail contains BGP, a classical serine-threonine phosphatase inhibitor and the caspase inhibitors zVAD-fmk and Z-Asp. It is used as a first step in heart perfusion to flush blood from the circulatory system and to inhibit the autoactivation of endogenous proteases and phosphatases.

2. Z-Fix, a ready to use zinc-buffered formalin solution, is a mild cross-linker compatible with blood pH. It is a convenient perfusion solution working excellently for the preservation of protein immunoreactivity and DNA/RNA complexity, allowing successful extraction of proteins for mass spectroscopy (Liquid Tissue®-SRM assay technology enables mass spectrometry measurement in ZFPE tissue of proteins; http://www.expressionpathology.com/contact_us.php?submenu-header=6 [20]), and DNA/RNA for PCR and next generation sequencing. Z-Fix is a relatively safe fixative, as overfixation does not produce formalin pigment precipitation, denaturation or masking of antigens, or degradation of DNA/RNA. It retains good antigen immunoreactivity, as tested by IMF assays in specimens that were processed through cryoprotection in 30 % sucrose buffer (FFCs) or by IHC after embedding into ZFPE blocks [7].

3. Tissue processors for paraffin embedding can be found at http://www.tradeindia.com/manufacturers/automatic-tissue-processor.html. Leica Microsystems supplies a complete paraffin embedding center and microtome (Reichert-Leica; http://www.leica-microsystems.com/company/this-is-leica-microsystems/).

4. The Aperio ePathology—Leica Biosystems consists of digital slide scanner (any model) and associated internet platform (Spectrum) with Imagescope analysis software (Aperio Technologies Inc., Vista, CA, http://www.leicabiosystems.com/pathology-imaging/epathology/).

5. Quality of paraffin sections obtained from ZFPE blocks is important, as variations in section thickness, wrinkles, and bubbles in the mounting media affect properties of digital images.

6. Although IMF methods permit registration of multicolor tags for multiple antigens generating spectacular images, their evaluation is rather qualitative than quantitative. Rapid development of imaging devices and associated software generating digital images offers more algorithm-based morphometric tools for measurement of cell events, launching the era of digital pathology. To quantify the cell death events in the brain, we focused on the development and evaluation of the bright-field digital pathology tools.

7. Both antigen retrieval buffers could be used in microwave or pressure cooker heat-induced epitope-retrieval (HIER). Routinely, for the majority of cytoplasmic epitopes, use high pH retrieval solution. For unmasking membrane and nuclear antigens use low pH retrieval solution. You may perform two types of retrieval procedures for proteins with dual expression and trafficking between cytoplasm and nucleus.

8. For better understanding the mechanistic background of IHC staining for tuning the variables of the method, read practical "Basic Immunochemistry" manual from Dako (http://www.dako.com/08002_03aug09_ihc_guidebook_5th_edition_chapter_2.pdf).

Acknowledgements

This work was supported by the National Institutes of Health (NIH) Grant NS-036821 and the Department of Defense (DoD) Grant W81XWH-10-1-0847 awarded to Stan Krajewski. We are greatly indebted to Dr. Allen Olsen from Aperio Technology, Inc. for collaborative development and beta-testing of Cell/Cytoplasmic v2 algorithm. We thank Xianshu Huang for high-quality histological preparations and immunocytochemistry, Tomi Omel and Francisco Beltran for excellent animal care assistance.

References

1. Galluzzi L, Maiuri MC, Vitale I et al (2007) Cell death modalities: classification and pathophysiological implications. Cell Death Differ 14:1237–1243

2. Vanden Berghe T, Grootjans S, Goossens V et al (2013) Determination of apoptotic and necrotic cell death in vitro and in vivo. Methods 61:117–129

3. Kroemer G, Galluzzi L, Vandenabeele P et al (2009) Classification of cell death: recommendations of the Nomenclature Committee on Cell Death 2009. Cell Death Differ 16:3–11

4. Rizzardi AE, Johnson AT, Vogel RI et al (2012) Quantitative comparison of immunohistochemical staining measured by digital image analysis versus pathologist visual scoring. Diagn Pathol 7:42

5. Krajewska M, Smith LH, Rong J et al (2009) Image analysis algorithms for immunohistochemical assessment of cell death events and fibrosis in tissue sections. J Histochem Cytochem 57:649–663

6. Weinstein RS (2009) Risks and rewards of pathology innovation: the academic pathology department as a business incubator. Arch Pathol Lab Med 133:580–586

7. Krajewska M, You Z, Rong J et al (2011) Neuronal deletion of caspase 8 protects against brain injury in mouse models of controlled cortical impact and kainic acid-induced excitotoxicity. PLoS One 6:e24341

8. Brazdziute E, Laurinavicius A (2011) Digital pathology evaluation of complement C4d component deposition in the kidney allograft biopsies is a useful tool to improve reproducibility of the scoring. Diagn Pathol 6 (Suppl 1):S5

9. Krajewska M, Xu L, Xu W et al (2011) Endoplasmic reticulum protein BI-1 modulates unfolded protein response signaling and protects against stroke and traumatic brain injury. Brain Res 1370:227–237

10. Okuno S, Saito A, Hayashi T et al (2004) The c-Jun N-terminal protein kinase signaling pathway mediates Bax activation and subsequent neuronal apoptosis through interaction with Bim after transient focal cerebral ischemia. J Neurosci 24:7879–7887

11. Silva RM, Ries V, Oo TF et al (2005) CHOP/ GADD153 is a mediator of apoptotic death in substantia nigra dopamine neurons in an in vivo neurotoxin model of parkinsonism. J Neurochem 95:974–986

12. Szegezdi E, Logue SE, Gorman AM, Samali A et al (2006) Mediators of endoplasmic reticulum stress-induced apoptosis. EMBO Rep 7: 880–885

13. Yuan J (2006) Divergence from a dedicated cellular suicide mechanism: exploring the evolution of cell death. Mol Cell 23:1–12

14. Blennow K, Hardy J, Zetterberg H (2012) The neuropathology and neurobiology of traumatic brain injury. Neuron 76:886–899

15. Krajewska M, Rosenthal RE, Mikolajczyk J et al (2004) Early processing of Bid and caspase-6, -8, -10, -14 in the canine brain during cardiac arrest and resuscitation. Exp Neurol 189:261–279

16. Ruifrok AC, Johnston DA (2001) Quantification of histochemical staining by color deconvolution. Anal Quant Cytol Histol 23: 291–299

17. Yu J, Wang Z, Kinzler KW et al (2003) PUMA mediates the apoptotic response to p53 in colorectal cancer cells. Proc Natl Acad Sci U S A 100:1931–1936

18. Paxinos G, Franklin K (2001) The mouse brain in stereotaxic coordinates. Academic, San Diego

19. Rosen GD, Harry JD (1990) Brain volume estimation from serial section measurements: a comparison of methodologies. J Neurosci Methods 35:115–124

20. Prieto DA, Hood BL, Darfler MM et al (2005) Liquid Tissue: proteomic profiling of formalin-fixed tissues. Biotechniques Suppl: 32–35

Chapter 15

In Vitro Oxygen-Glucose Deprivation to Study Ischemic Cell Death

Carla I. Tasca, Tharine Dal-Cim, and Helena Cimarosti

Abstract

Oxygen-glucose deprivation (OGD) is widely used as an in vitro model for stroke, showing similarities with the in vivo models of brain ischemia. In order to perform OGD, cell or tissue cultures, such as primary neurons or organotypic slices, and acutely prepared tissue slices are usually incubated in a glucose-free medium under a deoxygenated atmosphere, for example in a hypoxic chamber. Here, we describe the step-by-step procedure to expose cultures and acute slices to OGD, focusing on the most suitable methods for assessing cellular death and/or viability. OGD is a simple yet highly useful technique, not only for the elucidation of the role of key cellular and molecular mechanisms underlying brain ischemia, but also for the development of novel neuroprotective strategies.

Key words Acute brain slices, Cell death assays, Ischemia, Neuroprotection, Organotypic hippocampal slice cultures, Oxygen-glucose deprivation (OGD), Primary neuronal cultures

1 Introduction

In recent years, in vitro models of stroke, such as oxygen-glucose deprivation (OGD), have been increasingly used to better understand the cellular and molecular pathways associated with brain ischemia. In order to mimic the interruption of the supply of oxygen and nutrients to the brain occurring during an ischemic event, cell lines [1], primary cells [2] or organotypic slice cultures [3], as well as acute slices of brain tissue [4, 5] are typically incubated in a glucose-free medium under a deoxygenated atmosphere. The exposure to OGD may vary, depending on the preparation, from 15 min to over 36 h [4, 6]. Subsequently, the cultures or acute slices are frequently incubated in a glucose-containing medium under an oxygenated atmosphere. This re-oxygenation period simulates the reperfusion stage happening in transient ischemia in vivo, i.e. when the blood flow to the brain is re-established.

OGD in primary glial or neuronal cultures, which provide a monolayer of cells, allows excellent visualization of individual cells

Laura Lossi and Adalberto Merighi (eds.), *Neuronal Cell Death: Methods and Protocols*, Methods in Molecular Biology, vol. 1254, DOI 10.1007/978-1-4939-2152-2_15, © Springer Science+Business Media New York 2015

as well as detailed studies of particular molecular mechanisms and isolated cellular events [2, 7]. Organotypic cultures and acute slices offer the advantage of preserving the functional relationships and interactions between neighboring cells, such as neurons and astrocytes, keeping the intrinsic synaptic connections found in vivo [8]. OGD in organotypic hippocampal slices closely reproduces two important aspects of ischemia in vivo, namely selective vulnerability, i.e. CA1 pyramidal cells are more susceptible than dentate gyrus (DG) granule cells [9], and delayed neuronal death [10]. Conversely, OGD in acutely prepared brain slices is limited by the fact that slices can only be maintained ex vivo for several hours, excluding the possibility of studying delayed pathways and longer-term effects of neuroprotective agents. Despite this drawback, OGD in acute slices has been largely used for electrophysiological experiments to study synaptic plasticity and for the screening of novel compounds and neuroprotective strategies [4, 11].

We have been employing OGD in primary neurons and organotypic cultures, as well as acute brain slices, for over a decade and found it to be an extremely useful and reliable model, especially in order to study neuroprotection, both in the form of endogenous cytoprotective pathways [2, 4, 5, 11, 12] and pharmacological agents [13, 14] or lifestyle changes, e.g. exercise [15, 16]. This method represents an invaluable alternative to in vivo models, especially since OGD experiments are less complicated and require significantly less animals to produce sufficiently reliable data, which will allow informed decisions regarding the design of subsequent experimental stages. Here we describe a detailed protocol for the induction of OGD in primary neurons, organotypic cultures and acute slices, in order to mimic the cellular death observed in in vivo models of brain ischemia. The most suitable assays to evaluate cellular death following OGD in each of these preparations will also be described in detail.

2   Materials

All solutions are prepared using ultrapure water (prepared by purifying deionized water to attain a sensitivity of at least 18.2 MΩ-cm at room temperature) and analytical grade reagents. All procedures involving primary neurons and organotypic slices are carried out using sterile cell and tissue culture techniques. For these, all non-sterile glassware and plastic ware are autoclaved prior to use, and tools and equipment are sterilized by coating them in a 70 % ethanol/water solution (spraying) followed by exposure to UV light for 60 min. The materials for the preparation of primary neuronal cell and organotypic slice cultures are described elsewhere [17, 18]. The culture media used to maintain primary neurons and slice cultures and the OGD medium are prepared in a sterile laminar flow hood. The OGD medium can be stored at 4 °C for up to 2 months, whereas the culture media should be prepared freshly every 2 weeks.

2.1 Media for OGD in Primary Neurons and Organotypic Slices

1. Neurobasal medium (Life Technologies™, Invitrogen™, Carlsbad, CA).

2. B27 supplement (Life Technologies™, Invitrogen™).

3. Penicillin/streptomycin solution.

4. Minimal Essential Medium (MEM, Life Technologies™, Invitrogen™).

5. Hank's Balanced Salt Solution (HBSS, Life Technologies™, Invitrogen™).

6. Horse serum (Life Technologies™, Invitrogen™).

7. 0.2 filters.

8. Culture medium for primary neurons: 2 % B27, 1 % penicillin/ streptomycin in Neurobasal medium. Using sterile pipettes add to a 500 mL bottle of Neurobasal medium, 10 mL of B27 and 5 mL of penicillin/streptomycin. The solution is warmed to 37 °C in a water bath prior to use.

9. Culture medium for organotypic slices: 4 mM $NaHCO_3$, 25 mM HEPES (sodium salt), 36 mM glucose, 49.5 % MEM, 24.8 % HBSS, 24.8 % horse serum, 1 % penicillin/streptomycin, pH 7.3. For 100 mL: 0.01 g $NaHCO_3$, 0.65 g HEPES and 0.58 g glucose are added stepwise in a 100 mL glass bottle containing a magnetic stirrer bar. In the flow hood, add 49.5 mL MEM, 24.8 mL HBSS and 24.8 mL horse serum using sterile pipettes. All components are stirred well and the pH is adjusted to 7.3 with HCl. The solution is filtered (0.2 μm) in the flow hood before 1 mL of penicillin/streptomycin is added. Warm to 37 °C in a water bath, prior to use.

10. OGD medium (pH 7.3) (in): 1 mM $CaCl_2$, 5 mM KCl, 137 mM NaCl, 0.4 mM KH_2PO_4, 0.3 mM Na_2HPO_4, 0.5 $MgCl_2$, $0.4MgSO_4$, 25 mM HEPES (free acid), 4 mM $NaHCO_3$, 1 % penicillin/streptomycin. Add all reagents stepwise to a 250 mL glass bottle containing a magnetic stirrer bar. For 250 mL OGD medium (in g): $CaCl_2$ 0.04, KCl 0.10, NaCl 2, KH_2PO_4 0.02, Na_2HPO_4 0.01, $MgCl_2 \cdot 6H_2O$ 0.02, $MgSO_4 \cdot 6H_2O$ 0.02, HEPES (free acid) 1.49 and $NaHCO_3$ 0.08. Add approximately 200 mL of water and stir well the initially opaque solution until it becomes clear. Adjust pH to 7.3 with NaOH and add water to give a total volume of 250 mL. Filter (0.2 μm) in the flow hood into a sterile bottle and 2.5 mL of penicillin/streptomycin. Warm to 37 °C in a water bath prior to use.

2.2 Buffers for OGD in Acute Slices

1. Krebs-bicarbonate (KRB) buffer: 122 mM NaCl, 3 mM KCl, 1.2 mM $MgSO_4$, 1.3 mM $CaCl_2$, 25 mM $NaHCO_3$, 1.2 mM KH_2PO_4, 10 mM glucose, pH 7.4. For 100 mL: Thoroughly mix 10 mL of each of the following stock solutions (*see* **Note 1**):

   1.2 M NaCl.

   30 mM KCl.

   12 mM $MgSO_4$.

   250 mM $NaHCO_3$.

   4 mM KH_2PO_4.

   Then add 1 mL of 1 M glucose, 1 mL of 130 mM $CaCl_2$ and 48 mL of H_2O, before O_2 (95 %)/CO_2 (5 %) is bubbled through the solution for 15 min in order to adjust the pH (*see* **Note 2**).

2. OGD buffer: 1.3 mM $CaCl_2$, 137 mM NaCl, 5 mM KCl, 0.6 mM $MgSO_4$, 0.3 mM Na_2HPO_4, 1.1 mM KH_2PO_4, 25 mM HEPES (free acid), 10 mM 2-deoxy-glucose, pH 7.4 [19, 20]. Add all reagents stepwise to a 100 mL glass bottle containing a magnetic stirrer bar. For 100 mL: 0.02 g $CaCl_2$, 0.80 g NaCl, 0.04 g KCl, 0.02 g $MgSO_4$, 0.004 g Na_2HPO_4, 0.02 g KH_2PO_4, 1 mL of 2-deoxy-glucose, and 3.25 mL of 770 mM HEPES (free acid). Add approximately 80 mL water before the initially opaque solution is stirred well until a clear solution is obtained and the pH is adjusted to 7.4 with HEPES. The volume is increased to 100 mL with water and the solution is warmed to 37 °C in a water bath prior to use.

2.3 Equipment

2.3.1 General

1. Inverted fluorescence microscope (e.g. Nikon Eclipse TE 300, Tokyo, Japan) fitted with a standard rhodamine filter set, a low power objective (5×) and a CCD camera.

2. Modular incubator hypoxia chamber (e.g. Billups-Rothenberg MIC-101, Del Mar, CA). Chamber must be airtight sealed with inlet and outlet valves for gas exchange, allowing for prompt establishment of a constant hypoxic environment.

3. N_2 (95 %)/CO_2 (5 %) gas cylinder with an attached flow meter.

2.3.2 Acute Slices

1. McIlwain tissue chopper (e.g. Mickle Laboratory Engineering Co. Ltd., Guildford, UK) with a double-edged blade or custom-tailored blades from a chopper manufacturer. *See* **Note 3**.

2. Dissecting microscope and light source.

3. Sodium pentobarbital (60 mg/kg) or isoflurane.

4. Guillotine.

5. Instruments for dissection of the brain.

6. Semi-permeable nylon mesh.

<table>
<tr><td valign="top">

2.3.3 Primary Neurons and Organotypic Slices

</td><td valign="top">

1. Vertical or horizontal laminar flow hood.

2. Cell culture CO_2 incubator at 37 °C with an atmosphere of air (95 %) and CO_2 (5 %).

</td></tr>
<tr><td valign="top">

2.4 Cell Death and Viability Assays

</td><td valign="top">

1. 1 mM propidium iodide (PI, e.g. Sigma Chemicals, St. Louis, MO) stock solution (*see* **Note 4**): weigh 0.001 g of PI in an Eppendorf tube and add 1.49 mL water. Shake (vortex) the tube until all solids are completely dissolved. Divide into aliquots (100 μL), and store at 4 °C in individual foil-wrapped Eppendorf tubes in a dark box. *See* **Note 5**.

2. Hoechst 33342 (Sigma Chemicals) stock solution: prepare a 5 mg/mL solution in water by weighing 0.005 g of Hoechst 33342 in an Eppendorf tube and adding 1 mL water. Shake (vortex) until all solids are completely dissolved. Divide into aliquots (100 μL) and store at 4 °C in individual foil-wrapped Eppendorf tubes in a dark box. *See* **Note 6**.

3. Lactate dehydrogenase (LDH) assay kit (e.g. Sigma Chemicals TOX7): store at –20 °C. *See* **Note 7**.

4. 3-(4,5-Dimethyl-2-thiazolyl)-2,5-diphenyl-2H-tetrazolium bromide (MTT) assay (Sigma Chemicals) stock solution: prepare a 5 mg/mL stock solution of MTT in water by adding 0.05 g of MTT to 10 mL of water in a glass beaker containing a magnetic stirrer bar. Stir well, divide into aliquots (1 mL), and store at 4 °C in individual foil-wrapped Eppendorf tubes in a dark box. *See* **Note 8**.

5. Dimethyl sulfoxide (DMSO).

6. Microplate reader (e.g. Tecan Ultra 384, Tecan Group Ltd, Männedorf, Switzerland).

7. Optional: tissue-tearor (e.g. Bio Spec Products, Bartlesville, UK).

</td></tr>
</table>

---

3 Methods

The methods for preparing primary neuronal cell and organotypic hippocampal slice cultures are described elsewhere [17, 18]. The experimental design for the induction of OGD in primary neurons, organotypic slices, and acute slices is schematized in Fig. 1.

<table>
<tr><td valign="top">

3.1 OGD in Primary Neurons and Organotypic Slice Cultures

</td><td valign="top">

Primary neuronal cell cultures and organotypic slices are typically exposed to OGD on in vitro day 14 for the appropriate amounts of time. *See* **Note 9**.

1. Prior to OGD induction, check the viability of cultures by adding 7.5 μL of PI stock solution in 1 mL of culture medium (final concentration 7.5 μM) into the wells containing the primary neurons or the inserts with the organotypic slices.

</td></tr>
</table>

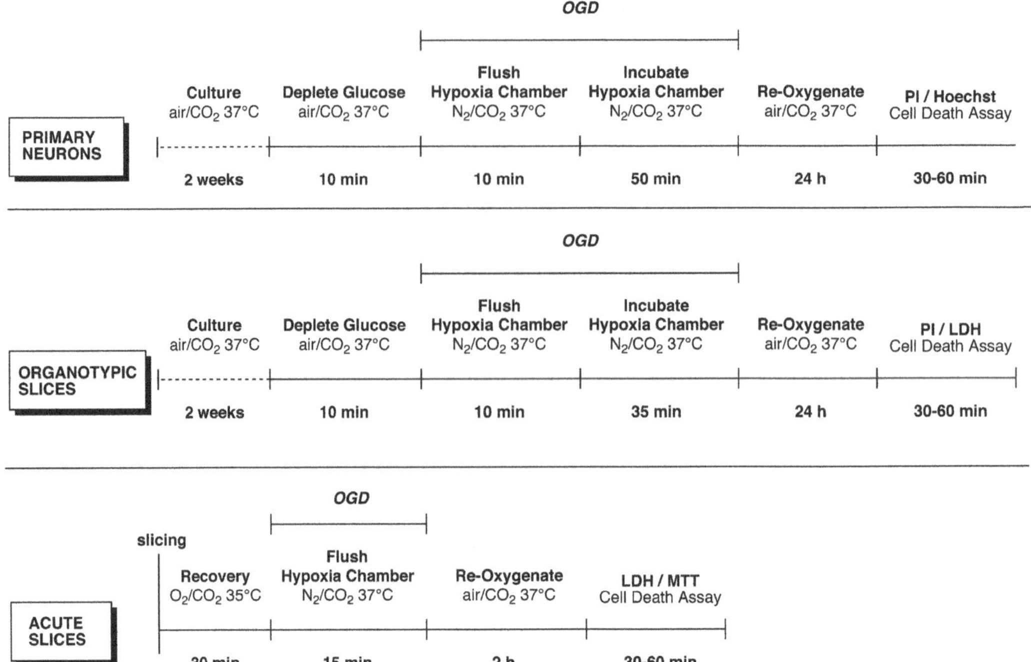

Fig. 1 A schematic timeline of the experimental design for the induction of OGD in primary neurons, organotypic slices, and acute slices. For each specific preparation, the most commonly used assays to measure OGD-induced cell death/viability are indicated

Gently mix the solution with the medium and observe results under the fluorescence microscope 30–60 min later. Discard cultures displaying distinct PI fluorescence above background levels. *See* **Note 10**.

2. Keep the hypoxia chamber inside the incubator at 37 °C for at least 1 h before starting the ODG (*see* **Note 11**).

3. To deplete glucose from intracellular stores and extracellular space, briefly rinse each well twice with pre-warmed OGD medium (primary cultures) or incubate for 10 min in 1 mL of OGD medium per well on a 6-well plate (organotypic slice cultures). Adjust the volume proportionally for smaller or bigger wells.

4. Exchange the medium for pre-warmed OGD medium, which has been bubbled with N_2 (95 %)/CO_2 (5 %) for 15 min prior to use.

5. Bring the hypoxia chamber into the flow hood.

6. Transfer the plates, without the closing lids, into the hypoxia chamber. Close the hypoxia chamber, place back in the incubator and attach the gas tube to the inlet valve.

7. Flush through the hypoxia chamber the N_2 (95 %)/CO_2 (5 %) gas mixture for 10 min at 8 L/min with an open outlet valve to achieve an N_2-enriched atmosphere.

8. Clamp the inlet and outlet valves and bring back the hypoxia chamber to the incubator at 37 °C for the remaining desired period of time (35–50 min).

9. After OGD exposure, remove cultures from the hypoxia chamber, briefly rinse twice with pre-warmed HBSS, return to culture medium and incubated for another 24 h.

10. Treat cultures in the control groups in parallel to those in the OGD group: incubate for 10 + 35–50 min, but wash and maintain in culture medium and expose to an atmosphere of air (95 %)/CO_2 (5 %) at 37 °C.

11. Where appropriate, incorporate pharmacological compounds in the culture medium and in the OGD medium during the periods indicated.

12. Evaluate cellular damage 24 h posterior to OGD exposure.

3.2   OGD in Acute Hippocampal Slices

The Krebs-bicarbonate (KRB) buffer used for dissection and slice preparation and the OGD buffer are both prepared on the bench, since they do not require sterile conditions. These buffers should be prepared freshly at the day of experimentation. All procedures are, unless otherwise specified, carried out at room temperature.

1. Prepare the slice chopper by placing a filter paper disc moisturized with KRB buffer on top of the polyethylene disc that covers the cutting table. Secure the razor blade and adjust settings in order to obtain slices of 0.4 mm thickness.

2. Deeply anesthetize the animal by intra-peritoneal injection of sodium pentobarbital (60 mg/kg) or isoflurane inhalation (0.96–0.75 MAC) before decapitation with a guillotine.

3. Rapidly remove brain from the skull with the help of a pair of large and a pair of small scissors and straight and curved tweezers. Fix the skin in position by the use of the straight tweezers and made an incision using the small scissors.

4. Cut the skull bone between the eyes using the large scissors, while with the small scissors cut along the sagittal middle line from back to front.

5. Open the skull bone with the help of the curved tweezers and place the brain on top of a Petri dish lid covered with a filter paper disc moisturized with ice-cold KRB buffer. The Petri dish lid should be kept on ice.

6. Remove the cerebellum with a scalpel or a spatula and separate the two hemispheres. Carefully dissect both hippocampi by inserting two brushes between the cortex and corpus callosum, and unfolding the hippocampus. Removed meninges and adhered tissue are eliminated with the help of spatulas and brushes.

7. Arrange both hippocampi side by side on the polyethylene disc, which is placed on top of the cutting table to cut the whole length of both hippocampi. Remove the polyethylene disc, turn it over and dip gently into a Petri dish containing ice-cold KRB buffer, releasing the slices into the dish.

8. Gently separate the slices using a pair of thin brushes. Select the best slices (*see* **Note 12**) and transfer them with a plastic Pasteur pipette into Eppendorf tubes (one slice/tube) containing KRB buffer (*see* **Note 13**). Allow slices to recover from slicing trauma before starting the OGD by bubbling O_2 (95 %)/CO_2 (5 %) through the medium for 30 min at 35 °C (equilibration period). Use a semi-permeable nylon mesh to completely submerge and protect the slices from the bubbles.

9. Aspirate the buffer with a Pasteur pipette and induce OGD by incubating the slices in pre-warmed OGD buffer, which has been previously bubbled with N_2 (95 %)/CO_2 (5 %) for 15 min at 37 °C. After OGD exposure, return the slices to an oxygenated pre-warmed KRB buffer and incubate under an atmosphere of air (95 %)/CO_2 (5 %) for 2 h at 37 °C (re-oxygenation period).

10. Treat slices in the control group in parallel to the slices in the OGD group but incubate in glucose-containing KRB buffer and expose to an atmosphere of air (95 %)/CO_2 (5 %) at 37 °C for the duration of the experiment.

11. Perform experiments using test drugs by adding the drugs either during the OGD or the re-oxygenation periods.

12. Evaluate cellular damage 2 h posterior to OGD exposure.

3.3 Assessment of Cellular Damage

Cellular death is preferably measured by PI uptake in primary neurons [2] and organotypic hippocampal slice cultures [3], as assessed by PI/Hoechst cell counts and levels of PI fluorescence in the CA1 region, respectively (*see* **Notes 14–16**). PI can also be used in green fluorescent protein (GFP) expressing cells. Depending on the extension of the damage caused by the slicing process in acute brain slices, PI uptake may also be a suitable method [21] (*see* **Note 17**) (Fig. 2). Alternatively, the LDH release assay (*see* **Note 18**) and the MTT reduction assay (*see* **Note 19**) can be used for acute brain slices, primary neurons and organotypic slice cultures [4, 22].

3.3.1 PI Uptake in Primary Neurons

1. Add 7.5 µL of PI stock solution into the wells containing the primary neurons or inserts with the organotypic slices in 1 mL of culture medium (final concentration 7.5 µM). Gently mix the plate contents by applying circular movements.

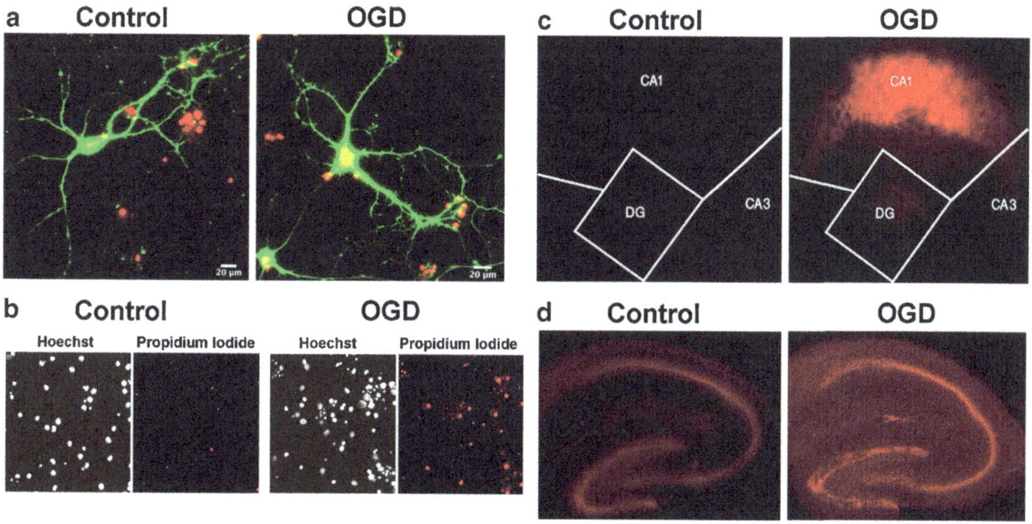

Fig. 2 Neuronal death induced by OGD in rat hippocampal primary neuronal cell cultures, organotypic, and acute slices. Cultured cells and slices were exposed to OGD for 45–60 min prior to incubation with propidium iodide (PI, 7.5 µM) 24 h later. Acute slices were exposed to OGD for 15 min prior to incubation with PI (10 µM) 2 h later. (**a**) Control and OGD neurons transfected with green-fluorescent protein (GFP) and stained with PI. The damaged OGD neuron shows a *yellow* nucleus due to the incorporation of PI (*overlay of green* from GFP with *red* from PI). (**b**) Control and OGD neurons stained with Hoechst 33342 (2.0 µg/mL) and PI. Manual cell counts provided both the number of Hoechst-positive (total neurons) and PI-positive (dead neurons) nuclei per field, which is indicative of the extent of OGD injury. (**c**) In organotypic hippocampal slice cultures, OGD causes a significantly increased PI uptake in the CA1 sub-region, relative to the CA3 and dentate gyrus (DG) sub-regions. Control organotypic hippocampal slices, kept under normoxic conditions, display only background PI fluorescence. (**d**) Due to the recent slicing trauma, control and OGD acute hippocampal slices display intense PI labeling not only in the CA1 sub-region, but also in CA2, CA3 and DG sub-region. The representative images shown in this figure clearly demonstrate the detectable difference in PI intensity between control and OGD slices. However, depending on the extension of the damage caused by the slicing process, PI uptake may however not be the most suitable method for the measurement of cellular death in acute slices

In case of primary neuronal cell cultures, Hoechst (2.0 µg/mL) can be added in combination with PI. Incubate the cultures under an atmosphere of air (95 %)/CO_2 (5 %) at 37 °C for 1 h.

2. Observe under the fluorescence microscope and capture digitalized images using a CCD camera.

3. Count the proportion of Hoechst-positive nuclei that are PI-positive across three fields of view. For each experiment, the mean of at least 20 images should be calculated per condition.

3.3.2 PI Uptake in Slices

1. Place slices are in a 96-well plate (1 slice/well) and incubate with 200 µL of a 10 µM PI solution in KRB buffer (*see* **Note 20**) at 37 °C for 30 min. Subsequently wash the slices twice with KRB buffer before microscopic analysis [23].

2. Observe under the fluorescence microscope and capture digitalized images using a CCD camera.

3. Analyze images using the Scion Image software (NIH). Use the intensity of PI fluorescence in the selected region of interest (CA1, CA2, CA3, or DG) as an index of cell death. Determine the pixel intensity and area in which PI fluorescence is detectable above background level using the "density slice" function of the software. Calculate the mean percentage values of fluorescence in the slices treated with test compounds (potential neuroprotective agents) and compare to standard damage, which is obtained as the mean of the intensity of PI fluorescence in the organotypic slices subjected to OGD with no added drug [13].

3.3.3 LDH Assay

1. Prepare the required solutions following all the instructions supplied in the TOX7 kit.

2. Transfer duplicate aliquots of 40 µL of media from each well with primary neurons, organotypic or acute slices (control and OGD) to a well of a 96-well plate. Keep the plates containing the samples on ice.

3. Optional: To obtain the total LDH activity add Triton X-100 to the medium to a final concentration of 10 % and scrape the cells or disrupt the slices by homogenization using a tissue-tearor.

4. Add 20 µL of LDH mixture per well. Protect the plate from light and incubate at room temperature for 20–30 min.

5. Measure absorbances spectrophotometrically at 490 nm and 690 nm (background) on the microplate reader.

6. Express results as a percentage relative to the total LDH activity. Data from different experiments may be pooled by normalizing relative to the amount of LDH released from disrupted slices.

3.3.4 MTT Assay

1. Prepare 5 mL of a 0.5 mg/mL solution of MTT by adding 500 µL of the 5 mg/mL stock solution of MTT in 4.5 mL of KRB buffer.

2. Place slices are in a 96-well plate (1 slice/well) and incubate with 200 µL of a 0.5 mg/mL MTT solution for 20 min in an incubator at 37 °C [24].

3. Replace the MTT solution with 200 µL of dimethyl sulfoxide (DMSO) and incubate for 30 min in an orbital shaker at room temperature or until complete solubilization of the formazan precipitate.

4. Measure absorbance spectrophotometrically at 540 nm on the microplate reader.

5. Express results usually as a percentage of the control group.

4   Notes

1. The following stock solutions can be prepared and stored at 4 °C for up to 2 months:

   12 mM $MgSO_4$: 0.7 g in 250 mL of H_2O

   1.2 M NaCl: 18 g in 250 mL of H_2O

   4 mM KH_2PO_4: 0.1 g in 250 mL of H_2O

   130 mM $CaCl_2$: 0.6 g in 30 mL of H_2O

   250 mM $NaHCO_3$: 5 g in 250 mL of H_2O

   30 mM KCl: 0.6 g in 250 mL of H_2O

   770 mM HEPES (free acid): 2 g in 10 mL of H_2O—pH is adjusted to 7.4 with 1 M NaOH

   1 M Glucose: 5 g in 30 mL of H_2O (store at –20 °C)

   1 M 2-Deoxy-glucose: 0.2 g in 1 mL of H_2O (store at –20 °C)

2. Only add the $CaCl_2$ solution to the buffer after pH 7.4 is reached.

3. The double-edged blade can be washed and reused for up to six different slice preparations (three on each side of the blade).

4. PI is a known mutagen and therefore gloves and a mask should be worn at any time when handling powders and solutions containing PI.

5. The photo-sensitivity of PI requires the use of foil-wrapped Eppendorf tubes when preparing the solutions and the storage of aliquots in the absence of light. The maximum absorption of PI is at 535 nm and the fluorescence emission maximum is at 617 nm.

6. Hoechst 33342 is also photo-sensitive, requiring the use of foil-wrapped Eppendorf tubes when preparing the solutions and storage of the aliquots in the absence of light. The maximum absorption of Hoechst 33342 is at 350 nm and the fluorescence emission maximum is at 461 nm.

7. In the TOX7 LDH assay kit an aliquot of the supernatant (cells or slices medium/buffer) is combined with a mixture of cofactor, substrate and dye solutions. The tetrazolium dye is converted to a soluble, colored compound by nicotinamide adenine dinucleotide (NADH), which is measured spectrophotometrically.

8. MTT is also photo-sensitive, requiring the use of foil-wrapped Eppendorf tubes when preparing the solutions and storage of the aliquots in the absence of light.

9. In order to achieve CA1-selective damage in the organotypic hippocampal slices and a significant amount of cellular death in hippocampal neurons, which can be attenuated by neuroprotective agents, our exposure periods of OGD are 45 and 60 min for organotypic and primary cultures, respectively.

10. Neurons and slice cultures can also be checked under phase-contrast microscope. Unhealthy neurons can be identified by vacuoles in their cell bodies or if the neurites have a beaded appearance, whereas damaged slices will appear brownish.

11. The temperature of 37 °C is a critical factor for OGD experiments to work, since hypothermia, as mild as 35 °C, is known to be neuroprotective [25].

12. Slices of same size should be chosen in order to maintain a high standard of reproducibility and minimize variability.

13. The tip of the plastic Pasteur should be cut large enough to allow the hippocampal slice to be sucked inside it without any damage being caused.

14. PI is a highly polar fluorescent dye, which can only enter cells after the loss of the cell membrane integrity. It then stains cell nuclei by binding to DNA and becoming highly fluorescent. In general, the PI uptake is a faster and more sensitive marker than the release of LDH into the medium.

15. In primary neurons, PI can be combined with Hoechst staining in order to provide information on the total number of (Hoechst positive) and the proportion of injured (PI positive) cells.

16. PI uptake is used as a marker for dead or dying cells, allowing direct identification of the main cell types affected and the most susceptible subfields in organotypic slice cultures.

17. Since acute slices cannot be allowed long periods of time to recover from the trauma caused by the slicing process, they can be covered by cells with damaged membranes, which stain positively with PI, making it an unsuitable method in order to analyze cell death caused by OGD.

18. The quantification of LDH released into the medium [26] provides a good indication of compromised cellular membrane and, since it does not interfere with the cells/tissue, it offers the advantage of allowing other cell death/viability assays, e.g. PI or MTT, to be carried out in combination. Moreover, it can be useful for measuring extensive cell death, since its maximum level is significantly beyond the saturation of PI uptake.

19. The MTT reduction assay [27], which measures mitochondrial function, can be used to quantify cell viability in primary neurons, organotypic and acute slices, but is typically used for acute slices. The tetrazolium ring of MTT can be cleaved by

active dehydrogenases in order to produce a precipitated formazan. The MTT reduction (formazan production) is proportional to cellular viability.

20. Approximately 5 mL of a 10 μM PI solution in KRB are prepared by adding 50 μL of the 1 mM PI stock solution in 4.95 mL of KRB buffer.

Acknowledgements

The authors would like to thank Dr. U.F.J. Mayer for his invaluable assistance with the manuscript.

References

1. Lee YJ, Castri P, Bembry J et al (2009) SUMOylation participates in induction of ischemic tolerance. J Neurochem 109: 257–267

2. Cimarosti H, Ashikaga E, Jaafari N et al (2012) Enhanced SUMOylation and SENP-1 protein levels following oxygen and glucose deprivation in neurones. J Cereb Blood Flow Metab 32:17–22

3. Cimarosti H, Kantamneni S, Henley JM (2009) Ischaemia differentially regulates GABA(B) receptor subunits in organotypic hippocampal slice cultures. Neuropharmacology 56:1088–1096

4. Dal-Cim T, Martins WC, Santos AR et al (2011) Guanosine is neuroprotective against oxygen/glucose deprivation in hippocampal slices via large conductance Ca^{2+}-activated K^+ channels, phosphatidilinositol-3 kinase/protein kinase B pathway activation and glutamate uptake. Neuroscience 183:212–220

5. Oleskovicz SP, Martins WC, Leal RB et al (2008) Mechanism of guanosine-induced neuroprotection in rat hippocampal slices submitted to oxygen-glucose deprivation. Neurochem Int 52:411–418

6. Colak G, Keillor JW, Johnson GV (2011) Cytosolic guanine nucleotide binding deficient form of transglutaminase 2 (R580a) potentiates cell death in oxygen glucose deprivation. PLoS One 6:e16665

7. Wang R, Zhang X, Zhang J et al (2012) Oxygen-glucose deprivation induced glial scar-like change in astrocytes. PLoS One 7:e37574

8. Gahwiler BH, Capogna M, Debanne D et al (1997) Organotypic slice cultures: a technique has come of age. Trends Neurosci 20:471–477

9. Newell DW, Malouf AT, Franck JE (1990) Glutamate-mediated selective vulnerability to ischemia is present in organotypic cultures of hippocampus. Neurosci Lett 116:325–330

10. Cho S, Liu D, Fairman D et al (2004) Spatiotemporal evidence of apoptosis-mediated ischemic injury in organotypic hippocampal slice cultures. Neurochem Int 45: 117–127

11. Dennis SH, Jaafari N, Cimarosti H et al (2011) Oxygen/glucose deprivation induces a reduction in synaptic AMPA receptors on hippocampal CA3 neurons mediated by mGluR1 and adenosine A3 receptors. J Neurosci 31: 11941–11952

12. Oliveira IJ, Molz S, Souza DO et al (2002) Neuroprotective effect of GMP in hippocampal slices submitted to an *in vitro* model of ischemia. Cell Mol Neurobiol 22:335–344

13. Cimarosti H, Rodnight R, Tavares A et al (2001) An investigation of the neuroprotective effect of lithium in organotypic slice cultures of rat hippocampus exposed to oxygen and glucose deprivation. Neurosci Lett 315: 33–36

14. Cimarosti H, Zamin LL, Frozza R et al (2005) Estradiol protects against oxygen and glucose deprivation in rat hippocampal organotypic cultures and activates Akt and inactivates GSK-3beta. Neurochem Res 30:191–199

15. Fontella FU, Cimarosti H, Crema LM et al (2005) Acute and repeated restraint stress influences cellular damage in rat hippocampal slices exposed to oxygen and glucose deprivation. Brain Res Bull 65:443–450

16. Scopel D, Fochesatto C, Cimarosti H et al (2006) Exercise intensity influences cell injury

in rat hippocampal slices exposed to oxygen and glucose deprivation. Brain Res Bull 71: 155–159

17. Gerace E, Landucci E, Scartabelli T et al (2012) Rat hippocampal slice culture models for the evaluation of neuroprotective agents. Methods Mol Biol 846:343–354

18. Giordano G, Costa LG (2011) Primary neurons in culture and neuronal cell lines for *in vitro* neurotoxicological studies. Methods Mol Biol 758:13–27

19. Pocock JM, Nicholls DG (1998) Exocytotic and nonexocytotic modes of glutamate release from cultured cerebellar granule cells during chemical ischaemia. J Neurochem 70:806–813

20. Strasser U, Fischer G (1995) Quantitative measurement of neuronal degeneration in organotypic hippocampal cultures after combined oxygen/glucose deprivation. J Neurosci Methods 57:177–186

21. Egea J, Martin-de-Saavedra MD, Parada E et al (2012) Galantamine elicits neuroprotection by inhibiting iNOS, NADPH oxidase and ROS in hippocampal slices stressed with anoxia/reoxygenation. Neuropharmacology 62:1082–1090

22. Parada E, Egea J, Buendia I et al (2013) The microglial alpha7-acetylcholine nicotinic receptor is a key element in promoting neuroprotection by inducing heme oxygenase-1 via nuclear factor erythroid-2-related factor 2. Antioxid Redox Signal 19:1135–1148. doi:10.1089/ars.2012.4671

23. Piermartiri TC, Vandresen-Filho S, de Araujo Herculano B et al (2009) Atorvastatin prevents hippocampal cell death due to quinolinic acid-induced seizures in mice by increasing Akt phosphorylation and glutamate uptake. Neurotox Res 16:106–115

24. Liu Y, Peterson DA, Kimura H et al (1997) Mechanism of cellular 3-(4,5-dimethylthiazol-2-yl)-2,5-diphenyltetrazolium bromide (MTT) reduction. J Neurochem 69:581–593

25. Feiner JR, Bickler PE, Estrada S et al (2005) Mild hypothermia, but not propofol, is neuroprotective in organotypic hippocampal cultures. Anesth Analg 100:215–225

26. Koh JY, Choi DW (1987) Quantitative determination of glutamate mediated cortical neuronal injury in cell culture by lactate dehydrogenase efflux assay. J Neurosci Methods 20:83–90

27. Mosmann T (1983) Rapid colorimetric assay for cellular growth and survival: application to proliferation and cytotoxicity assays. J Immunol Methods 65:55–63

Chapter 16

Laser Microbeam Targeting of Single Nerve Axons in Cell Culture

Nicholas Hyun, Linda Z. Shi, and Michael W. Berns

Abstract

By focusing a laser with short pulses to a diffraction-limited spot, single nerve axons can be precisely targeted and injured. Subsequent repair can be analyzed using various imaging and biochemical techniques to understand the repair process. In this chapter, we will describe a robotic laser microscope system used to injure nerve axons while simultaneously observing repair using phase and fluorescence microscopy. We provide procedures for controlled laser targeting and an experimental approach for studying axonal repair in embryonic rat hippocampus neurons.

Key words Ablation, Axotomy, Hippocampus, Laser, Neurons, Repair, Targeting

---

1 Introduction

Neuronal growth cones are specialized motile structures that react to environmental cues to guide nerve growth. The repair and reassembly of the growth cone after injury is crucial in the process of nerve regeneration and is an important step in driving axonal growth to reconnect with its target. Nerves with incomplete growth cone regeneration often exhibit dystrophic end bulbs which are markers for degenerating axons [1, 2]. The regeneration of a new growth cone involves many intracellular processes including proteolytic events, cytoskeletal rearrangement, and regulated transport of repair materials [3]. Although progress has been made in identifying causes for abnormal regeneration, additional novel ways to study nerve repair and regeneration following injury can have a significant impact on our understanding of the process.

To investigate the role of growth cones in nerve repair and regeneration, we have developed a system that can locally damage axons while simultaneously observing the repair process in real time using phase and fluorescence microscopy. By focusing a short-pulsed picosecond (ps) laser beam to a diffraction-limited spot, individual axons in cell cultures can be manipulated without dam-

Laura Lossi and Adalberto Merighi (eds.), *Neuronal Cell Death: Methods and Protocols*, Methods in Molecular Biology, vol. 1254, DOI 10.1007/978-1-4939-2152-2_16, © Springer Science+Business Media New York 2015

aging adjacent cells. The laser microscope system uses a 532 nm ps laser to partially damage (sub-axotomy) hippocampus neurons. We can then study the growth cone response to the damage. Previous preliminary studies on goldfish retinal ganglion cells in culture have shown this approach to be feasible [4, 5]. In the following sections, we will highlight key steps in using a laser ablation approach for neuron repair studies in embryonic rat hippocampus. This approach can be applied to nerve cells from other sources as well.

2   Materials

2.1   Cell Culture

1. 35 mm imaging dishes (MatTek, Ashland, MA).
2. 45 × 50 mm microscope cover glass (Fisherbrand, Waltham, MA).
3. 70× 50 microscope cover glass (Fisherbrand, Waltham, MA).
4. Poly-L-Lysine (Sigma-Aldrich, St. Louis, MO).
5. Culture grade water (Sigma-Aldrich).
6. Embryonic day (E) 17–18 rat embryos.
7. Fetal Bovine Serum (FBS, Life Technologies™).
8. Hanks' Balanced Salt Solution (HBSS, Life Technologies™).
9. HEPES (Life Technologies™).
10. Penicillin Streptomycin (Life Technologies).
11. 1× Neurobasal (Life Technologies™).
12. 50× B27 (Life Technologies™).
13. Glutamax (Life Technologies™).
14. OkoLab stage incubator (OkoLab, Napoli, Italy).

2.2   RoboLase: Optical Design and Hardware

Overview: Neuronal repair studies were performed using a robotic laser microscope system (RoboLase) consisting of an ablation laser, external optics to direct the laser path into the microscope, an inverted microscope with a motorized stage, CCD cameras, and Labview-based software to control the optical components as well as phase and fluorescence imaging [6]. Various optical designs can achieve similar results. We will describe our optical system and outline key components necessary for controlled laser ablation (sub-axotomy or axotomy) (Fig. 1).

2.2.1   Individual Components

1. 532 nm picosecond laser (VDGD2000-80-HM532 Spectra-Physics, Mountain View, CA).
2. Fast scanning mirror (FSM-200-01, Newport Corp, Irvine, CA).
3. Polarizers (CLPA-12.0-425-575, CVI laser, LLC, Albuquerque, NM).
4. Mechanical shutters (Vincent Associates, Rochester, NY).

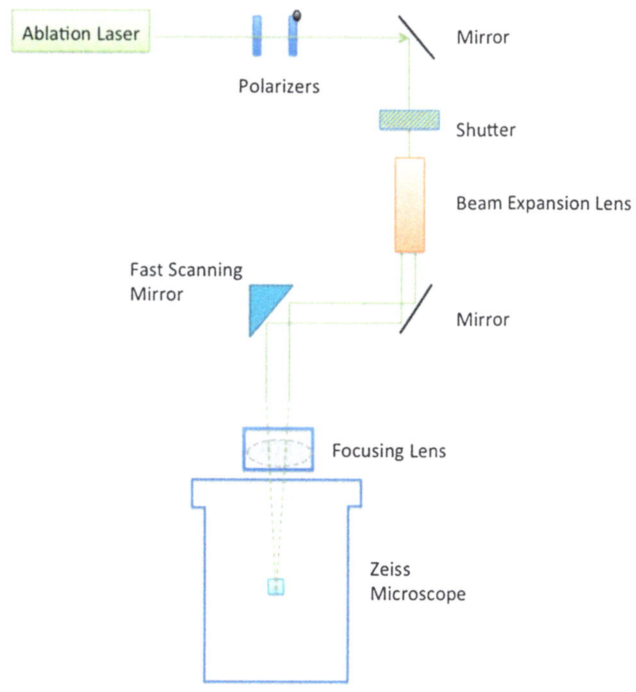

Fig. 1 Schematic of the optical hardware to direct the laser into the microscope system

5. 63×, phase III, Na1.4 oil emersion Plan-Apochromat objective (Carl Zeiss Microscopy GmbH, Jena, Germany) for neuronal ablation.

6. Beam expander lens (2–8×, 633/780/803 nm correction, Rodenstock, Germany).

7. Zeiss axiovert 200 M microscope (Carl Zeiss) with motorized objective turret, reflector turret, fluorescence filter cubes, condenser turret, halogen lamp shuttering with intensity control, mercury arc lamp shutter, camera port selection, objective focus with parfocal switching between objectives.

8. Hamamatsu camera (c4742-80-12AG, Hamamatsu Photonics, K.K., Hamamtsu, Japan).

9. Mirrors (Y2-1025-45-S, CVI Laser LLC).

10. Rotary mount (Newport Corp).

11. Photometer (Newport Corp).

12. X-Y stepper stage (Ludl Electronic Products Hawthorne, NY).

13. Stepper motion controller (National Instruments, Austin, TX).

14. MID-7604 power drive (National Instruments)

15. Computer (AlienWare Corp, Intel® Core™ 2CPU, 2.67 GHz, 2.75 GB of RAM).

16. Labview software (National Instruments).

<table>
<tr><td>

2.2.2 External Laser
Optics and Hardware

</td><td>

1. The ablation laser (**item 1** in Subheading 2.2.1) is a diode-pumped Vanguard Nd:YVO$_4$ second harmonic generator (SHG) 532 nm laser light linearly polarized with 100:1 purity, 76 MHz repetition rate, 12 ps pulse duration, and 2 W average power. *See* **Note 1**.

2. The glan linear polarizer (**item 3** in Subheading 2.2.1) is positioned after the ablation laser to increase the laser beam polarization purity.

3. The laser beam next passes through a second polarizer mounted in a rotary mount, which controls the amount of laser power and energy entering the microscope. The motorized mount (**item 10** in Subheading 2.2.1) can rotate the second polarizer to its vertical position for maximum transmission (95 %) or to its horizontal position for minimum transmission below the damage threshold of biological samples. The rotary mount is controlled through the motion controller (**item 13** in Subheading 2.2.1) in the PXI chassis [8].

4. A mechanical shutter (**item 4** in Subheading 2.2.1) with a 30 ms duty cycle gates the main laser beam resulting in a 30 ms burst of pulses that enters the microscope. The number of pulses is calculated based on the pulse rate of the laser [6, 7].

5. The laser beam passes through an adjustable-beam expander (2–8×, 633/780/803 nm correction) (**item 6** in Subheading 2.2.1) and is lowered to a height just above the optical table by using additional mirrors and mirror mounts.

6. A dual-axis fast scanning mirror (**item 2** in Subheading 2.2.1) is used to steer the laser beam at an imaged plane conjugate to the back focal plane of the microscope objective.

7. Neurons are mounted on a X-Y stepper stage (**item 12** in Subheading 2.2.1) controlled with a stepper motion controller (**item 13** in Subheading 2.2.1) and a power drive (**item 14** in Subheading 2.2.1).

8. The ORCA camera (**item 8** in Subheading 2.2.1) is linked to the system with a controller and a DCAMPI driver with Hammatsu's video Capture library for LabView.

</td></tr>
<tr><td>

2.3 RoboLase: Software

</td><td>

The software used to control the microscope, cameras, and external light paths is programmed in the LabVIEW language. The RoboLase software also manages image and measurement file storage. The "Front panel" software was designed to provide a user-friendly interface capable of meeting the demands of single cell manipulation (Fig. 2). A detailed description of the RoboLase Software has been published [8].

</td></tr>
</table>

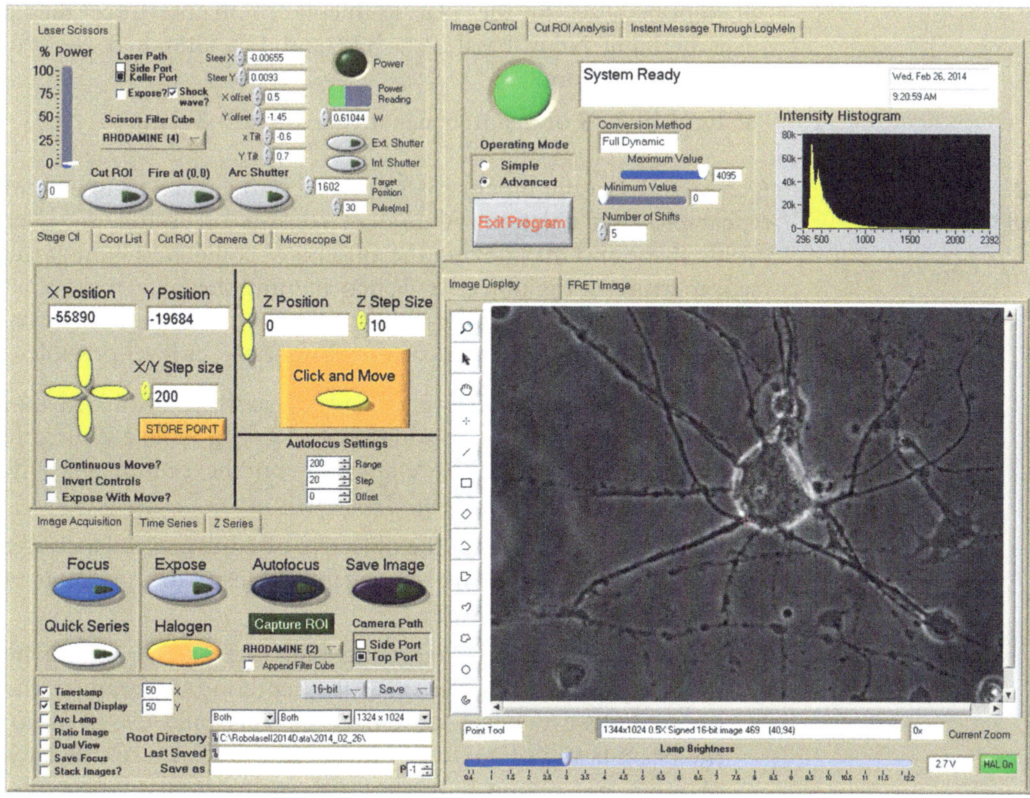

Fig. 2 Screen shot of the LabView RoboLase interface system to control laser ablation

3 Methods

3.1 Primary Nerve Cell Preparation

3.1.1 Coating Coverslips

1. Coat 35 mm imaging cell culture dishes with 250 µL of 0.2 mg/mL Poly-L-Lysine. Set dishes in incubator for 3 h or let dishes sit overnight at room temperature.

2. Aspirate Poly-L-Lysine solution and washout coating solution using culture grade water. Allow water to sit for 10 min before aspirating. Repeat 3 times. Let coated dishes dry in sterile hood before plating cells.

3.1.2 Dissection

1. Dissect hippocampal neurons from E17 to 18 rats in HBSS containing 10 mM HEPES and 1 % penicillin/streptomycin.

2. Plate them onto coated Poly-L-Lysine dishes into plating media containing Neurobasal, 2 % B27, 1 % Glutamax and 5 % FBS. *See* **Note 2**.

3. Replace 2/3 of the plating media to maintenance media containing Neurobasal, 2 % B27 and 1 % Glutamax the day after dissection.

4. Replace 2/3 of maintenance media every 2–3 days.

3.2 Laser Alignment and Targeting

3.2.1 Laser Alignment

Overview: Laser alignment is a critical step to ensure accurate and repeatable laser ablation. Alignment consists of optimizing laser profile shape and maximizing the amount of collimated light at the back aperture of the objective. When aligning the laser, remove all unnecessary personnel from the room and wear appropriate eyewear. Perform alignment tasks at the lowest power level possible.

1. Place a targeting bulls-eye at the back aperture of objective in an open turret. Place an additional targeting bulls-eye directly perpendicular to the back aperture on the ceiling 5–10 ft. above the microscope (Fig. 3).

2. Fill the back aperture of the objective by adjusting the focusing lenses and mirrors to maximize laser intensity to the center of the bulls-eye. *See* **Note 3**.

3. Align the laser to the targeting bulls-eye on the ceiling by adjusting the steering mirrors. After centering the beam on the ceiling bulls-eye, recheck the back aperture bulls-eye to insure the beam is still centered. If not, realign the beam on the back aperture and ceiling bulls-eyes with incremental adjustments. Repeat the process until both are centered. The back aperture alignment is more critical than the ceiling targeting. This alignment procedure insures that the laser beam is coming through the microscope at a perpendicular angle rather than at an angle that will result in uneven energy distribution in the focal spot and an irregular shaped focal spot in the optical field.

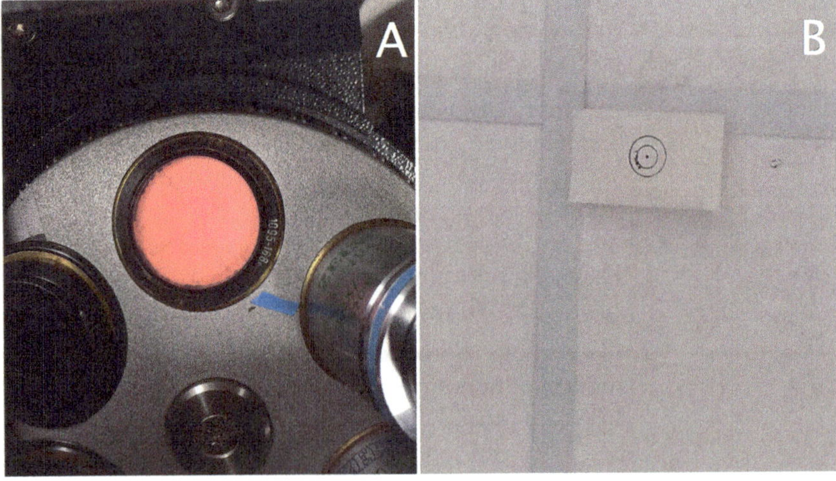

Fig. 3 Bulls-eye targets placed at the (**a**) back aperture of the objective and the (**b**) ceiling for laser alignment

3.2.2 Laser Power Measurement

Overview: The amount of laser power entering the microscope is controlled using a polarizer in a rotatory mount. By rotating the angle of the polarizer, the transmission of the laser passing through the polarizer and into the microscope can be precisely controlled. Calibrating the laser power versus polarizer position is critical for controlled and repeatable laser ablation experiments.

Laser power measurements are also important for calculating the energy and power in the focal spot used to damage axons. As the laser passes through the optical hardware and objective lens, some of the beam power is attenuated. To determine the laser power at the focus spot of the objective, the transmittance of the objective is multiplied by the laser power measured at the back aperture. A modified dual objective method is used to calculate the transmission of the objective (*see below and* 9).

1. Manually open mechanical shutter controller switch in laser path and turn on laser. Make sure the microscope port is changed to the base port such that the laser can enter the back aperture of the objective.

2. Measure the laser power at the back aperture of the objective with a photometer. This value is the power entering the objective (P_{in}). Make sure the photometer is set to the correct wavelength and is aligned to provide maximum intensity.

3. Place two objectives (A and B) coaxially such that the lenses are facing each other with emersion oil and a glass coverslip in-between the objectives (Fig. 4).

4. Align the combined objectives in the X, Y, Z planes to maximize the amount of laser light exiting the back aperture of objective B (P_{out}) and measure the power using a photometer.

5. The transmission of the combined objectives is calculated using the following equations:

$$T_1 = \frac{P_{out}}{P_{in}} \tag{1}$$

$$T_1 = T_A \times T_B \tag{2}$$

where the T_A and T_B equal the transmission of objective A and B respectively.

6. To solve for T_A and T_B a third objective is needed. Using Eqs. (1) and (2), the following equations can be used to solve the transmittance for three objectives (A, B, and C)

$$T_A^2 \times T_B^2 \times T_C^2 = T_1 \times T_2 \times T_3 \tag{3}$$

$$T_A \times T_B \times T_C = \sqrt{T_1 \times T_2 \times T_3} \tag{4}$$

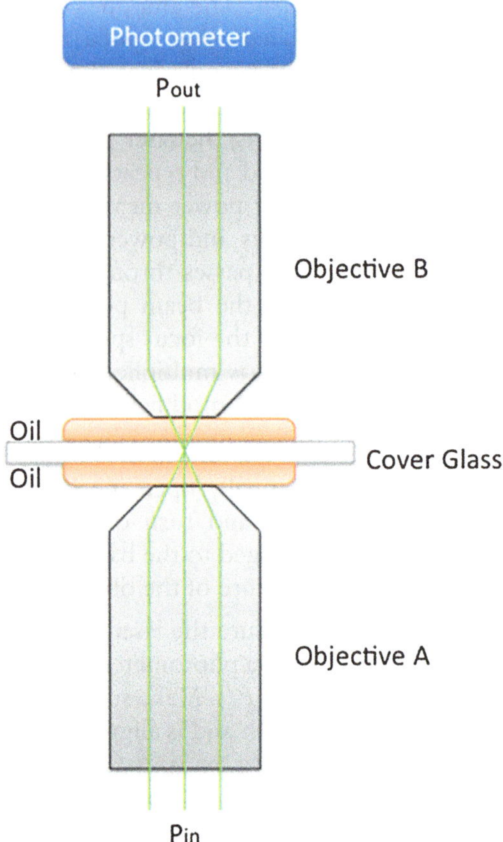

Fig. 4 Schematic of the dual-objective method

$$T_A = \frac{\sqrt{T_1 \times T_2 \times T_3}}{T_2} \tag{5}$$

$$T_B = \frac{\sqrt{T_1 \times T_2 \times T_3}}{T_3} \tag{6}$$

$$T_C = \frac{\sqrt{T_1 \times T_2 \times T_3}}{T_1} \tag{7}$$

The focus spot power equals (P_{in}) times the transmittance of the objective. *See* **Note 4** [10].

7. Repeat **step 2** at different polarizer positions to calibrate the laser power vs. polarizer position. *See* **Note 5**.

3.2.3 Laser Dosage Calculation

1. Assuming X is the power reading that creates the expected laser damage, the energy per pulse is equal to

$$\frac{X \times \text{Objective transmission}}{\text{Laser repetition rate}}.$$

2. The total laser dosage is energy per pulse $\times$ number of pulses per shot $\times$ number of spots in the damage.

3.3 Laser Targeting

Overview: After laser alignment and power measurements, the accuracy of the laser must be calibrated by firing at a uniform pigmented target. We use dried human red blood cells (RBC) that have been spread on a 45×50 mm cover glass. *See* **Note 6**.

A good RBC smear (*see* the cell images in Fig. 5) mounted and inverted on a 70×50 mm microscope slide can be used for over a year.

1. Open and run the RoboLase software. Load prepared RBC cells on microscope stage. *See* **Note 7**.

2. Set the power at a relatively high output by moving the slider for laser power in the upper left "Laser Scissors".

3. Click the "point" cursor on the side bar of the image and draw a point underneath the crosshair in the acquired image as shown in Fig. 2.

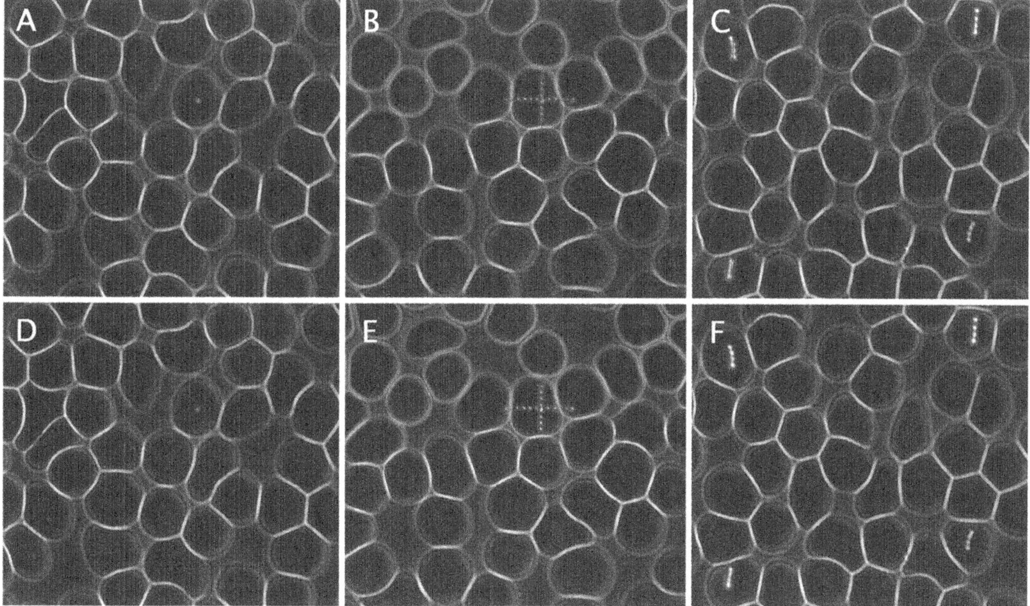

Fig. 5 (a) Point spot ablations on RBCs to calibrate targeting. **(b)** Intersected line cuts used to calibrate tilt. **(c)** Four line cuts at each corner to calibrate steering. Figures **(d)**–**(f)** show the ablated RBCs without line cut overlay

4. Fire the laser by clicking "CUT ROI" button under the "Laser Scissors" tab and acquire a new image by clicking "Expose" button under the "Image Acquisition" tab.

5. There will be a new bright dot displayed in the acquired image. It may not be underneath the crosshair.

6. Click "Advanced" under "Operation Mode" tab to access the laser offset parameters in the "Laser Scissors" tab.

7. Adjust the "x offset" and "y offset" parameters to redirect the laser to the crosshair position. Fire at the RBC cells with new offset parameters and acquire image. Repeat process until the bright dot is shown under the crosshair (Fig. 5a).

8. Draw one long horizontal and one vertical line passing through the crosshair, and acquire image to see if there is a tilt horizontally or vertically. If yes, change the values in "x Tilt" and "y Tilt" and repeat process until the two lines are straight (Fig. 5b).

9. Draw lines in the four corners and see if the bright lines overlie with the lines. If not, change the value in "Steer X" and "Steer Y" and repeat process until the targeting is accurate (Fig. 5c).

10. Reduce the laser power to a level that only a very small damage spot can be detected in the RBC. Move the z focus up and down using the yellow button under "Stage Ctl" tab. The difference between the firing z position to the position with the sharpest dot is the z offset. The value can be saved under "Cut ROI."

3.4 Nerve Regeneration: Live Cell Imaging

To perform live imaging of neuronal cells for extended periods of time, the temperature, humidity, and CO_2 concentration of the environment should be controlled. We use a microscope incubator that maintains the proper environment (37 °C, 5 % CO_2) on an inverted microscope.

1. Load the cells onto the microscope and determine the laser threshold for axotomy and sub-axotomy for neuron cells (*see* **Note 8**). The amount of power necessary to create sub-axotomy will vary depending on the laser used. Start by targeting a nerve axon at a low laser power and gradually increase the laser dose until visible thinning of the axon can be seen within a second after laser exposure (Fig. 6). Repeat on several axons to determine the laser dose thresholds for sub-axotomy lesions. Cells may be observed under phase contrast, differential interference contrast (DIC), or fluorescence microscopy.

2. Once the sub-axotomy threshold is determined, look for established growth cones that have active filopodia and lamellipodia. Typically, we use cell cultures 3–5 days post dissection. During this period nerve cultures are actively growing and growth cones are easily identifiable.

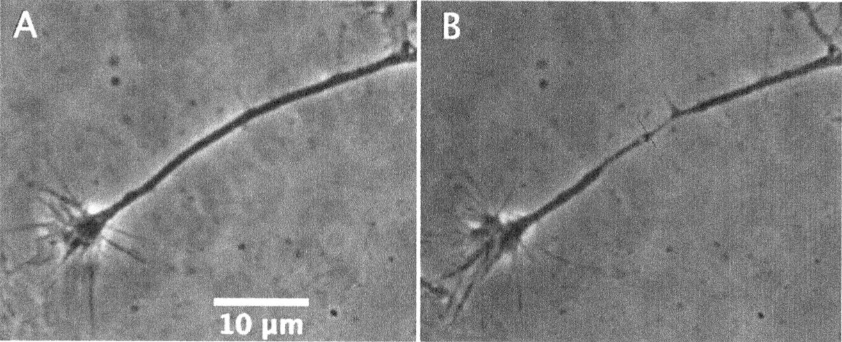

Fig. 6 Hippocampus thinning after laser damage. Growth cone can be found by looking for structures at the end of nerve axons: lamellipodia can be distinguished by the actin network formed at the leading edge. The slender protrusions that extend outward past the lamellipodia are the filopodia

3. Follow and record phase contrast images (or using other forms of microscopy) of the neuron prior to laser ablation to characterize pre-irradiation morphology and behavior. This will be used for the analysis of the growth cone dynamics before laser axotomy.

4. Measure the distance from the growth cone and cell body to the targeted damaged site. The relative position of the cut may affect the dynamics of the repair process. *See* **Note 9**.

5. Damage axon and follow cells by phase contrast images (or other microscopic imaging methods) immediately after damage for up to 60 min. *See* **Note 10**.

6. Compile and analyze phase contrast images from pre and post laser exposure using ImageJ. After laser ablation, phase contrast images showed a visible thinning of the nerve axon followed by a distinct repair process involving cytoskeletal remodeling. The growth cone retracted toward the damage site, possibly providing cytoskeletal material to repair the injured region of the axon. In severely damaged axons, the growth cone retracted past the damaged site and did not recover. The observed retraction may be initiated through the release of chemotrophic factors from the damage site, and which initiate axonal repair [4]. In some neurons, lamellipodia formed at the damage site, or near the cell body traveling toward the damage site probably to assist in the repair of the injured axon (Fig. 7).

   This repair process is consistent with results seen by Difata et al. using a UVA laser to damage hippocampus neurons [11]. After axonal recovery, the nerve cell proceeds to form an apparently normal new growth.

7. Qualitatively score nerve repair based on axonal recovery and growth cone reformation. A "++" score is given to nerve cells that exhibited both axonal repair and growth cone reformation,

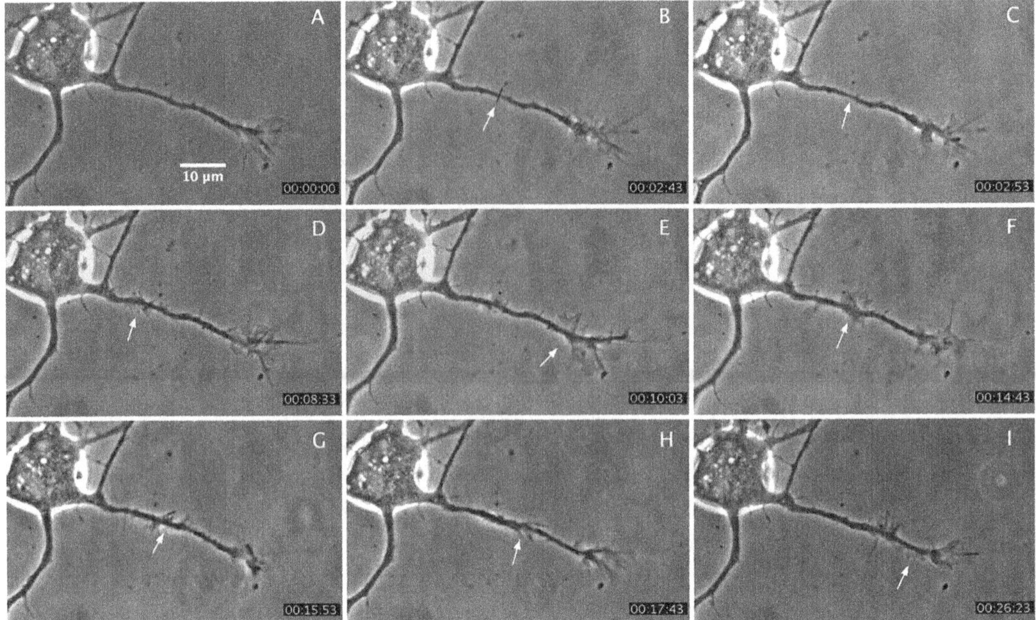

Fig. 7 (**a**) Growth cone before laser sub-axotomy. (**b**) The axon is damaged along the *red line* causing a thinning of the axon. (**c, d**) Axon is slightly thinner at ablation site and actin accumulates at the base of the cell body. (**e**) Growth cone retracts slightly and sends repair material toward damage site. (**f**) The growth cone is reduced in size and actin is accumulated at the damage site. (**g, h**) Nerve axon thickens and is repaired. (**i**) Lamellipodia progress forward and combines with the growth cone

"+" is given to nerve cells that exhibited axonal repair with partial or no reformation of growth cone, and "–" is given to nerve cells that did not recover from laser damage. In our experiments with hippocampus neurons 24 % of the cells exhibited a "++" response, 29 % showed a "+" response, and 47 % of the neurons did not recover from the laser damage (Table 1).

The laser damage mechanism has not been fully characterized, and may be a combination of single or two photon absorption, or possibly the creation of a plasma and shock wave, all of which likely affect the axon cytoskeletal complex [4]. The axonal thinning produced with the 532 nm ps green laser beam was similar to that observed using a 532 nm ns laser on goldfish retina neurons. Electron micrographs of the retina cells ablated with 532 ns laser showed that the damage did not rupture the cell membrane (Fig. 8).

It appears that the thinning is a result of loss of cytoskeletal structure in the axon even though there was evidence of microtubules in the damage area. However, it is not clear whether or not the microtubules were normal, as they appeared somewhat compressed as viewed in the electron micrographs.

8. To visualize microtubule damage after laser sub-axotomy, transduce nerve cells with red fluorescent protein (RFP) tubulin.

Table 1
Hippocampus neurons

Response	Number	Percent
+	5	29
++	4	24
–	8	47
Total	17	100

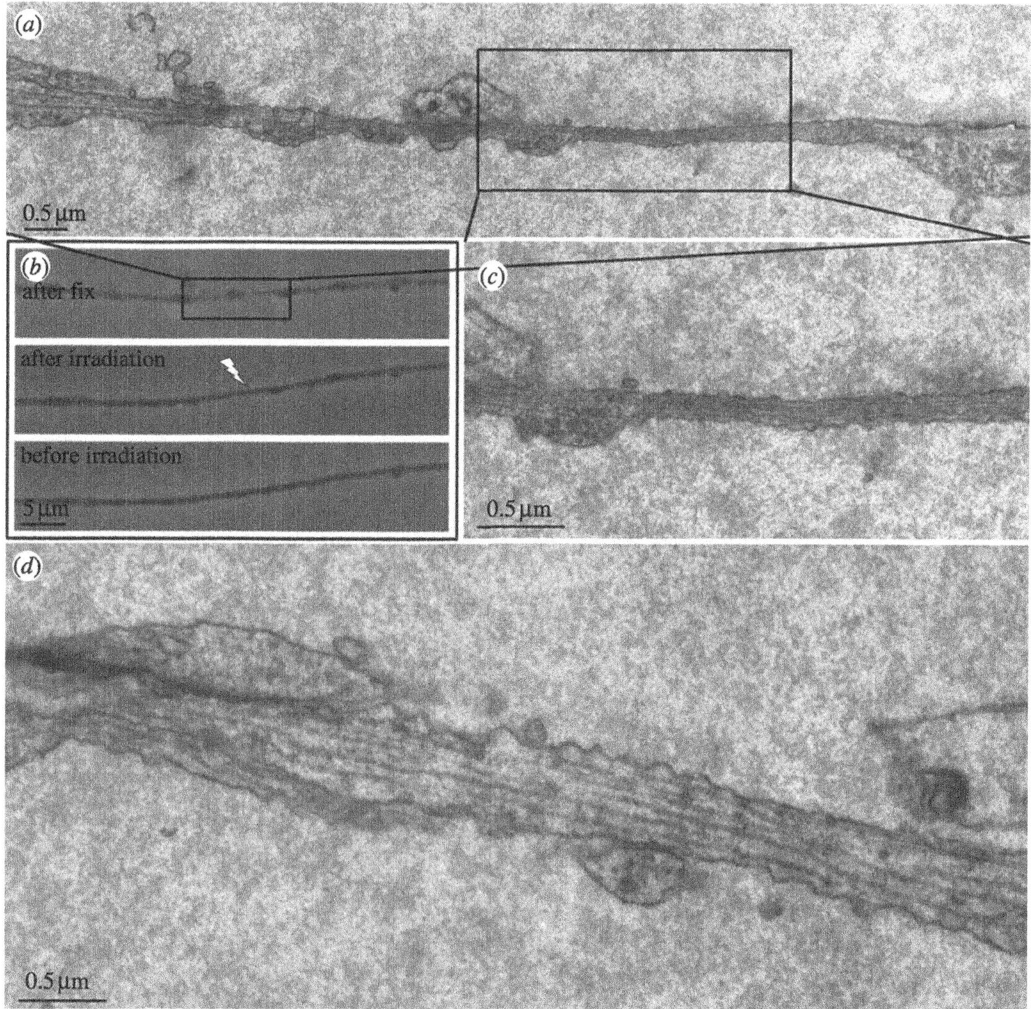

Fig. 8 TEM and phase contrast images of damaged axon: (**a**) reconstructed collage of multiple TEM images of an axon fixed 30 s after laser irradiation. (**b**) Live phase contrast images taken before (*bottom*) and after irradiation (*middle*) and after fixation (*top*). Images are matched with the electron microscope images in (**a**). (**c**) Electron micrograph of the region in the center of the "thinned" zone. Note the intact cell membrane and the presence of contiguous microtubules. (**d**) Non-irradiated region 36 mm away from the laser-irradiated region (Wu et al. [4]). Reprinted from [4] with permission from the Royal Society Interface

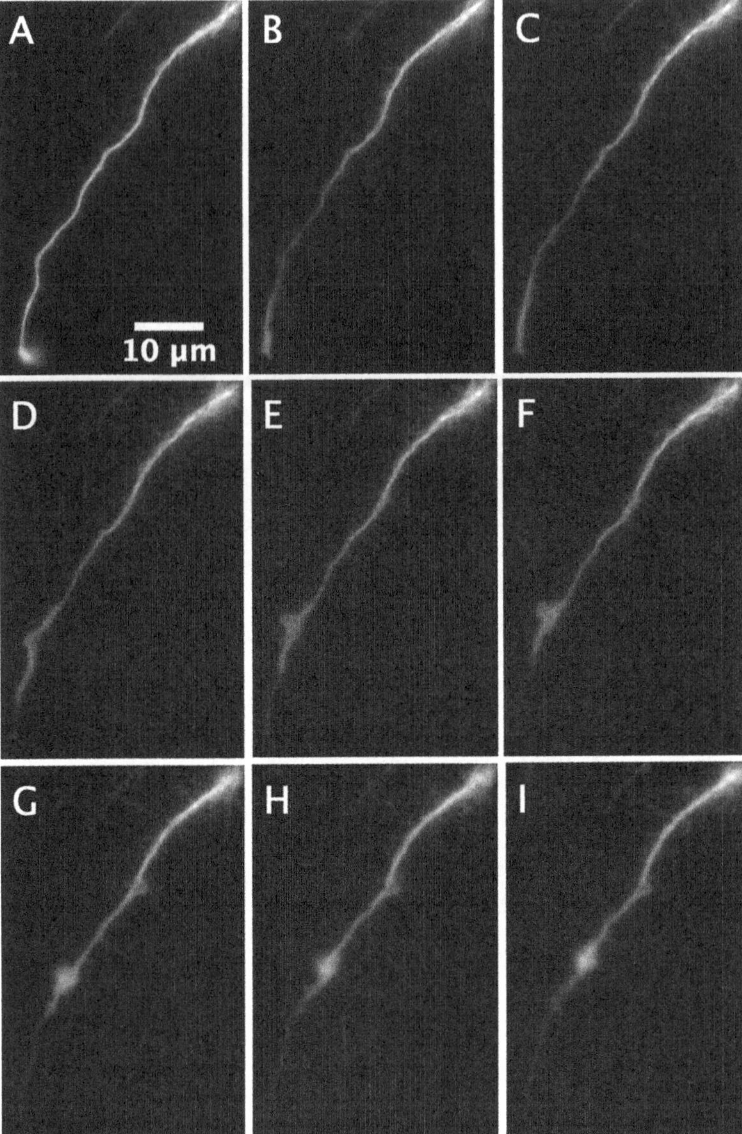

Fig. 9 Hippocampus neuron transduced with GFP tubulin before and after laser radiation

Fluorescence time series analysis showed a visible retraction and accumulation of tubulin at the damage site consistent with the response seen in phase contrast images (Fig. 9).

RFP tubulin signal decreased in intensity at the ablation site, but recovered in intensity as axonal repair proceeded. The fluorescent signal at the damage site indicates microtubule structure was partially preserved after laser sub-axotomy. The decrease in signal is likely due to the microtubule damage seen in electron micrograph.

In summary, we have described a robotic laser microscope that can be used to target axons and create sub-axotomy lesions. The repair process can be studied at the level of individual axons. Additional studies should be done to study the biochemical pathways involved in growth cone sensing and cytoskeletal remodeling. Lasers provide a novel way to study growth cones and their possible role in neuronal repair and regeneration.

4 Notes

1. Selection of the laser for the RoboLase, or any other system should be based on the desire to produce cellular damage using multiphoton process produced by short-pulsed lasers.

2. We have found that some nerve cultures do not grow well on 35 mm dishes. Nerve cell cultures can also be prepared on glass coverslips and placed in culture dishes for imaging.

3. Most laser-related eye injuries occur during laser alignment. When designing optical hardware, try to direct the laser path in such a way that it is not at eye level. Follow all safety procedures and always wear protective eyewear.

4. The transmission factor provided by the manufacturer may vary slightly between lenses and can degrade with usage and time. Thus it is important to measure transmission through the entire system periodically, preferably weekly, but certainly monthly. If multiple people use the system, transmission and alignment should be checked daily and even between different users of the system.

5. Measure power at incremental changes to the polarizer position. This power reading should be done periodically since laser power may fluctuate between experiments.

6. The blood smear is prepared by holding a standard microscope slide at a 45° angle to the cover glass, and while exerting a slight pressure down, sliding the edge of a standard microscope slide through a small drop of human blood. This is a standard "smear" method used in pathology/hematology labs and can easily be mastered.

7. The calibration method shown on the RoboLase system serves as an example. Target calibration can be done differently depending on the software used to control the laser ablation.

8. Start at the RBC threshold and gradually increase power. We have found line cuts work better than point cuts to create laser sub-axotomy.

9. In our experience, we make laser ablations 10–20 μm from the growth cone to get the most consistent response.

10. The repair process can range from 10 min to over 40 min. A frame rate of 10 s was used to capture the growth cone dynamics. Image processing was used to quantify the behavior of the growth cone.

References

1. Tom VJ, Steinmetz MP, Miller JH et al (2004) Studies on the development and behavior of the dystrophic growth cone, the hallmark of regeneration failure, in an in vitro model of the glial scar and after spinal cord injury. J Neurosci 24:6531–6539

2. Erturk A, Hellal F, Enes J et al (2007) Disorganized microtubules underlie the formation of retraction bulbs and the failure of axonal regeneration. J Neurosci 27:9169–9180

3. Bradke F, Fawcett JW, Spira ES (2012) Assembly of a new growth cone after axotomy: the precursor to axon regeneration. Nat Rev Neurosci 13:183–193

4. Wu T, Mohanty S, Gomez-Godinez V et al (2011) Neuronal growth cones respond to laser-induced axonal damage. J R Soc Interface. doi:10.1098/rsif.2011.0351

5. Wu T, Nieminen TA, Mohanty S et al (2012) A photon-driven micromotor can direct nerve fibre growth. Nat Photonics 6:62–67

6. Botvinick EL, Berns MW (2005) Internet-based robotic laser scissors and tweezers microscopy. Microsc Res Tech 68:65–74

7. Berns MW, Botvinick EL, Liaw L et al (2005) Micromanipulation of chromosomes and the mitotic spindle using laser microsurgery (laser scissors) and laser-induced optical forces (laser tweezers). In: Cell biology: a laboratory handbook. Elsevier Press, Burlington, MA

8. Shi L, Berns MW, Botvinick EL (2008) RoboLase: internet-accessible robotic laser scissor and laser tweezers microscope system. Medical Robotics, I-Tech Education and Publishing. ISBN 978-3-902613-18-9

9. Misawa H, Koshioka M, Sasaki K et al (1991) Three-dimensional optical trapping and laser ablation of a single polymer latex particle in water. J Appl Phys 70:3829–3836

10. Gomez-Godinez V, Wu T, Sherman AJ (2010) Analysis of DNA double-strand break response and chromatin structure in mitosis using laser microirradiation. Nucleic Acids Res 38:e202

11. Difato F, Tsushima H, Pesce M et al (2011) The formation of actin waves during regeneration after axonal lesion is enhanced by BDNF. Sci Rep 1:183

Real-Time Imaging of Retinal Cell Apoptosis by Confocal Scanning Laser Ophthalmoscopy

Eduardo M. Normando, Mohammad H. Dehabadi, Li Guo, Lisa A. Turner, Gaia Pollorsi, and M. Francesca Cordeiro

Abstract

Retinal cell apoptosis occurs in many eye conditions, including glaucoma, diabetic retinopathy and Alzheimer's disease. Real-time detection of retinal cell apoptosis has potential clinical value in early disease detection, as well as evaluating disease progression and treatment efficacy. Here, we describe our novel imaging technology DARC (Detection of Apoptosing Retinal Cells), which can be used to visualize single retinal neurons undergoing apoptosis in real time, by using fluorescently labeled Annexin A5 and confocal scanning laser ophthalmoscopy (cSLO). Clinical trials of DARC in glaucoma patients are due to start shortly, but in this chapter, we describe this technique in experimental animal models.

Key words Apoptosis, Retinal neurodegeneration, Glaucoma, Detection of apoptosing retinal cells, Scanning laser ophthalmoscope

1 Introduction

The process of apoptosis (which means "to fall off" in Greek) is a physiological phenomenon which occurs in all tissues resulting in a controlled program of cellular scavenging. Apoptosis is also implicated in many pathological conditions including AIDS [1], hematological diseases [2, 3], myocardial infarction [4], ischemic renal damage [5], cerebrovascular accident [4], neurodegenerative diseases [6–10], and retinal diseases [11]. In retinal neurodegenerative conditions, apoptosis accounts for the earliest cell loss and eventual onset of blindness [12–14]. Cells undergoing apoptosis translocate phosphatidylserine molecules from the inner to the outer membrane [15]. Annexin A5 has a strong affinity for this molecule [16] and this characteristic has been used in cell biology for many years to identify cells undergoing apoptosis [17]. Clinically, radiolabeled Annexin has been used frequently to detect apoptosing cells [18]. However, this technique had never previously been used in the eye until a novel imaging technology called

Laura Lossi and Adalberto Merighi (eds.), *Neuronal Cell Death: Methods and Protocols*, Methods in Molecular Biology, vol. 1254, DOI 10.1007/978-1-4939-2152-2_17, © Springer Science+Business Media New York 2015

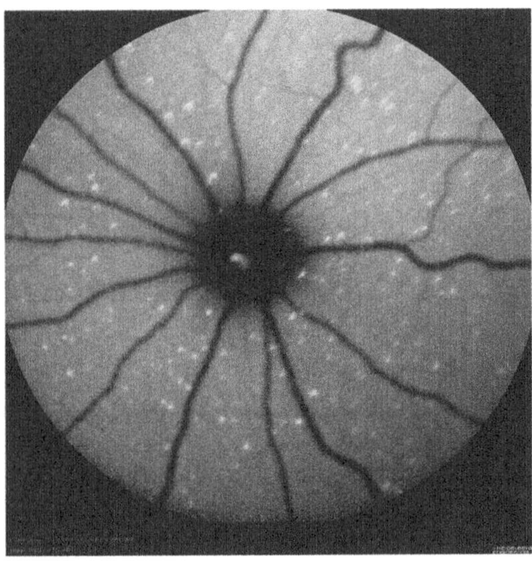

Fig. 1 Detection of apoptosing retinal cells (DARC). This DARC image shows apoptotic retinal ganglion cells (RGCs) labeled by fluorescent Annexin V (*white spots*) in a rat model of staurosporine-induced RGC apoptosis

DARC (Detection of Apoptosing Retinal Cells) was developed in 2004 [19]. Due to the unique transparent optical media of the eye, it is possible to directly visualize a single retinal cell undergoing apoptosis in real time using DARC [19]. This technique uses fluorescently labeled Annexin A5 to bind apoptosing retinal cells, and a fluorescence-enabled confocal scanning laser ophthalmoscope (cSLO) (Fig. 1) to detect the dying cells over a period of time. Recently, this technique has been used to assess different phases of cell death using different fluorescent probes [20].

As mentioned above, the instrument central to DARC is the cSLO. Invented in 1980 [21], the cSLO uses the confocal principle to enhance image resolution by utilizing a narrow beam of light from a point source laser at a defined wavelength. Light is rapidly swept over the retina, reducing both exposure and light scattering, while improving patient compliance and producing a better quality image (Fig. 2).

Since its invention, several versions of cSLO have been commercially available (Table 1), such as the Rodenstock SLO-101 [22], the Optos SLO [23], the Zeiss SLO [24], and the Heidelberg Engineering HRT and the HRA [25].

DARC has been used experimentally with a number of these devices (Table 2) [26].

Although clinical trials of DARC in glaucoma patients are due to start shortly, in this chapter, we describe this technique in experimental animal models.

Confocal mode

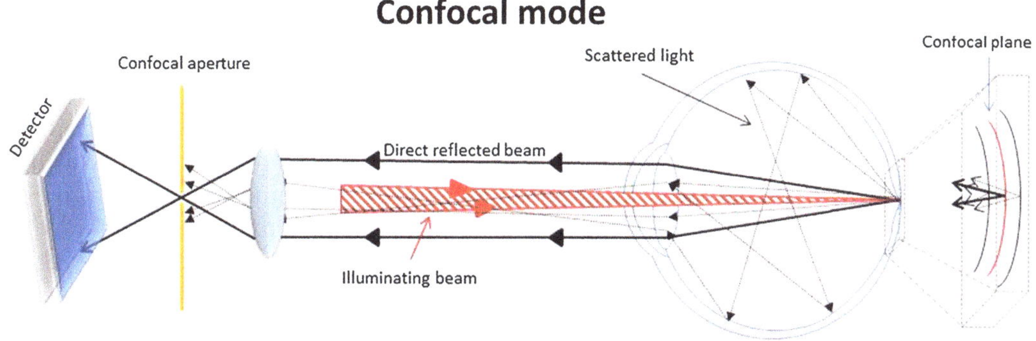

Fig. 2 Confocal mode: only the light from a specific layer of the retina is collected by the photodetector giving an image with a high signal-to-noise ratio

Table 1
Commercially available cSLO

Make/model	FAF	FA	ICGA	OCT	Tracking	Field (°)
NIDEK F10 (Gamagori, Japan)	X	X	X			40
OPTOS® P200 (Dunfermline, UK)	X	X	X			200
OPTOS® Daytona		X				200
I-Optics Easyscan ('S-Gravenhage, Netherlands)	X		X			30
OTI/OPKO OCT/SLO™ (OPKO Health, Inc., Miami, FL)	X	X	X	X	X	40
HE Spectralis® (Heidelberg, Germany)	X	X	X	X	X	30, 55, 190

FA fluorescein angiography, *FAF* Fundus AutoFluorescence, *ICGA* indocyanine green angiography, *OCT* optical coherence tomography

Table 2
Technical characteristics of cSLOs which have been used experimentally

	HRA Spectralis® (Heidelberg, Germany)	Rodenstock GMBH Model 101 SLO (Munich, Germany)	Zeiss SM 304024 (Oberkochen, Germany)
Laser type	486 nm Diode 786/815 nm Diode	488 nm Argon	488 nm Argon
Laser output	275 µW	250 µW	300 µW
Detector	Avalanche photodiode	Photodiode	Photomultiplier cathode tube
Field of view (°)	15/30/55/180	20/40	20/40
Barrier filter	500 nm/800 nm	520 nm	520 nm

2 Materials

2.1 cSLO Devices

Commercially available cSLO devices allow retinal imaging at different wavelengths according to the requirements of commonly used fluorescent dyes in clinical diagnostic techniques such as fluorescein angiography (FA) and indocyanine green angiography (ICGA). A cSLO can be specifically customized for small animal imaging, as well as high contrast and high resolution imaging of the retina. For example, a cSLO can be used in the following ways: infrared reflection (820 nm), infrared autofluorescence (ICGA, 790 nm), red free reflection (488 nm), and blue laser autofluorescence (FA, 488 nm). Images can be acquired as single images, time sequence images or depth sequence images (Z-stack) in any of the above modalities. It is also possible to simultaneously acquire images using two laser sources. For example, it is feasible to capture concurrent FA and ICGA images of the same retinal area [27]. This allows, when necessary, a direct comparison of results in two modes.

A recent, advantageous feature of imaging is the application of automatic, real-time averaging, which can detect and correct for eye movements using a secondary camera dedicated to motion tracking. When this system is initiated, the software detects any activity in real time and compensates for translational and rotational changes by comparing each image with a reference (real-time eye tracking). The resulting image is greatly enhanced, with reduced background noise and movement artifacts.

The depth resolution (the minimum distance where two objects can be seen as separate) is determined by the numerical aperture of the objective lens. The larger the pupil size the lower the depth resolution, secondary to optical aberrations. Previous studies have shown that in order to exploit the optical properties of the human eye, without taking into account higher order aberrations, a maximum pupil diameter of 3 mm is required [28], producing a true depth resolution of about 300 μm [29]. However, to allow the passage of a greater amount of light so that a greater number of photons can reach the cSLO detector, the optical aperture of the cSLO has a larger diameter [30].

When the pupil is dilated to a diameter greater than 3 mm, cSLO image quality is reduced secondary to greater aberrations (astigmatism, coma, and higher order aberrations) produced in the peripheral areas of the lens. This reduction in image quality is not offset by the increased amount of light entering the eye. For this reason, when the instrument is properly positioned, the diameter of the cSLO laser beam is limited to a circle of 3 mm at the level of the pupil [31].

The field of view of commercially available SLO is between 15° and 200° (Table 2). The standard optics of a SLO for small

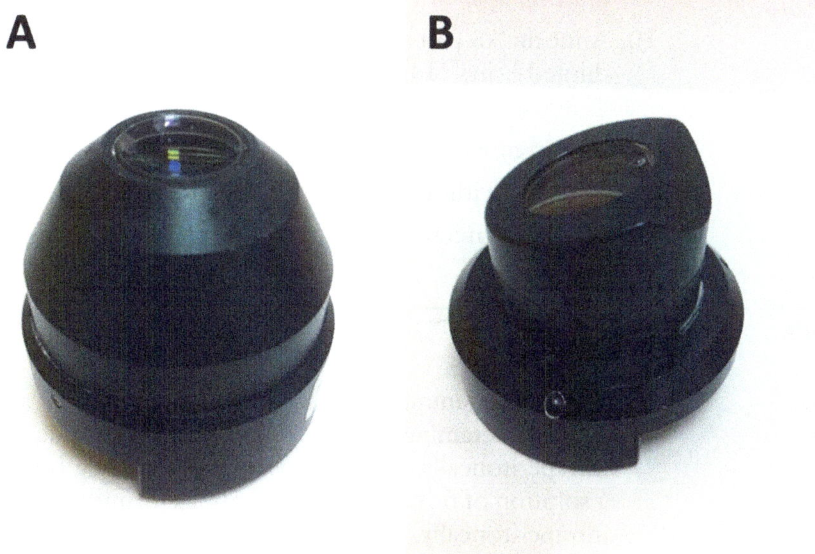

Fig. 3 Lenses used for DARC imaging. (**a**) A 55° wide field lens, (**b**) a 30° standard lens

laboratory rodents has a typical 30° field of view (Fig. 3). Using software magnification, a 15° field can be achieved; wide-angle objectives are also available for use with cSLOs, allowing fields of view up to 200° [32–35].

2.2 Drugs and Other Materials

For general anesthesia:

1. Ketamine (Ketaset; Fort Dodge Animal Health, Fort Dodge, IA).
2. Medetomidine (Dormitor; Pfizer, New York, NY).
3. Sterile water.

For local anesthesia:

4. Proxymetacaine Hydrochloride 0.5 % w/v: 1–2 eye drops.
5. Lidocaine.

For mydriasis:

6. 2.5 % Phenylephrine hydrochloride (Bausch&Lomb, Rochester, NY), and 1 % Cyclopentolate hydrochloride (Bausch&Lomb): 1–2 eye drops (rats).
7. 1.0 % Tropicamide (Alcon, Camberly, UK): 1–2 eye drops to avoid formation of cataract (mice).

For moistening the cornea:

8. Hypromellose 0.3 % eye drops.
9. Viscotears Liquid Gel (Carbomer, polyacrylic acid, Novartis, Basel, Switzerland) 2 mg/g.

For labeling apoptotic cells:

10. Annexin A5 (An-F). For intravitreal dosing, use 0.2–3 μg/mL diluted in 1× PBS. For intravenous dosing, use 4–300 μg/mL diluted in 1× PBS [36]. *See* **Note 1**.

11. Heating mat.

12. Platform with mouth/nose holder.

13. Operating microscope.

3   Methods

3.1 Animal Preparation

1. Anesthetize animal: For rats use 0.2 mL/100 g of a solution of 37.5 % Ketamine and 25 % Dormitor in sterile water [37], intraperitoneally. For mice use 0.5 mL/100 g body weight of a solution of 6 % Ketamine and 10 % Dormitor in sterile water, intraperitoneally. *See* **Note 2**.

2. Keep animals on a heating mat to maintain body temperature.

3. Apply the mydriatic eye drops to dilate the pupils. Complete pupil dilation is achieved within 5 min. *See* **Note 3**.

4. Wipe the eyelids carefully with outward movements to remove eyelashes from the corneal surface.

3.2 Administration of Fluorescent Annexin A5

3.2.1 Intravitreal Injection (See Note 4)

1. Place animal on a platform equipped with a mouth/nose holder (customized for rat and mouse).

2. Apply a drop of Proxymetacaine hydrochloride 0.5 % to the eye.

3. Apply Carbomer 2 mg/g to the eye to cover the cornea and negate the refractive power of the air corneal interface.

4. Place a cover slide on top of the Carbomer, allowing direct visualization of the retina under the operating microscope. Perform an intravitreal injection of fluorescently labeled Annexin A5 under an operating microscope.

5. Leave the eye covered with Carbomer for 2 h before imaging.

3.2.2 Intravenous Injection (See Note 4)

1. Place the anesthetized animal on a flat surface.

2. Apply 1 drop of lidocaine on animal's tail allowing 2 min to act.

3. Carry out intravenous injection into the tail vein. For intravenous injection, a total volume of 500 μL and 50 μL should be used for rat and mouse respectively.

3.3 Adaptation of the cSLO for Small Animal Imaging

1. Replace the chinrest with a metal platform to accommodate laboratory animals (Fig. 4).

Fig. 4 cSLO ready to be used for small animal imaging

3.4   Imaging Using a cSLO

1. Choose the appropriate lens for your SLO (Fig. 3).

2. Switch on the power supply, the computer and the laser box.

3. Start the cSLO Manufacturing Software.

4. Create or open a new database entry for the animal.

5. Place anesthetized animal in front of objective, and adjust the head to align the longitudinal axis of the eye with the device's optical path.

6. Apply Hypromellose eye drops every 5 min. *See* **Note 5**.

7. Adjust the camera head, first by directing it toward the center of the eye from a distance of about 5–10 cm, then by slowly moving the camera closer to the eye focusing on parts of the retina discerned on the screen. Continue to realign the camera head and readjust the focus until the distance from the eye is reduced to about 5 mm, and a clear retinal image can be seen.

8. Initially, use reflective modes to focus (IR or Red Free). A frame with a live image appears on the monitor.

9. Pan and tilt the camera head until the optic nerve disc is in the center of the frame.

10. Fine-tune the focus to obtain the best image.

3.5   Image Acquisition (Fig. 5)

1. Select IR or Red Free mode on the control panel to focus on the desired retinal layer.

2. Adjust the brightness manually using the manufacturer's instructions.

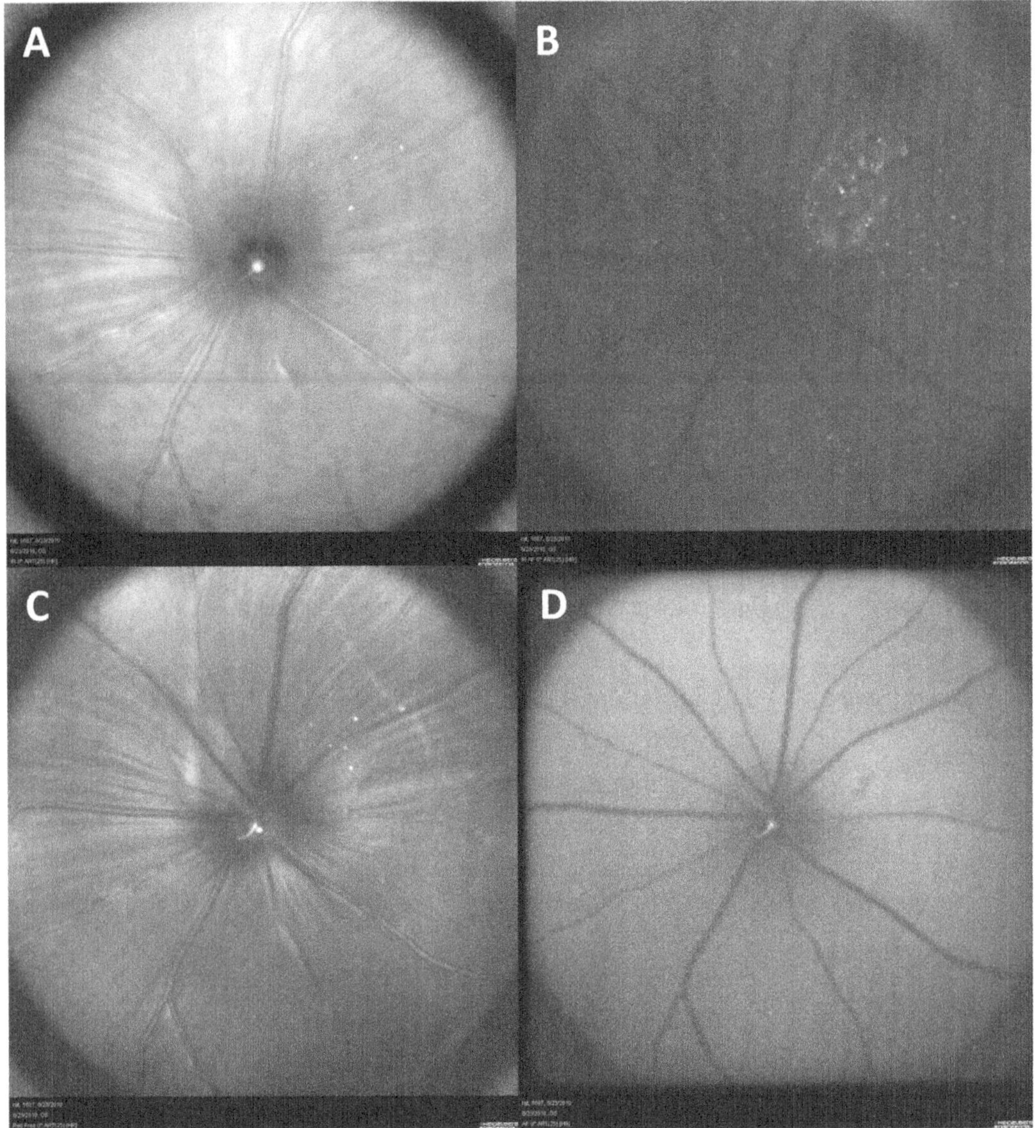

Fig. 5 Experimental SLO images: (**a**) IR reflectance (IR mode), (**b**) IR autofluorescence (ICGA mode), (**c**) Blue Laser reflectance (Red Free mode), (**d**) Blue Laser autofluorescence (FA mode)

3. When the focus is optimal, "acquire" image.

4. To achieve recordings with improved signal-to-noise ratio, use averaging software. This summates a defined number of individual fundus images. When the image quality is not adequate to create an averaged image, a movie should be acquired so that an averaged image can be constructed later [19, 38].

5. After a reflective image is acquired, proceed to acquiring a fluorescent image (FA or ICGA) by selecting the correct mode on the control panel and repeating **steps 2–4** above.

3.6 Data Analysis

Detailed analysis may be limited when using the software provided. However, a number of third-party software programs are available for post hoc analysis, such as the recently validated MatLab script, an automated technique of counting apoptosing retinal cells imaged with DARC [39].

4 Notes

1. Fluorescent Annexin A5 is not readily available commercially, and we recommend in-house manufacturing to be sure of final concentrations. The optimal concentration of freshly formulated Annexin A5 should be titrated before use, as it is dependent on the fluorescent probe, its quantum yield and the inherent retinal autofluorescence at the appropriate wavelength. The concentration of An-F for intravitreal dosing is 0.2–3 µg/mL. The concentration of An-F for intravenous dosing is 4–300 µg/mL. The volume for intravitreal injection is 5 µL for rat or 1 µL for mouse. For intravenous injection, a total volume of 500 µL and 50 µL should be used for rat and mouse respectively.

2. DARC imaging is conducted in experimental animals under general anesthesia, with each species requiring a different anesthetic regime. Careful selection of species-specific mydriatic drops must be considered to maintain a clear lens, required for DARC imaging.

3. To achieve the best results, it is important to fully dilate the pupils of the animal. Both rats and mice have a dilated pupil size smaller than 3 mm. In this case, the larger the pupil the greater the amount of light entering and exiting the rodent's eye, giving better illumination despite a lower resolution.

4. Fluorescent materials should be kept in the dark at all times.

5. While the animal is anesthetized, it is essential to keep the cornea lubricated. To ensure good picture quality, it is also important to clean the cornea carefully with Hypromellose eye drops and moistened cotton buds prior to imaging.

Conflict of Interest

M. Francesca Cordeiro is an inventor on patent applications owned by UCL and pertaining to Detection of Apoptosing Retinal Cells.

References

1. Ascher MS, Sheppard HW, Krowka JF et al (1995) AIDS as immune system activation. Key questions that remain. Adv Exp Med Biol 374:203–210

2. Selleri C, Maciejewski JP, Sato T et al (1996) Interferon-gamma constitutively expressed in the stromal microenvironment of human marrow cultures mediates potent hematopoietic inhibition. Blood 87:4149–4157

3. Raza A, Mundle S, Iftikhar A et al (1995) Simultaneous assessment of cell kinetics and programmed cell death in bone marrow biopsies of myelodysplastics reveals extensive apoptosis as the probable basis for ineffective hematopoiesis. Am J Hematol 48:143–154

4. Thompson CB (1995) Apoptosis in the pathogenesis and treatment of disease. Science 267:1456–1462

5. Shimizu A, Yamanaka N (1993) Apoptosis and cell desquamation in repair process of ischemic tubular necrosis. Virchows Arch B Cell Pathol Incl Mol Pathol 64:171–180

6. Gschwind M, Huber G (1995) Apoptotic cell death induced by beta-amyloid 1-42 peptide is cell type dependent. J Neurochem 65:292–300

7. Festoff BW (1996) Amyotrophic lateral sclerosis: current and future treatment strategies. Drugs 51:28–44

8. Walkinshaw G, Waters CM (1995) Induction of apoptosis in catecholaminergic PC12 cells by L-DOPA. Implications for the treatment of Parkinson's disease. J Clin Invest 95: 2458–2464

9. Portera-Cailliau C, Sung CH, Nathans J et al (1994) Apoptotic photoreceptor cell death in mouse models of retinitis pigmentosa. Proc Natl Acad Sci U S A 91:974–978

10. Pollard H, Cantagrel S, Charriaut-Marlangue C et al (1994) Apoptosis associated DNA fragmentation in epileptic brain damage. Neuroreport 5:1053–1055

11. Reme CE, Grimm C, Hafezi F et al (1998) Apoptotic cell death in retinal degenerations. Prog Retin Eye Res 17:443–464

12. Raynal P, Pollard HB (1994) Annexins: the problem of assessing the biological role for a gene family of multifunctional calcium- and phospholipid-binding proteins. Biochim Biophys Acta 1197:63–93

13. Vermes I, Haanen C, Steffens-Nakken H et al (1995) A novel assay for apoptosis. Flow cytometric detection of phosphatidylserine expression on early apoptotic cells using fluorescein labelled Annexin V. J Immunol Methods 184:39–51

14. Baskic D, Popovic S, Ristic P et al (2006) Analysis of cycloheximide-induced apoptosis in human leukocytes: fluorescence microscopy using annexin V/propidium iodide versus acridin orange/ethidium bromide. Cell Biol Int 30(11):924–932

15. Fadok VA, Bratton DL, Frasch SC et al (1998) The role of phosphatidylserine in recognition of apoptotic cells by phagocytes. Cell Death Differ 5:551–562

16. Reutelingsperger CP, van Heerde WL (1997) Annexin V, the regulator of phosphatidylserine-catalyzed inflammation and coagulation during apoptosis. Cell Mol Life Sci 53:527–532

17. Coxon KM, Duggan J, Cordeiro MF et al (2011) Purification of annexin V and its use in the detection of apoptotic cells. Methods Mol Biol 731:293–308

18. Blankenberg FG, Tait J, Ohtsuki K et al (2000) Apoptosis: the importance of nuclear medicine. Nucl Med Commun 21:241–250

19. Cordeiro MF, Guo L, Luong V et al (2004) Real-time imaging of single nerve cell apoptosis in retinal neurodegeneration. Proc Natl Acad Sci U S A 101:13352–13356

20. Cordeiro MF, Guo L, Luong V et al (2010) Imaging multiple phases of neurodegeneration: a novel approach to assessing cell death in vivo. Cell Death Dis 1:e3. doi:10.1038/cddis.2009.3

21. Webb RH, Hughes GW, Pomerantzeff O (1980) Flying spot TV ophthalmoscope. Appl Opt 19:2991–2997

22. Seth R, Gouras P (2004) Assessing macular pigment from SLO images. Doc Ophthalmol 108:197–202

23. Kernt M, Schaller UC, Stumpf C et al (2010) Choroidal pigmented lesions imaged by ultra-wide-field scanning laser ophthalmoscopy with two laser wavelengths (Optomap). Clin Ophthalmol 4:829–836

24. Rudnicka AR, Burk RO, Edgar DF et al (1998) Magnification characteristics of fundus imaging systems. Ophthalmology 105:2186–2192

25. Eter N (2010) Molecular imaging in the eye. Br J Ophthalmol 94:1420–1426

26. Maass A, Lundh von Leithner P, Luong V et al (2007) Assessment of rat and mouse RGC apoptosis imaging in vivo with different scanning laser ophthalmoscopes. Curr Eye Res 32:851–861

27. Hassenstein A, Meyer CH (2009) Clinical use and research applications of Heidelberg retinal angiography and spectral-domain optical coherence tomography—a review. Clin Experiment Ophthalmol 37:130–143

28. Hartwig A, Atchison DA (2012) Analysis of higher-order aberrations in a large clinical population. Invest Ophthalmol Vis Sci 53: 7862–7870

29. Hecht E (1987) Optics/Eugene Hecht/with contributions by Alfred Zajac

30. Donnelly WJ 3rd, Roorda A (2003) Optimal pupil size in the human eye for axial resolution. J Opt Soc Am A Opt Image Sci Vis 20: 2010–2015

31. Holz FG, Bellmann C, Rohrschneider K et al (1998) Simultaneous confocal scanning laser fluorescein and indocyanine green angiography. Am J Ophthalmol 125:227–236

32. Soliman AZ, Silva PS, Aiello LP et al (2012) Ultra-wide field retinal imaging in detection, classification, and management of diabetic retinopathy. Semin Ophthalmol 27:226–232

33. Campbell JP, Leder HA, Sepah YJ et al (2012) Wide-field retinal imaging in the management of noninfectious posterior uveitis. Am J Ophthalmol 154:908–911

34. Tsui I, Franco-Cardenas V, Hubschman JP et al (2012) Ultra wide field fluorescein angiography can detect macular pathology in central retinal vein occlusion. Ophthalmic Surg Lasers Imaging 43:257–262

35. Witmer MT, Kozbial A, Daniel S et al (2012) Peripheral autofluorescence findings in age-related macular degeneration. Acta Ophthalmol 90:e428–e433

36. Galvao J, Davis B, Tilley M et al (2013) Unexpected low-dose toxicity of the universal solvent DMSO. FASEB J. doi:10.1096/fj. 13-235440

37. Guo L, Salt TE, Maass A et al (2006) Assessment of neuroprotective effects of glutamate modulation on glaucoma-related retinal ganglion cell apoptosis in vivo. Invest Ophthalmol Vis Sci 47:626–633

38. Guo L, Salt TE, Luong V et al (2007) Targeting amyloid-beta in glaucoma treatment. Proc Natl Acad Sci U S A 104:13444–13449

39. Bizrah M, Dakin SC, Guo L, Rahman F, Parnell M, Normando E, Nizari S, Davis B, Younis A, Cordeiro MF (2014) A semi-automated technique for labeling and counting of apoptosing retinal cells. BMC Bioinform 15:169. doi:10.1186/1471-2105-15-169

Chapter 18

Targeted Toxicants to Dopaminergic Neuronal Cell Death

Huajun Jin, Arthi Kanthasamy, Dilshan S. Harischandra, Vellareddy Anantharam, Ajay Rana, and Anumantha Kanthasamy

Abstract

Parkinson's disease (PD) is mainly characterized by a progressive degeneration of dopaminergic neurons in the substantia nigra resulting in chronic deficits in motor functions. Administration of the neurotoxin 1-methyl-4-phenyl-1,2,3,6-tetrahydropyridine (MPTP) produces PD symptoms and recapitulates the main features of PD in human and animal models. MPTP is converted to 1-methyl-4-phenylpyridine (MPP^+), which is the active toxic compound that selectively destroys dopaminergic neurons. Here, we describe methods and protocols to evaluate $MPTP/MPP^+$-induced dopaminergic neurodegeneration in both murine primary mesencephalic cultures and animal models. The ability of $MPTP/MPP^+$ to cause dopaminergic neuronal cell death is assessed by immunostaining of tyrosine hydroxylase (TH).

Key words Parkinson's disease, Dopaminergic neurons, Neurodegeneration, MPTP, MPP^+, Primary mesencephalic cultures, Animal model, Tyrosine hydroxylase, Immunostaining

Abbreviations

BSA	Bovine serum albumin
HBSS	Hank's balanced salt solution
IACUC	Institutional Animal Care and Use Committee
MPP	1-Methyl-4-phenylpyridine
MPTP	1-Methyl-4-phenyl-1,2,3,6-tetrahydropyridine
OCT	Optimum cutting temperature
PBS	Phosphate-buffered saline
PD	Parkinson's disease
PDL	Poly-D-lysine
PFA	Paraformaldehyde
SNc	Substantia nigra pars compacta
TH	Tyrosine hydroxylase

Laura Lossi and Adalberto Merighi (eds.), *Neuronal Cell Death: Methods and Protocols*, Methods in Molecular Biology, vol. 1254, DOI 10.1007/978-1-4939-2152-2_18, © Springer Science+Business Media New York 2015

1 Introduction

Parkinson's disease (PD) is the second most common neurodegenerative disease, which affects several million people worldwide [1, 2]. PD results primarily from the progressive and selective loss of dopaminergic neurons within the substantia nigra pars compacta (SNc), leading to the marked depletions of dopamine, its biosynthetic enzyme tyrosine hydroxylase (TH), and the dopamine transporter in the striatum, as well as in the SNc [3, 4]. The cardinal clinical features of PD include bradykinesia, resting tremors, rigidity and postural instability [5]. Those behavioral symptoms manifest by a threshold effect, when 50–60 % of SNc dopaminergic neurons and 70–80 % of striatal dopamine are lost [5, 6]. Symptoms that do not involve movement, coordination, or mobility, such as constipation, loss of smell, fatigue, depression, and cardiovascular dysfunction are also noted during the very early stages of PD [7, 8]. Thus far, the actual pathogenic mechanisms underlying dopaminergic neuron loss are still not fully understood. Current treatments for PD are all symptomatic; none slow or prevent neuronal death progression in the dopaminergic system [9, 10]. The most potent treatment remains dopamine replacement therapy by administration of the dopamine precursor, L-dopa [11]. Although dopamine replacement with L-dopa is initially effective for most patients to improve PD symptoms, long-term manipulation of L-dopa can lead to disabling side effects such as wearing-off, dyskinesias, and dystonia. Moreover, the clinical efficacy often declines as the disease advances [12]. Therefore, there is an urgent need to elucidate the cascade of events underlying the dopaminergic neurodegenerative process in PD.

Over the past several decades, a number of neurotoxin-based PD models of selective cell death in dopaminergic neuronal populations that mimic PD have been developed, although none of these models displays all features of PD [13, 14]. Among the different neurotoxin-based PD models, the 1-methyl-4-phenyl-1,2,3,6-tetrahydropyridine (MPTP) model has been extensively used in various mammalian species for the investigation of degenerative mechanisms and development of therapeutic strategies for PD. MPTP was originally recognized as a neurotoxin in 1982 when several drug users developed sub-acute onset of severe parkinsonism after the intravenous use of a synthetic heroin that was contaminated with MPTP [15]. It is now well established that systemic injection of MPTP induces in humans, non-human primates, and mice a severe parkinsonian syndrome characterized by most of the clinical features of PD. MPTP passes through the blood–brain barrier and accumulates in the brain after systemic administration. Once in the brain, MPTP undergoes a biotransformation process in glia and serotonergic neurons to convert to the active toxic metabolite, 1-methyl-4-phenylpyridine (MPP^+) [16, 17], by mono amine oxidase B. MPP^+ is later selectively taken up by dopaminergic neurons via dopamine transporters, where it accumulates in mitochondria and causes the complex I defect [18–21]. This chapter summarizes step-by-step procedures of immunostaining of TH-positive neurons in primary mesencephalic cultures and animals, which can be used as a simple and efficient assay for measuring $MPTP/MPP^+$-induced dopaminergic neuronal cell death. The methods described in this chapter are used successfully in our laboratory and routinely produce reliable results [22–25].

2 Materials

All protocols using live animals must be first approved and supervised by an Institutional Animal Care and Use Committee (IACUC). Appropriate safety procedures for handling and use of $MPTP/MPP^+$ should be followed as described in [20].

2.1 Treatment of Primary Mouse Mesencephalic Cultures with MPP+

1. E14 (*see* **Note 1**) pregnant C57Bl/6 mice (Charles River Laboratories, Wilmington, MA).

2. Dissection tools (dissecting scissors, microdissection forceps, 35- and 100-mm sterile Petri dishes, autoclaved Whatman paper filters). *See* **Note 2**.

3. Dissecting microscope.

4. Calcium- and magnesium-free Hank's balanced salt solution (HBSS) (Life Technologies™, Carlsbad, CA). *See* **Note 3**.

5. B-27 serum-free supplement (Life Technologies™). *See* **Note 4**.

6. Neurobasal medium (Life Technologies™).

7. Dissection medium: To 490 mL of calcium- and magnesium-free HBSS add 5 mL of penicillin/streptomycin (final concentrations: 50 units of penicillin and 50 μg of streptomycin per mL) and 5 mL of L-glutamine (final concentration 2 mM).

8. Neurobasal culture medium: 4 mL of B-27 serum-free supplement, 500 μL of L-glutamine (final concentration 500 μM), 400 μL of L-glutamate (final concentration 25 μM) and 8 mL of penicillin/streptomycin into 187 mL of Neurobasal medium. Pre-warm to 37 °C before use.

9. Trypsin solution: 2.5 % trypsin/EDTA (Life Technologies™) in HBSS. Warm to 37 °C before use.

10. Trypsin inhibition solution: DMEM medium supplemented with 10 % fetal bovine serum and 5 % penicillin/streptomycin. Pre-warm to 37 °C before use.

11. Poly-D-lysine stock solution (PDL, Sigma Chemicals, St. Louis, MO): dissolve 1 mg PDL in 1 mL of sterile double-distilled water. Divide stock in aliquots and store at −20 °C. Dilute to a fresh 0.1 mg/mL PDL solution with sterile double-distilled water for use. *See* **Note 5**.

12. 15-mm glass coverslips (sterilized).

13. 24-Well cell culture plate (BD Falcon, BD Biosciences, San Jose, CA).

14. 50- and 15-mL conical-bottom polystyrene tubes (BD Falcon).

15. Vi-CELL Cell Viability Analyzer (Beckman Coulter, Inc, Brea, CA) for counting cells. *See* **Note 6**.

16. 10 μM cytosine β-D-arabinofuranoside (Ara-c) in Neurobasal culture medium.

17. MPP+ (Sigma Chemicals).

18. MPP+ stock solution: prepare a fresh 10 mM solution in sterile double-distilled water. Dilute to 10 μM with culture medium before use.

19. Water bath (37 °C).

20. Cell culture incubator at 37 °C with 5 % CO_2 in a humidified environment.

21. BD cell strainer, 40 μm nylon filter (BD Biosciences).

2.2 Tyrosine Hydroxylase (TH) Immunocytochemistry of Primary Cultures

1. 4 % Paraformaldehyde (PFA) as fixative. Prepare fresh on day of use or the night before. *See* **Notes** 7 and **8**.

2. Blocking solution: 2 % Bovine serum albumin (BSA), 0.5 % Triton X-100, 0.05 % Tween-20 in PBS.

3. Primary antibody solution: dilute mouse monoclonal antibody anti TH (Chemicon) in PBS containing 1 % BSA at 1:1,500.

4. Secondary antibody solution: dilute Alexa Fluor 488 goat anti-mouse antibody (Life Technologies™) in PBS containing 1 % BSA at 1:1,500.

5. Poly-L-lysine-coated slides (Sigma Chemicals).

6. Belly dancer or shaker.

7. 1× Phosphate-buffered saline (PBS), pH 7.4.

8. Fluorescence-free mounting medium (e.g. Sigma Chemicals).

9. Fluorescence microscope with appropriate filter set.

2.3 Treatment of Intact Animals with MPTP

1. Male C57Bl/6 mice, 8- to 10-week-old (24–28 g) (Charles River). *See* **Note 9**.

2. Sterile syringes and needles for injection.

3. 1× sterile PBS.

4. MPTP-HCl solution: dissolve MPTP-HCl (Sigma Chemicals) in sterile PBS 2.5 mg/mL. Make a fresh MPTP solution every time. *See* **Note 10**.

5. Protective clothing for handling MPTP (disposable nitrile gloves, lab coats, safety glasses with side-shields or safety goggles, and N95 masks).

6. Chemical fume hood.

7. Scale (for weighing mice).

2.4 TH Immunohisto-chemistry of Intact Animals

1. Surgical tools for transcardial perfusion (scissors, forceps, and clamps).

2. Perfusion pump with tubing.

3. Butterfly catheter with needle.

4. 1× PBS.

5. 4 % Paraformaldehyde (PFA) as fixative. Prepare fresh on day of use or the night before. *See* **Notes 7** and **8**.

6. Anesthetic: Ketamine/Xylazine mixture 2:1 (v/v).

7. 50-mL conical-bottom polystyrene tubes.

8. Safety goggle and gloves.

9. Cryoprotectant: 30 % sucrose in PBS.

10. OCT™(optimum cutting temperature) embedding compound (Tissue-Tek®, Sakura Finetek Europe BV, Alphen aan den Rijn, The Netherlands). *See* **Note 11**.

11. Brush (camel hair).

12. 6-Well plates.

13. Vibratome or cryostat.

14. Antifreezing solution for cryostat sections: 30 % ethylene glycol, 30 % sucrose in PBS.

15. Antigen retrieval solution (sodium citrate buffer, pH 8.5): Add 2.94 g trisodium citrate powder into 1,000 mL sterile double-distilled water and adjust pH to 8.5.

16. Blocking solution: 2 % BSA, 0.5 % Triton X-100, 0.05 % Tween-20 in PBS.

17. Primary antibody solution: dilute mouse monoclonal antibody anti-TH (Chemicon) in PBS containing 1 % BSA at 1:1,500.

18. Secondary antibody solution: dilute Alexa Fluor 488 goat anti-mouse antibody (Life Technologies™) in PBS containing 1 % BSA at 1: 1,500.

19. Poly-L-lysine-coated slides (Sigma Chemicals).

20. Xylene and ethanol (100 %, 95 %, 70 % each).

21. DPX mounting medium (Electron Microscopy Science, Hatfield, PA).

22. Belly dancer or shaker.

23. Water bath (80 °C).

24. Fluorescence microscope with appropriate filter set.

3 Methods

3.1 Treatment of Primary Mouse Mesencephalic Cultures with MPP+

1. One day before starting the culture preparation, place sterilized 15-mm glass coverslips into the wells of a 24-well cell culture plate. Coat the coverslips with 500 μL of 0.1 mg/mL sterile PDL solution overnight in the incubator at 37 °C.

2. After overnight incubation, wash the wells and coverslips three times with double-distilled water and once with HBSS. Air-dry the coverslips completely (lids on) in a culture hood until use. *See* **Note 12**.

3. Place 3 mL of dissection medium in a 35-mm sterile Petri dish and keep it on ice.

4. Asphyxiate E14 pregnant mouse with CO_2. *See* **Note 13**.

5. After wiping the abdomen with 70 % ethanol, use scissors to cut through abdominal cavity to expose abdominal contents. Lift up the string of embryos with forceps and separate yolk sack from embryonic chain. Then use small scissors to remove each embryo from the amniotic sac and pinch off placenta. The brains from each embryo are carefully dissected out and collected into the 35-mm sterile Petri dish filled with dissection medium.

6. Put an autoclaved Whatman filter paper on the bottom of a 100-mm sterile Petri dish, rinse the filter paper with 2 mL of dissection medium, and place this dish under a dissecting microscope. Transfer a brain into the dish and use a fine forceps to cut out the ventral mesencephalic region. *See* **Note 14**.

7. Repeat with all brains and collect all ventral mesencephali obtained into a 15-mL polypropylene conical tube filled with cold dissection medium. Keep the 15-mL conical tube on ice during the dissection procedures. *See* **Note 15**.

8. Place the 15-mL conical tube into a cell culture hood and wash three times with cold dissection medium. *See* **Note 16**.

9. Add 5 mL of trypsin solution to the 15-mL tube and digest tissues in a water bath at 37 °C for 30 min.

10. At the end of digestion, add 5 mL of trypsin inhibition solution to stop the trypsin activity. Carefully remove the supernatant containing trypsin and wash the digested tissues three times with trypsin inhibition solution.

11. Resuspend the tissue with 5 mL Neurobasal culture medium and pass suspension ten times through a wide mouth pipette to dissociate large aggregates.

12. Remove and filter the cell suspension through a 40 μm filter (BD cell strainer) into a 50 mL sterile conical tube.

13. Count cells using the Vi-CELL Cell Viability Analyzer, and plate the mesencephalic cells in 24-well plates with PDL-coated coverslips at a density of 0.5×10^6 cells per well in Neurobasal culture medium. *See* **Notes 17** and **18**.

14. Incubate the cells in a humidified culture incubator at 37 °C in 5 % CO_2.

15. After 2 days of culture, add 10 μM cytosine β-d-arabinofuranoside (Ara-c) to the Neurobasal culture medium to reduce the population of glial cells (*see* **Note 19**). One day later, replace the Ara-c-containing medium with Neurobasal culture medium.

16. Change half of the medium at least twice a week. *See* **Note 20**.

17. Six-day-old cultures are incubated in the presence or absence of 10 μM MPP for 24 h. *See* **Note 21**.

3.2 TH Immunocytochemistry of Primary Mesencephalic Cultures

1. Remove the culture medium from the wells. Add 700 μL of freshly prepared 4 % PFA to each coverslip and fix for 15 min. *See* **Note 22**.

2. Rinse the cells with 1× PBS (4 × 5 min).

3. Block the cells in blocking solution for 45 min at room temperature.

4. Incubate the cells in primary antibody solution overnight at 4 °C.

5. Wash the cells with 1× PBS (6 × 5 min).

6. Incubate in secondary antibody solution for 1.5 h at room temperature.

7. Wash the cells with 1× PBS (6 × 5 min).

8. Mount coverslip on Poly-L-lysine-coated slides with fluorescence-free mounting medium.

9. Examine the TH-positive cells under a fluorescence microscope.

Figure 1 is an example of an in vitro model of mesencephalic neuron-enriched cultures used to evaluate MPP+-induced dopaminergic neurodegeneration. It clearly shows the MPP+-treated primary TH-positive neurons with significant neurite loss compared to untreated cultures.

3.3 Treatment of Intact Animals with MPTP

1. Prior to injection, determine the live weight of each mouse and calculate the volume of MPTP solution based on the desired dose. In this protocol, we adopt a sub-acute dosing regimen of MPTP (25 mg/kg body weight single injection each day for 5 consecutive days). *See* **Notes 23** and **24**.

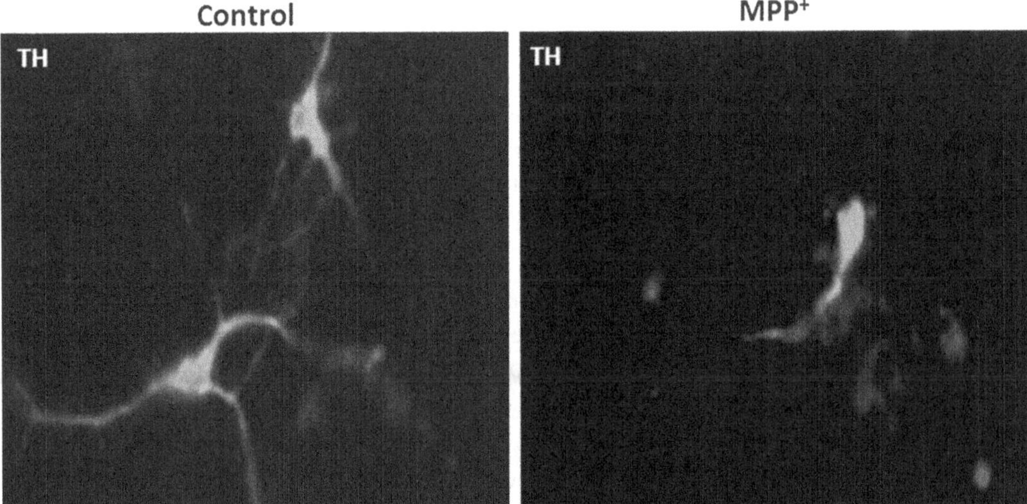

Fig. 1 MPP⁺ induced dopaminergic neurotoxicity in primary mouse mesencephalic cultures. Mesencephalic tissues from E14 mouse embryos were cultured and treated with 10 μM of MPP⁺ for 24 h. After treatment, the cultures were immunostained for tyrosine hydroxylase (TH, *green*). Images were obtained using a Nikon TE2000 fluorescence microscope (magnification 60×). Representative immunofluorescent cells are shown (Color figure online)

2. In a chemical fume hood, load the appropriate amount of MPTP in syringe and administer intraperitoneally to each mouse. Control mice receive equal volumes of sterile PBS.

3. Return animal to cage and observe that it recovers from injection.

4. Perform studies 7 days post-MPTP injection.

3.4 TH Immunohisto-chemistry of Intact Animals

1. Set up perfusion pump and adjust the flow rate to approximately 10 mL/min.

2. Anesthetize mouse with an intraperitoneal injection of 60 μL of Ketamine/Xylazine mixture.

3. Once mouse is sedated (*see* **Note 25**), grab skin with forceps at the level of the diaphragm and make an incision with scissors. Continue cutting until the entire heart is visible.

4. With a sharp scissor, make a small incision on the right atrium (*see* **Note 26**). Then pierce the butterfly catheter needle into the left ventricle, and release valve to perfuse with 1× PBS for 3–5 min. *See* **Notes 27** and **28**.

5. When blood has been cleared from body, switch to 4 % PFA and perfuse for 7–10 min.

6. Turn off pump, decapitate the animal and quickly remove the brain. The brain should be post-fixed by immersing it in 4 % PFA/PBS overnight at 4 °C.

7. Cryoprotect brains with 30 % sucrose for 24 h at 4 °C. Provide enough 30 % sucrose to completely submerge samples.

8. Cover the entire brain with OCT embedding compound and freeze the tissue block at –80 °C until ready for sectioning. *See* **Note 29**.

9. Cut the tissue block into sections of the desired thickness (e.g. 30 μm) using a cryostat at –20 °C and use a brush to move the emerging sections away from the knife blade. Keep the SNc sections in the 6-well plate filled with antifreezing solution at –20 °C. *See* **Note 30**.

10. Rinse the free floating brain sections with 1× PBS (3 × 15 min).

11. For antigen retrieval, incubate sections with antigen retrieval solution pre-heated to 80 °C in a water bath for 30 min. Cool the solution to room temperature and then wash the sections 3 × 5 min in 1× PBS. *See* **Notes 31** and **32**.

12. Block sections for 1 h in blocking solution at room temperature.

13. Incubate in primary antibody solution overnight at room temperature.

14. Rinse sections in 1× PBS (6 × 5 min).

15. Incubate with secondary antibody solution for 1.5 h at room temperature.

16. Wash sections in 1× PBS (6 × 5 min).

17. Carefully mount sections on Poly-L-lysine-coated slides.

18. Dry the slides for 24 h at room temperature. Sections can be stored in a sealed slide box at –80 °C for later use.

19. Dehydrate sections as follows: 1 min in distilled water, 1 min in 70 % ethanol, 1 min in 95 % ethanol, 1 min in 100 % ethanol, and dip in xylene twice. Coverslips are adhered with DPX mounting medium.

20. Examine the TH-positive cells under a fluorescence microscope.

Figure 2 shows that sub-acute treatment of MPTP led to a marked decrease in TH-positive neurons in ventral midbrain sections compared to saline-treated mice.

4   Notes

1. The choice of the appropriate age of the tissue from which the cells are obtained is critical in this culture. The optimal ages for ventral mesencephalic cultures are embryonic days 10–14 (E10–E14), a time when midbrain dopaminergic neurogenesis takes place at the ventral midbrain floor plate in the mouse [26–28].

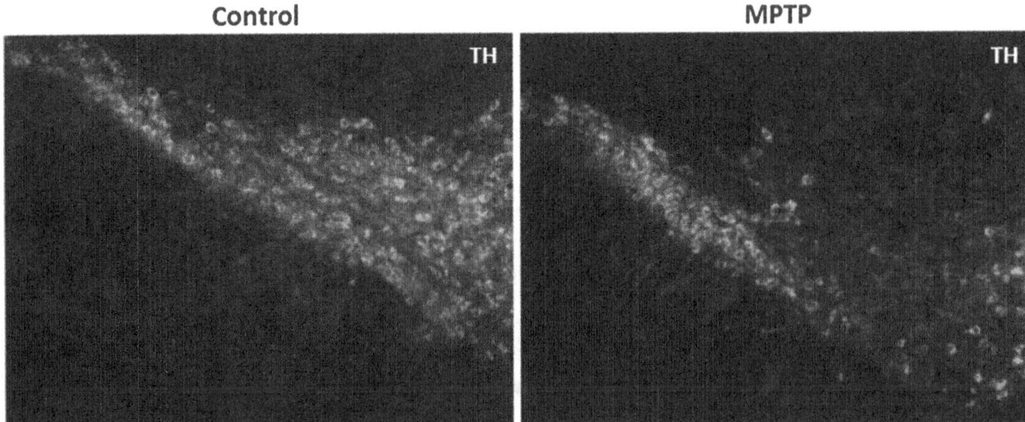

Fig. 2 MPTP-induced dopaminergic neurotoxicity in animals. Mice received single intraperitoneal doses of MPTP (25 mg/kg) each day for 5 consecutive days and control mice received an equal volume of saline. Seven days after MPTP treatment, substantia nigra sections from control and MPTP-treated mice were subjected to TH immunohistochemistry (*green*) analysis. Images were obtained using a Nikon TE2000 fluorescence microscope (magnification 20×). Representative immunofluorescent cells are shown (Color figure online)

2. All surgical instruments must be sterile (autoclaved) before use, and during dissection they should be cleaned with 70 % ethanol as needed.

3. The dissection medium containing calcium and magnesium should not be used, because their presence leads to immediate and widespread cell death [29].

4. B-27 serum-free supplement is formulated with cortical antioxidants and minimizes glial cell proliferation [30, 31]. B27 supplement minus AO (without antioxidants) has the same composition as B27 but without cortical antioxidants, thereby making it ideal for studies under oxidative stress conditions.

5. Poly-lysine is applied to glass coverslips to enhance the electrostatic attraction between the glass surface and the cells, thereby improving cell attachment. The D rather than L form of poly-lysine is favored since it is less prone to break down to lysine, and thus is less toxic. The use of poly-ornithine or collagen as alternative substrates to poly-D-lysine also has been successfully tested in the culture of substantia nigra neurons [32, 33].

6. The Vi-CELL system is a viability analyzer and cell counter that automates the widely accepted trypan blue cell exclusion method [34]. Alternatively, the amount of cells can be manually assessed with a hemocytometer and trypan blue.

7. PFA is toxic. Gloves and safety goggles should be worn, and the solution should be prepared inside a chemical fume hood.

8. For making 1 L of PFA in PBS, dissolve 40 g of PFA powder in 1× PBS and heat to approximately 60 °C while stirring. The powder generally takes 20–30 min to completely dissolve into

solution. If suspended PFA is still visible, a few drops of 1 N NaOH can be added until the solution clears. The solution is then cooled, filtered, and stored at 4 °C.

9. It should be noted that strain, vendor, gender, age, and body weight are all significant factors that influence the sensitivity and reproducibility of the lesion caused by MPTP treatment and the degree of dopaminergic neurodegeneration. In our and others' experience [20], male C57Bl/6 mice, 8–10 weeks old and weighing 24–28 g, produce optimal reproducibility of MPTP neurotoxicity.

10. The inappropriate handling of MPTP may cause neurological damage to research staff. Thus, its storage, handling, and disposal must follow a strict safety protocol. Always operate under a chemical fume hood.

11. OCT compound is a formulation of water-soluble glycols and resins and can be used to surround and cryoprotect the fresh tissue for cryostat sectioning at temperatures below –10 °C.

12. After air-drying, the PDL-coated cell culture plates and glass coverslips can be stored for up to 6 months at 4 °C covered with parafilm.

13. Anesthesia of mice via intraperitoneal injection of anesthetic agents is not recommended, as it may damage the embryos.

14. The yield of nigral dopaminergic neurons from the newborn mouse brain is very low, about 5–10 % of total cells. Thus, the presence of sufficient dopaminergic neurons in this culture requires proper dissection of mid-brain sections.

15. Keeping the 15-mL polypropylene conical tube on ice retards the action of proteolytic enzymes. The polypropylene tube is used because cells do not adhere readily to its surface.

16. To wash the tissues, let the tissue settle for 1–2 min, and then aspirate the liquid.

17. By using the Vi-CELL Cell Viability Analyzer, one can determine the viability of the cells. Empirically, this protocol will obtain cell suspension with viability of 90 % or higher. Excessively long dissection time may cause poor cell viability.

18. Seeding cell density is an important factor for good survival. Generally, high density conditions enhance the survival of dopaminergic neurons in dissociated cell cultures [35].

19. To maintain viability of the postmitotic neurons, we use the antimitotic agent Ara-c to reduce the population of dividing glia and other non-neuronal cells capable of DNA synthesis [36].

20. We usually exchange half of the medium because the primary cultured cells seem to be sensitive to air exposure. Changing the medium too frequently is not recommended, as it results in the loss of paracrine and autocrine factors produced within mesencephalic cultures [37].

21. The mesencephalic cells can be used 4 days post-dissection since very mature dopaminergic neurons usually appear at 3–6 days in vitro.

22. Using a belly dancer is recommended for all washing and incubation steps, as noted in Subheadings 3.2 and 3.4.

23. Various routes have been used for administering MPTP, such as gavage and stereotaxic injection into the brain; however, systemic administration (intraperitoneal, subcutaneous, and intravenous) is still the most commonly used and reproducible means [20, 38, 39].

24. MPTP can be administered to mice by a variety of dosing regimens. The sub-acute dosing regimen used in this protocol was originally developed by Heikkila et al. [40], who administered mice with 30 mg/kg/day MPTP for 5–10 consecutive days. With this treatment paradigm, MPTP induces significant dopaminergic degeneration within a few to 20 days after MPTP treatment is terminated [25, 41].

25. After anesthetics administration, the animal should be immobile within a few minutes. Do not conduct surgery until the toe-pinch response is completely gone.

26. Watch for a flush of blood in the body cavity as a superficial indicator of efficient cut on the right atrium. This cut also can be made immediately after turning on a pump to perfuse PBS.

27. Usually 100 mL of each solution is sufficient to perfuse one mouse.

28. The needle should not be extended too far in (no more than ¼ in.), as it may come out the other side and compromise circulation of the solution.

29. Some soft tissues, like brain tissue, can also be frozen in M-1 medium at −3 °C.

30. For the SNc sections, coronal sections staring at ~2.6 mm posterior to the bregma are collected. The location of the brain slice and identification of brain regions are based on atlas images [42].

31. We use a heat treatment procedure to effectively retrieve the antigens that are masked by PFA fixation and embedding procedures from free floating sections. There is also an enzymatic antigen retrieval method available (proteinase K- or trypsin-based).

32. During the boil, use a sufficient volume of antigen retrieval solution to cover the sections if boiling in a non-sealed vessel.

Acknowledgement

This work was supported by the National Institutes of Health grants, NS65167 and NS078237 to AK, and ES10586 and NS074443 to AGK. W. Eugene and Linda Lloyd Endowed Chair to AGK is also acknowledged.

References

1. Jankovic J (1988) Parkinson's disease: recent advances in therapy. South Med J 81:1021–1027

2. Tanner CM (1992) Epidemiology of Parkinson's disease. Neurol Clin 10:317–329

3. Przedborski S, Jackson-Lewis V, Djaldetti R et al (2000) The parkinsonian toxin mptp: action and mechanism. Restor Neurol Neurosci 16:135–142

4. Przedborski S, Dawson TM (2001) The role of nitric oxide in Parkinson's disease. Meth Mol Med 62:113–136

5. Dauer W, Przedborski S (2003) Parkinson's disease: mechanisms and models. Neuron 39:889–909

6. Abou-Sleiman PM, Muqit MM, Wood NW (2006) Expanding insights of mitochondrial dysfunction in Parkinson's disease. Nat Rev Neurosci 7:207–219

7. Chaudhuri KR, Healy DG, Schapira AH (2006) Non-motor symptoms of Parkinson's disease: diagnosis and management. Lancet Neurol 5:235–245

8. Magerkurth C, Schnitzer R, Braune S (2005) Symptoms of autonomic failure in Parkinson's disease: prevalence and impact on daily life. Clin Auton Res 15:76–82

9. Olanow CW (2004) The scientific basis for the current treatment of Parkinson's disease. Annu Rev Met 55:41–60

10. Obeso JA, Rodriguez-Oroz MC, Goetz CG et al (2010) Missing pieces in the Parkinson's disease puzzle. Nature Med 16:653–661

11. Carlsson A, Lindqvist M, Magnusson T (1957) 3,4-Dihydroxyphenylalanine and 5-hydroxytryptophan as reserpine antagonists. Nature 180:1200

12. Lewitt PA (2008) Levodopa for the treatment of Parkinson's disease. New Engl J Med 359:2468–2476

13. Manning-Bog AB, Langston JW (2007) Model fusion, the next phase in developing animal models for Parkinson's disease. Neurotoxicity Res 11:219–240

14. Jenner P (2008) Functional models of Parkinson's disease: a valuable tool in the development of novel therapies. Ann Neurol 64(Suppl 2):S16–S29

15. Langston JW, Ballard P, Tetrud JW et al (1983) Chronic parkinsonism in humans due to a product of meperidine-analog synthesis. Science 219:979–980

16. Hartley A, Stone JM, Heron C et al (1994) Complex i inhibitors induce dose-dependent apoptosis in pc12 cells: relevance to Parkinson's disease. J Neurochem 63:1987–1990

17. Schapira AH (1993) Mitochondrial complex i deficiency in Parkinson's disease. Adv Neurol 60:288–291

18. Ramsay RR, Salach JI, Dadgar J et al (1986) Inhibition of mitochondrial nadh dehydrogenase by pyridine derivatives and its possible relation to experimental and idiopathic parkinsonism. Biochem Biophys Res Commun 135:269–275

19. Schober A (2004) Classic toxin-induced animal models of Parkinson's disease: 6-ohda and mptp. Cell Tiss Res 318:215–224

20. Przedborski S, Jackson-Lewis V, Naini AB et al (2001) The parkinsonian toxin 1-methyl-4-phenyl-1,2,3,6-tetrahydropyridine (MPTP): a technical review of its utility and safety. J Neurochem 76:1265–1274

21. McGeer PL, McGeer EG (2008) Glial reactions in Parkinson's disease. Mov Disord 23:474–483

22. Jin H, Kanthasamy A, Ghosh A et al (2011) Alpha-synuclein negatively regulates protein kinase cdelta expression to suppress apoptosis in dopaminergic neurons by reducing p300 histone acetyltransferase activity. J Neurosci 31:2035–2051

23. Jin H, Kanthasamy A, Anantharam V et al (2011) Transcriptional regulation of proapoptotic protein kinase cdelta: implications for oxidative stress-induced neuronal cell death. J Biol Chem 286:19840–19859

24. Kanthasamy AG, Anantharam V, Zhang D et al (2006) A novel peptide inhibitor targeted to caspase-3 cleavage site of a proapoptotic

kinase protein kinase C delta (PKCdelta) protects against dopaminergic neuronal degeneration in Parkinson's disease models. Free Radic Biol Med 41:1578–1589

25. Ghosh A, Chandran K, Kalivendi SV et al (2010) Neuroprotection by a mitochondria-targeted drug in a Parkinson's disease model. Free Radical Biol Med 49:1674–1684

26. Bayer SA, Wills KV, Triarhou LC et al (1995) Time of neuron origin and gradients of neurogenesis in midbrain dopaminergic neurons in the mouse. Exp Brain Res 105:191–199

27. Ang SL (2006) Transcriptional control of midbrain dopaminergic neuron development. Development 133:3499–3506

28. Jonsson ME, Ono Y, Bjorklund A et al (2009) Identification of transplantable dopamine neuron precursors at different stages of midbrain neurogenesis. Exp Neurol 219:341–354

29. Takeshima T, Shimoda K, Johnston JM et al (1996) Standardized methods to bioassay neurotrophic factors for dopaminergic neurons. J Neurosci Meth 67:27–41

30. Brewer GJ, Torricelli JR, Evege EK et al (1993) Optimized survival of hippocampal neurons in b27-supplemented neurobasal, a new serum-free medium combination. J Neurosci Res 35:567–576

31. Brewer GJ (1995) Serum-free b27/neurobasal medium supports differentiated growth of neurons from the striatum, substantia nigra, septum, cerebral cortex, cerebellum, and dentate gyrus. J Neurosci Res 42: 674–683

32. Kittappa R, Chang WW, Awatramani RB et al (2007) The foxa2 gene controls the birth and spontaneous degeneration of dopamine neurons in old age. PLoS Biol 5:e325

33. Blakely BD, Bye CR, Fernando CV et al (2011) Wnt5a regulates midbrain dopaminergic axon growth and guidance. PloS One 6:e18373

34. Szabo SE, Monroe SL, Fiorino S et al (2004) Evaluation of an automated instrument for viability and concentration measurements of cryopreserved hematopoietic cells. Lab Hematol 10:109–111

35. O'Malley EK, Black IB, Dreyfus CF (1991) Local support cells promote survival of substantia nigra dopaminergic neurons in culture. Exp Neurol 112:40–48

36. Liu B, Linley JE, Du X et al (2010) The acute nociceptive signals induced by bradykinin in rat sensory neurons are mediated by inhibition of m-type K^+ channels and activation of Ca^{2+}-activated Cl^- channels. J Clin Invest 120: 1240–1252

37. Seroogy KB, Gall CM (1993) Expression of neurotrophins by midbrain dopaminergic neurons. Exp Neurol 124:119–128

38. Blesa J, Phani S, Jackson-Lewis V et al (2012) Classic and new animal models of Parkinson's disease. J Biomed Biotechnol 2012:845618

39. Ghosh A, Saminathan H, Kanthasamy A et al (2013) The peptidyl-prolyl isomerase Pin1 up-regulation and proapoptotic function in dopaminergic neurons: relevance to the pathogenesis of Parkinson disease. J Biol Chem 288:21955–21971

40. Heikkila RE, Hess A, Duvoisin RC (1984) Dopaminergic neurotoxicity of 1-methyl-4-phenyl-1,2,5,6-tetrahydropyridine in mice. Science 224:1451–1453

41. Seniuk NA, Tatton WG, Greenwood CE (1990) Dose-dependent destruction of the coeruleus-cortical and nigral-striatal projections by MPTP. Brain Res 527:7–20

42. Paxinos G, Franklin KBJ (2001) The mouse brain in stereotaxic coordinates, 2nd edn. Academic, San Diego

Part III

Neural Stem Cells, Progenitors, and Gene Therapy Strategies

Chapter 19

Stem Cells, Neural Progenitors, and Engineered Stem Cells

Raj R. Rao and Shilpa Iyer

Abstract

Human pluripotent stem cells (hPSCs) have the unique potential to form every cell type in the body. This potential provides opportunities for generating human progenitors and other differentiated cell types for understanding human development and for use in cell type-specific therapies. Equally important is the ability to engineer stem cells and their derived progenitors to mimic specific disease models. This chapter will focus on the propagation and characterization of human neural progenitors (hNPs) derived from hPSCs with a particular focus on engineering hNPs to generate in vitro disease models for human neuro-mitochondrial disorders. We will discuss the methodologies for culturing and characterizing hPSCs and hNPs; and protocols for engineering hNPs by using a novel mitochondrial transfection technology.

Key words Human pluripotent stem cells, Neural progenitors, Mitochondria, Protofection, Genetic engineering

1 Introduction

Human pluripotent stem cells (hPSCs) that include human embryonic stem cells (hESCs) and human induced pluripotent stem cells (hiP-SCs) are expected to serve as useful models for studying human embryonic development in addition to their proposed potential for regenerative biomedical therapies [1–3]. Under specific culture conditions, hPSCs can be uniformly differentiated into multipotent human neural progenitors (hNPs) [4, 5]. The hNPs can also be differentiated into neurons and glial cells of the neural lineage [6–9]. These studies represent a first step toward generating a homogeneous, self-renewable and easy to propagate hNP cell population, committed to differentiation toward major cell types of the neural lineage. These cells not only provide healthy cell types for biomedical therapies; but serve as cell models for understanding the mechanisms of diseases progression in early human development.

We have developed methods for directed differentiation of hPSCs to generate uniform populations of hNPs with the potential

Laura Lossi and Adalberto Merighi (eds.), *Neuronal Cell Death: Methods and Protocols*, Methods in Molecular Biology, vol. 1254, DOI 10.1007/978-1-4939-2152-2_19, © Springer Science+Business Media New York 2015

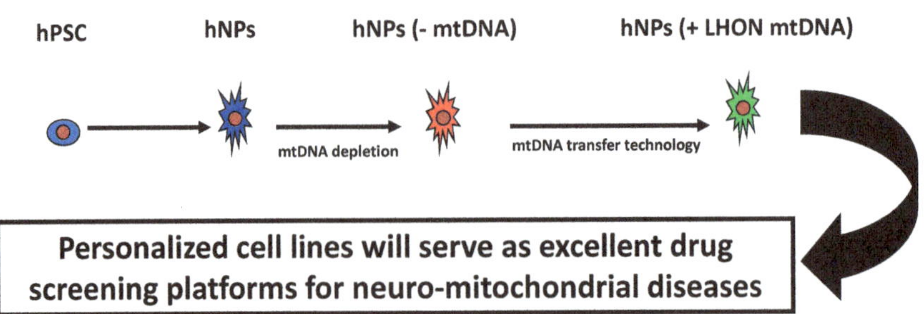

Fig. 1 Sequence of steps involved in creation of neural progenitor cell model based on mitochondrial gene replacement

for developing novel in vitro models to understand the mitochondrial genetics in many devastating childhood mitochondrial disorders. Recent work demonstrated the potential for use of a mitochondrial transduction technology to successfully introduce and express exogenous pathogenic mitochondrial DNA (obtained from a patient afflicted with Leber's hereditary mitochondrial neuropathy); into hNPs devoid of their own mitochondrial DNA. This technology termed "protofection" requires a recombinant engineered protein referred to as TFAM or mitochondrial transcription factor A for introduction of full-length human mtDNA into mitochondria [10, 11]. The "protofection" protein TFAM has been engineered to contain an N-terminal protein transduction domain (PTD)—stretch of 11 arginines—a strong mitochondrial localization sequence (MLS) derived from mitochondrial superoxide dismutase 2 (*SOD2*) gene sequence, yielding the final recombinant protein rhTFAM [11–13]. Our encouraging findings demonstrated that the human mitochondria could be manipulated from outside the cell, and provide the basis for engineering the human neural progenitors to understand the role and effects of pathogenic mtDNA during human development and for providing cell type-specific drug screening platforms in specific childhood and adult neurodegenerative disorders affecting the mitochondria (Fig. 1).

2   Materials

2.1   Human Pluripotent Stem Cell Passaging and Subculture

1. NIH approved WA09 human embryonic stem cell line (http://www.stemcells.nih.gov/research/registry).

2. Human pluripotent stem cell (hPSC) medium (all chemicals from Life Technologies™, Grand Island, NY, unless otherwise labeled): Dulbecco's Modified Eagle's Medium (DMEM/F12) supplemented with 15 % fetal bovine serum (FBS, HyClone, Ogden, UT), 5 % knockout serum replacer (KSR), 1× non-essential amino acids (NEAA), 20 mM l-glutamine, 0.5 U/mL penicillin, 0.5 U/mL streptomycin, 0.1 mM β-mercaptoethanol

(Sigma Chemicals, St. Louis, MO), 4 ng/mL fibroblast growth factor-2 (FGF-2) (Sigma Chemicals). Store at 4 °C and use within 1 week. Protect from light and wide swings in temperature.

3. Mouse embryonic fibroblast (MEF) medium (all chemicals from Life Technologies™, unless otherwise labeled): DMEM-High Glucose (DMEM-HiGlu) supplemented with 10 % FBS, (HyClone), 2 mMl-glutamine, 0.5 U/mL penicillin, 0.5 U/mL streptomycin. Store at 4 °C and use within 2 weeks.

4. 1 mg/mL collagenase: weigh out 10 mg collagenase type IV (Life Technologies™) and dissolve in 10 mL of DMEM/F12 medium supplemented with 15 % FBS, 5 % KSR at 37 °C. Filter sterilize and store at 4 °C and use within 1 week.

5. 0.05 % trypsin solution (Life Technologies™).

6. 1 mM thylenediamine tetraacetic acid (EDTA) (Life Technologies™).

7. Trypan blue (Sigma Chemicals).

2.2 Derivation of Human Neural Progenitors from Human Pluripotent Stem Cells

1. 0.25 % trypsin (Life Technologies™).

2. 1 mM thylenediamine tetraacetic acid (EDTA) (Life Technologies™).

3. Laminin-coated dishes: dissolve 5 µg/mL laminin (Sigma Chemicals) in cold, sterile deionized water. Add corresponding volume (96 well plate: 100 µL/well; 35 mm dish: 2 mL/dish; 6-well plate: 2 mL/well) of laminin solution to the plate size being used. Swirl to ensure the entire surface of the plate or well is covered with the laminin solution. Allow dishes to incubate at room temperature for at least 30 min, but no longer than 1 h before use. Store coated dishes at 4 °C and use within 1 month. *See* **Note 1**.

4. Medium 1 (all chemicals from Life Technologies™): DMEM/F12 supplemented with 20 % KSR, 1 % NEAAs, 1 mMl-glutamine, 0.5 U/mL penicillin, 0.5 U/mL streptomycin. Store at 4 °C and use within 1 month. Protect from light and wide swings in temperature.

5. Medium 2 (all chemicals from Life Technologies™): DMEM/F12 with 1× N2 supplement, 4 ng/mL FGF-2, 2 mMl-glutamine, 0.5 U/mL penicillin, 0.5 U/mL streptomycin. Store at 4 °C and use within 1 month. Protect from light and wide swings in temperature.

6. Medium 3 (all chemicals from Life Technologies™, unless otherwise labeled): Neurobasal medium supplemented with 1× Pen-Strep and 1× B27, 2 mMl-glutamine, 20 ng/mL FGF-2 (Sigma Chemicals), 10 ng/mL LIF (Millipore, Billerica, MA). Store at 4 °C and use within 1 month. Protect from light and wide swings in temperature.

2.3 Passage and Subculture of Human Neural Progenitors

1. hPSC-derived hNPs. *See* **Note 2**.

2. hNP propagation medium (all chemicals from Life Technologies™, unless otherwise labeled): Neurobasal medium supplemented with 1× Pen-Strep and 1× B27, 2 mMl-glutamine, 20 ng/mL FGF-2 (Sigma Chemicals), 10 ng/mL LIF (Millipore). Store at 4 °C and use within 1 month. Protect from light and wide swings in temperature.

3. Poly-ornithine/Laminin (POLA)-coated dishes: Dissolve 20 μg/mL POLA (Sigma Chemicals) 5 μg/mL laminin (Sigma Chemicals) in cold, sterile deionized water. Use in sequential steps for coating dishes. Store coated dishes at 4 °C and use within 1 month. *See* **Note 3**.

2.4 Dideoxycytidine (ddC) Treatment of Human Neural Progenitors

1. 2′,3′-Dideoxycytidine (ddC) (Sigma Chemicals). Store at –20 °C.

2. hNP medium during ddC treatment (all chemicals from Life Technologies™, unless otherwise labeled): Neurobasal medium supplemented with 1× Pen-Strep and 1× B27, 2 mMl-glutamine, 20 ng/mL FGF-2 (Sigma Chemicals), 0.1 mg/mL pyruvate (Sigma Chemicals), 0.05 mg/mL uridine (Sigma Chemicals), 10 ng/mL LIF (Millipore). Store at 4 °C and use within 1 month. Protect from light and wide swings in temperature. *See* **Note 4**.

2.5 Isolation and Introduction of Pathogenic mtDNA into ddC-Human Neural Progenitors

1. Qproteome Mitochondria Isolation Kit (Qiagen, Valencia, CA). *See* **Note 5**.

2. Mitochondrial DNA Extraction Kit (PromoKine, Heidelberg, Germany). *See* **Note 6**.

3. Plasmid-Safe ATP-dependent DNase (Epicentre Biotechnologies, Madison, WI). *See* **Note 7**.

4. UltraClean GelSpin DNA purification kit (MO BIO Laboratories, Carlsbad, CA). *See* **Note 8**.

5. DNA Quant-iT assay kit (Life Technologies™). *See* **Note 9**.

6. E-Gel® Precast Agarose Gels (Life Technologies™). Store at room temperature and protect from light.

7. Recombinant human TFAM (Gencia Corporation, Charlottesville, VA). *See* **Note 10**.

8. Roche Expand long template buffer 3 containing high Mg²⁺ (Roche Applied Sciences, Indianapolis, IN). Store at –20 °C.

2.6 Labeling Pathogenic mtDNA and rhTFAM

1. Rabbit anti-nestin antibody (Millipore).

2. Mitotracker Red or Mitotracker Green (Mitosciences, Eugene, OR, USA). *See* **Note 11**.

3. Appropriate AlexaFluor-488 secondary antibodies (Molecular Probes™, Invitrogen™, Carlsbad, CA).

4. Cy3 dye (Mirus, Madison, WI).

**2.7 Immuno-
cytochemistry**

1. Chamber slides (ThermoFisher Scientific, Rochestor, NY).

2. 10× Phosphate-buffered saline (PBS)$^{++}$ with Ca^{2+} Mg$^+$ (Millipore).

3. 4 % (w/v) paraformaldehyde (PFA, e.g. Sigma Chemicals): Prepare fresh for each experiment. Work in a fume hood and wear gloves, as paraformaldehyde is toxic. Weigh out 4 g PFA and add to glass beaker. Add 75 mL distilled water and place on heated stirrer to dissolve. The solution needs to be carefully heated (use a stirring hot-plate at a temperature ~56 °C) to dissolve the PFA. Add 2 drops of 1 M sodium hydroxide and once all has gone into solution, add 10 mL of 10× PBS^{++}. Make sure that pH is between 7.2 and 7.4. Make up volume to 100 mL with distilled water. Store at 4 °C and use within 1 week.

4. Blocking solution: 3 % (w/v) goat serum (Hyclone Labs) in PBS^{++}. Store at 4 °C and use within 48 h.

5. High salt buffer (HSB): 50 mL 1 M Tris, pH 7.4, 950 mL deionized water, 0.25 M sodium chloride (14.61 g). Store at room temperature and use within 2 months.

6. Permeabilization buffer: add 25 μL of Tween 20 (EMD Chemicals, Darmstadt, Germany) to 50 mL HSB.

7. DAPI.

8. Fluorescence-free mounting medium.

9. Nail varnish.

**2.8 Restriction
Enzyme Digestion**

1. AllPrep DNA/RNA Mini Kit (Qiagen, Valencia, CA). *See* **Note 12**.

2. Quant-IT DNA and RNA assays (Life Technologies™). *See* **Note 13**.

3. iScript™ cDNA Synthesis Kit (BioRad, Hercules, CA). *See* **Note 14**.

4. SfaN1 restriction enzyme (New England BioLabs, Ipswich, MA). Store at –20 °C.

5. PCR primers for ND4 region of the mitochondrial genome. To detect the LHON mutation at G11778A: where G gets replaced by A:

   (a) Forward Primer: 11632-11651 (CAGCCACATAGCCC TCGTAG).

   (b) Reverse Primer: 11843-11862 (GCGAGGTTAGCGAGGC TTGC).

**2.9 Differentiation
of Engineered Human
Neural Progenitors
into Neurons**

1. hNP differentiation medium (all chemicals from Life Technologies™): Neurobasal medium supplemented with 1× Pen-Strep and 1× B27, 2 mM l-glutamine. Store at 4 °C and use within 1 month. Protect from light and wide swings in temperature.

2. Anti-β-tubulin III antibody (e.g. R&D Systems, Minneapolis, MN).

2.10 Equipment	1. Dissection microscope.

2.10 Equipment

1. Dissection microscope.
2. Cell culture incubator.
3. Centrifuge.
4. Hemocytometer.
5. Water bath.
6. Light microscope with phase contrast optics.
7. Aspirator for fluids.
8. DNA gel electrophoresis apparatus (an automated apparatus is preferable, e.g. Experion, BioRad).
9. Fluorescence microscope with confocal capabilities (e.g. Olympus IX70, Olympus, Center Valley, PA).
10. PCR thermocycler.

---

3 Methods

3.1 Human Pluripotent Stem Cell Passaging and Subculture

1. Use at least 3-day-old MEF plates (1.2×10^5 cells/cm^2), aspirate off medium and replace with 2 mL of hPSC medium. Place dish at 37 °C until ready to plate out cells.

2. Add 1 mL of collagenase solution per 35 mm dish containing hESC colonies. Place on 37 °C stage for 2–3 min. Colonies can be observed rounding under dissection microscope.

3. Aspirate off collagenase solution and add 1 mL 0.05 % trypsin solution.

4. Allow trypsin to contact cells for no more than 40 s, then aspirate off the trypsin solution.

5. Add 1 mL of 10 % FBS in DMEM/F12 to 35 mm dish and begin to gently pipette up and down to dislodge or knock off and break up cell clumps, while continuously observing cells under microscope.

6. Place harvested cells in 15 mL tube containing 8 mL hPSC medium.

7. Add fresh medium to the dish and wash off and collect any remaining trypsinized cells, add to 15 mL tube.

8. Spin harvested cells for 4 min at $200 \times g$ at room temperature.

9. Resuspend trypsinized pellet in 2 mL of hPSC medium per 35 mm dish used.

10. Count cells using hemocytometer by taking 10 µL cell suspension and mixing with 10 µL Trypan blue.

11. Aspirate medium from pre-equilibrated dishes and plate cells at 150,000 cells per 35 mm dish in 2 mL medium. Place at 37 °C in a 5 % CO$_2$ incubator. Evenly distribute cells on the plate via a uni-directional quick movement of the plate and place it back on the shelf in the incubator (do not swirl cells in the plate).

12. Feed everyday with a 50 % medium change until ready to passage again in 3–4 days. The cells at this stage should exhibit distinct colony morphology (high nuclear to cytoplasmic ratio) characteristic of hPSCs.

13. hPSCs can be maintained in suspension prior to use in immunocytochemical and/or flow cytometry analysis.

14. For immunocytochemical analysis, hPSCs propagated under routine conditions are passaged and transferred into MEF-seeded chamber slides. Medium is changed every day prior to fixing for immunocytochemical analysis (*see* Subheading 3.5).

3.2 Derivation of Human Neural Progenitors from Human Pluripotent Stem Cells

1. Trypsinize hPSCs using 0.25 % trypsin in EDTA solution for 1 min.

2. Count cells using hemocytometer and transfer 3×10^5 hPSCs in fresh hPSC medium to a 35 mm laminin-coated plate.

3. After 24 h, replace hPSC medium with Medium 1, with medium change every day for a period of 7 days.

4. After 7 days in Medium 1, replace with Medium 2, with medium change every day for a period of 7 days.

5. After 7 days in Medium 2, replace with Medium 3, with medium change every day for a period of 7 days.

6. After 3 weeks of differentiation, hNPs can be sub-cultured in hNP propagation medium as detailed below.

3.3 Passage and Subculture of Human Neural Progenitors

1. Thaw hNPs by rotating the cryovial in a 37 °C water bath.

2. Swirling slowly, add 10 mL of pre-warmed hNP propagation medium to the cell suspension in a sterile 15 mL tube.

3. Spin cells for 4 min at $200 \times g$ at room temperature.

4. Aspirate supernatant and resuspend cells in 2 mL of hNP propagation medium.

5. Plate cells at 1×10^6 cells per 35 mm POLA-coated dish. Place at 37 °C in a 5 % CO_2 incubator. Replace medium everyday.

6. Once the hNPs reach 90 % confluence, carefully remove the media.

7. Add pre-warmed hNP propagation medium to the cells. Triturate the cells with a pipette to manually detach them from the dish. Alternatively, the cells can be detached using a cell scraper.

8. Observe the dishes under a bright field or a phase contrast microscope to ensure that all cells have been removed. If necessary, the cells can be centrifuged at $200 \times g$ for 4 min in order to resuspend the pellet at a specific concentration.

9. It is recommended to subculture the cells by passaging 1:2 or 1:3 to maintain a high density.

10. hNP cells can be maintained in suspension, prior to use in immunocytochemical analysis.

11. For immunocytochemical analysis, propagate hNPs in chamber slides coated with POLA to provide a similar culture condition as on regular dishes.

3.4 Dideoxycytidine (ddC) Treatment of Human Neural Progenitors

1. Treat hNPs with 10 μM 2′,3′ ddC solution for reduction of endogenous mtDNA, for a period of 6 days. *See* **Note 15**.

2. Use immunocytochemical analyses to determine expression of hNP markers in ddC-treated hNPs (Subheading 3.5).

3.5 Immunocytochemistry

3.5.1 Fixation

1. Work in a fume hood. Wash once cells obtained after completing Subheadings 3.1 or 3.2, in PBS⁺⁺. Use aspirator to remove and a transfer pipette to add PBS.

2. Add enough PFA solution (~0.8 mL/1.8 cm²) to cover bottom of well of chamber slides or dish.

3. Let sit at room temperature for 15–20 min.

4. Wash cells 3 times in 1 mL of PBS⁺⁺.

5. Store fixed cells at 4 °C until ready to stain with chosen markers.

3.5.2 Localization of Specific Markers in Human Neural Progenitors by Immunocytochemical Analysis

1. Wash once fixed cells in wells of chamber slides (from Subheading 3.3) with 1 mL of PBS⁺⁺. Use aspirator to remove and transfer pipette to add PBS.

2. If staining for intracellular markers, wash cells with permeabilization buffer 3 times for 5 min each.

3. Add 0.8 mL blocking solution to each well. Incubate at room temperature for 45 min.

4. Dilute the primary antibody in 1 mL of blocking solution at supplier recommended dilution (for anti-nestin antibody 1:200).

5. Aspirate off blocking solution from wells and add 300 μL of diluted primary antibody at optimal titer. Cover to prevent exposure to light and incubate 1 h at room temperature. This can be extended to overnight at 4 °C, if necessary.

6. Wash cells 4 times in HSB (5 min each) if staining for intracellular markers. Wash cells 4 times in PBS⁺⁺ (5 min each) if staining for cell surface markers.

7. Whilst completing washes in step 6, prepare the secondary antibody in blocking solution at supplier recommended dilution (for AlexaFluor 1:1,000).

8. Aspirate off last wash from wells and add 300 μL of diluted secondary antibody at optimal titer. Incubate 1 h at room temperature. During incubation, cover sample with aluminum foil to prevent fluorescence bleaching.

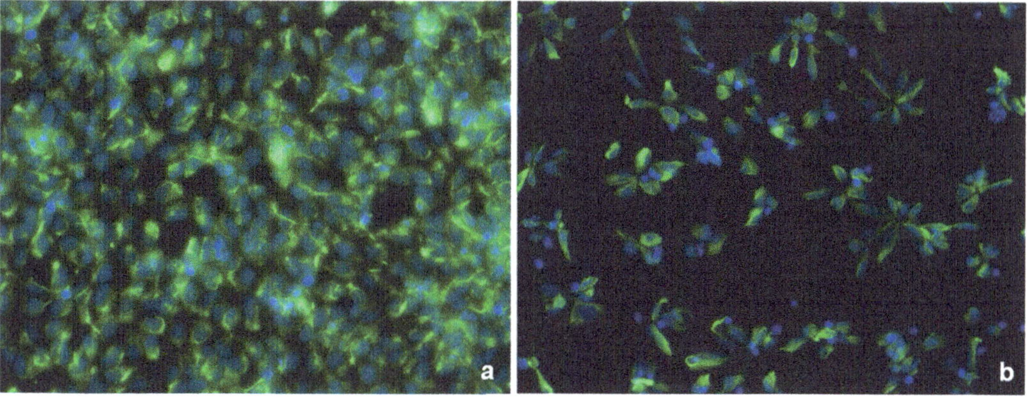

Fig. 2 Immunostained images of (**a**) hNPs and (**b**) ddC-treated hNPs. The cultures were highly homogeneous and exhibit neural rosette morphology, that stain positive for Nestin (*green*). DAPI (*blue*) was used for staining the nuclei. Magnification 20×

9. Wash wells 4 times in 1 mL of PBS^{++} for 5 min each wash.

10. Add 0.8 mL of a 1:10,000 dilution of DAPI in distilled H$_2$O to each well. Incubate for 5 min at room temperature. Cover with foil during incubation, to prevent exposure to light.

11. Wash cells 3 times in 1 mL of PBS^{++}.

12. Verify that staining has taken place under fluorescence microscope before mounting.

13. Gently remove sides of chamber and aspirate excess surrounding PBS.

14. Place one drop of mounting media directly in the center of each well area.

15. At an angle gently lower a cover slip onto the slide trying to avoid air bubbles where possible. Remove excess mounting media from slide and seal with nail varnish on all 4 sides.

16. Keep in dark storage until results are observed/documented. It is recommended to document the same day. Examples of nestin expression in hNPs and ddC-treated hNPs are shown in Fig. 2.

3.6 Preparation of LHON Pathogenic mtDNA and Introduction of LHON Pathogenic mtDNA into ddC-Treated Human Neural Progenitors

1. Propagate LHON cybrid cells containing a >96 % homoplasmic G11778A mutation in the ND4 region of mitochondrial genome in DMEM containing 10 % FCS.

2. Isolate mitochondrial DNA from the LHON cybrid cells using Qproteome Mitochondria Isolation Kit and the Mitochondrial DNA Extraction Kit according to the manufacturer's instructions.

3. Treat the isolated mitochondrial DNA with limiting amounts of Plasmid-Safe ATP-dependent DNase, which selectively digests all forms of DNA but does not affect closed circular or nicked circular double-stranded DNA.

4. Purify the isolated mitochondrial DNA according to protocols for the UltraClean GelSpin DNA purification kit. *See* **Note 16**.

5. Quantify the DNA solution with DNA Quant-iT assay kit and visualize the linear band on a 0.8 % agarose gel.

6. For introduction into ddC-hNPs, the rhTFAM (+/− LHON mtDNA) prepare mix as follows: incubate at 37 °C for 30 min the purified rhTFAM to 3 mg of LHON mtDNA in PBS containing high salt Mg^{2+} buffer. *See* **Note 17**.

7. Incubate ddC-hNPs in growth medium containing rhTFAM complexed with LHON pathogenic mtDNA for 2 h at 37 °C. Controls involve incubating the hNPs in growth medium containing only rhTFAM or PBS.

8. After the incubation period, replace the medium with regular growth medium and use cells in different experimental analyses as outlined below.

3.7 Imaging of LHON mtDNA Entry into ddC-Treated Human Neural Progenitors

1. For real-time imaging of LHON mtDNA entry into ddC-treated hNPs, propagate cells to 80 % confluence on laminin-coated 35 mm glass bottom dishes.

2. Stain Mitochondria in hNPs using either Mitotracker Red or Mitotracker Green depending on whether tracking is being conducted for Alexa488-rhTFAM or Cy3 labeled LHON-mtDNA complexed with unlabeled rhTFAM, respectively.

3. Purify rhTFAM by dialyzing with 1× PBS buffer and label with Alexa 488 dye according to the manufacturer's instructions. *See* **Note 18**.

4. Label circular mtDNA carrying the LHON G11778A mutation using Cy3 dye according to the manufacturer's instruction.

5. Mix the labeled rhTFAM or the labeled LHON mtDNA complexed with MTD-TFAM in 2 mL DMEM and add to independent dishes of hNPs with labeled mitochondria. *See* **Note 19**.

6. Obtain real-time single plane-time lapse images every 5 min using a confocal microscope.

3.8 Restriction Enzyme Digestion Analysis

1. Amplify a PCR product in the ND4 gene spanning the SfaN1 site removed by the G11778A mutation from the cDNA of hNPs, and digest with SfaN1 to check for the presence or absence of the mutation.

2. Digest cDNA PCR products and analyze using an automated electrophoresis apparatus.

3.9 Differentiation Potential of Engineered hNP Cells into Neurons

1. Allow the engineered hNPs to attain 100 % confluency in regular growth medium.

2. Replace the propagation medium with differentiation medium (*see* Subheading 2.8) for a period of 14 days.

3. For immunocytochemical analysis, propagate hNPs in chamber slides coated with POLA to provide a similar culture condition as on regular dishes. After differentiation for 14 days.

4. Subject the neurons obtained to immunocytochemical analysis for β-tubulin-III expression.

4 Notes

1. Laminin-coated dishes provide the appropriate surface for derivation of hNPs.

2. It is recommended to use WA09-derived [STEMEZ hNP™] human neural progenitors (http://www.arunabiomedical.com) during initial experimentation, and for comparison when deriving hNPs from hPSCs.

3. POLA-coated plates provide the appropriate surface for propagation of hNPs.

4. The pyruvate/uridine mixture is normally made at 1,000× in PBS and aliquoted as 500 μL samples and stored at –20 °C.

5. The Qproteome Mitochondria Isolation kit contains buffers and reagents for isolation of high-purity mitochondria preparations from cells. Lysis Buffer should be stored at –20 °C upon arrival. All other buffers and Protease Inhibitor Solution (100×) should be stored at 2–8 °C.

6. The Mitochondrial DNA Extraction kit contains 5× Cytosol Extraction Buffer, Mitochondrial Lysis Buffer, Enzyme B Mix (lyophilized), TE Buffer. After opening the kit, store Enzyme B Mix at –70 °C. Store all other buffers at 4 °C. Make 1× Cytosolic Extraction Buffer by mixing the 1 mL of the 5× buffer with 4 mL ddH$_2$O. Add 275 μL of TE buffer to Enzyme B Mix. Mix well, aliquot and refreeze immediately at –70 °C. Stable for up to 3 months at –70 °C. All buffers should be placed on ice at all times during the experiment.

7. Plasmid-Safe DNase is supplied in a 50 % glycerol solution containing 50 mM Tris–HCl pH 7.5, 0.1 M NaCl, 0.1 mM EDTA, 1 mM dithiothreitol (DTT), 0.1 % Triton® X-100. One unit degrades 1 nmol of deoxynucleotides in linear dsDNA in 30 min at 37 °C in 1× Plasmid-Safe Reaction Buffer and 1 mM ATP. Plasmid-Safe 10× Reaction Buffer contains: 330 mM Tris-acetate (pH 7.5), 660 mM potassium acetate, 100 mM magnesium acetate, 5.0 mM DTT. ATP is required for Plasmid-Safe DNase activity and should be added to a final concentration of 1 mM. Store only at –20 °C in a freezer without a defrost cycle.

8. The UltraClean GelSpin DNA purification kit uses an economical silica binding particle method to purify DNA from agarose gels. Kit reagents and components should be stored at room temperature (15–30 °C).

9. The DNA Quant-iT assay kit provides concentrated assay reagent, dilution buffer, and pre-diluted DNA standards. Store in refrigerator (2–8 °C) and protect from light.

10. Store the purified protein in Tris-buffered 50 % glycerol at –20 °C.

11. Use Mitotracker Red for staining and tracking mitochondria as Alexa488 is used for tracking MTD-TFAM. Use Mitotracker Green for tracking Cy3-labeled pathogenic mtDNA complexed with unlabeled MTD-TFAM.

12. The AllPrep DNA/RNA Mini Kit contains buffers and reagents and should be stored at room temperature (15–25 °C). It is stable for at least 9 months under these conditions.

13. The Quant-IT DNA and RNA assay kit provides concentrated assay reagent, dilution buffer, and pre-diluted DNA standards. Store in refrigerator (2–8 °C) and protect from light.

14. The iScript™ cDNA Synthesis Kit is a modified MMLV-derived reverse transcriptase, optimized for reliable cDNA synthesis over a wide dynamic range of input RNA. Store the enzyme at –20 °C. Nuclease-free water can be stored at room temperature.

15. ddC is a potent inhibitor of mitochondrial DNA polymerase γ. Loss of cell viability is expected during ddC treatment of hNPs. Care should be taken in replacing growth medium supplemented with pyruvate and uridine once every 24 h.

16. Treated DNA can be further purified by ethanol precipitation, spin columns, or organic extraction.

17. The presence of the high salt Mg^{2+} buffer is critical for facilitating protein binding to the DNA.

18. It is recommended to assess DNA-binding capacity of rhT-FAM by electrophoretic mobility shift assay (EMSA) that is based on maximum retardation of circular DNA.

19. A ratio of 500 ng of circular DNA complexed with 1 µg of rhTFAM is recommended.

Acknowledgements

This work was supported in part by start-up funds provided to Dr. Iyer by the Center for the Study of Biological Complexity at Virginia Commonwealth University and a grant from the Presidential Research Innovation Program (PRIP) at Virginia

Commonwealth University. The authors would also like to thank Gencia Corporation, Charlottesville for provision of the recombinant human TFAM. Mention of vendor names is for information only and does not imply endorsement.

References

1. Thomson JA, Itskovitz-Eldor J, Shapiro SS et al (1998) Embryonic stem cell lines derived from human blastocysts. Science 282:1145–1147

2. Yu J, Vodyanik MA, Smuga-Otto K et al (2007) Induced pluripotent stem cell lines derived from human somatic cells. Science 318:1917–1920

3. Takahashi K, Tanabe K, Ohnuki M et al (2007) Induction of pluripotent stem cells from adult human fibroblasts by defined factors. Cell 131: 861–872

4. Shin S, Mitalipova M, Noggle S et al (2006) Long-term proliferation of human embryonic stem cell-derived neuroepithelial cells using defined adherent culture conditions. Stem Cells 24:125–138

5. Dhara SK, Hasneen K, Machacek DW et al (2008) Human neural progenitor cells derived from embryonic stem cells in feeder-free cultures. Differentiation 76:454–464

6. Shin S, Dalton S, Stice SL (2005) Human motor neuron differentiation from human embryonic stem cells. Stem Cells Dev 14:266–269

7. Wilczynska KM, Singh SK, Adams B et al (2009) Nuclear factor I isoforms regulate gene expression during the differentiation of human neural progenitors to astrocytes. Stem Cells 27:1173–1181

8. Iyer S, Alsayegh K, Abraham S et al (2009) Stem cell-based models and therapies for neurodegenerative diseases. Crit Rev Biomed Eng 37:321–353

9. Wilson PG, Cherry JJ, Schwamberger S et al (2007) An SMA project report: neural cell-based assays derived from human embryonic stem cells. Stem Cells Dev 16:1027–1041

10. Khan SM, Bennett JP Jr (2004) Development of mitochondrial gene therapy. J Bioenerg Biomembr 36:387–393

11. Iyer S, Thomas RR, Portell FR et al (2009) Recombinant mitochondrial transcription factor A with N-terminal mitochondrial transduction domain increases respiration and mitochondrial gene expression. Mitochondrion 9:196–203

12. Iyer S, Bergquist K, Young K et al (2012) Mitochondrial gene therapy improves respiration, biogenesis, and transcription in G11778A Leber's hereditary optic neuropathy and T8993G Leigh's syndrome cells. Hum Gene Ther 23:647–657

13. Iyer S, Xiao E, Alsayegh K et al (2012) Mitochondrial gene replacement in human pluripotent stem cell-derived neural progenitors. Gene Ther 19:469–475

Chapter 20

Herpes Simplex Virus Type 1 (HSV-1)-Derived Recombinant Vectors for Gene Transfer and Gene Therapy

Peggy Marconi, Cornel Fraefel, and Alberto L. Epstein

Abstract

Herpes simplex virus type 1 (HSV-1) is a human pathogen whose lifestyle is based on a long-term dual interaction with the infected host, being able to establish both lytic and latent infections. The virus genome is a 153-kilobase pair (kbp) double-stranded DNA molecule encoding more than 80 genes. The interest of HSV-1 as gene transfer vector stems from its ability to infect many different cell types, both quiescent and proliferating cells, the very high packaging capacity of the virus capsid, the outstanding neurotropic adaptations that this virus has evolved, and the fact that it never integrates into the cellular chromosomes, thus avoiding the risk of insertional mutagenesis. Two types of vectors can be derived from HSV-1, recombinant vectors and amplicon vectors, and different methodologies have been developed to prepare large stocks of each type of vector. This chapter summarizes the approach most commonly used to prepare recombinant HSV-1 vectors through homologous recombination, either in eukaryotic cells or in bacteria.

Key words HSV-1, Recombinant vectors, Homologous recombination, Transfection, Virus, Plasmid

1 Introduction

1.1 Herpes Simplex Virus Type 1 and Its Derived Vectors

1.1.1 HSV-1

HSV-1 is a major human pathogen whose lifestyle is based on a long-term dual interaction with the infected host. After initial infection and lytic multiplication at the body periphery, generally in oral or genital epithelial cells, the virus enters the sensory neurons that innervate the infected epithelia and, following retrograde transport of the capsids to the cell bodies, establishes a lifelong latent infection in sensory ganglia. Periodic reactivation from latency usually leads to the return of the virus to epithelial cells, where it produces secondary lytic infections (recurrences) resulting in mild illness symptoms, such as cold sores. Often, infectious virus can be detected in the saliva of people presenting no clinical symptoms of disease. In rare cases, HSV-1 can spread

Laura Lossi and Adalberto Merighi (eds.), *Neuronal Cell Death: Methods and Protocols*, Methods in Molecular Biology, vol. 1254, DOI 10.1007/978-1-4939-2152-2_20, © Springer Science+Business Media New York 2015

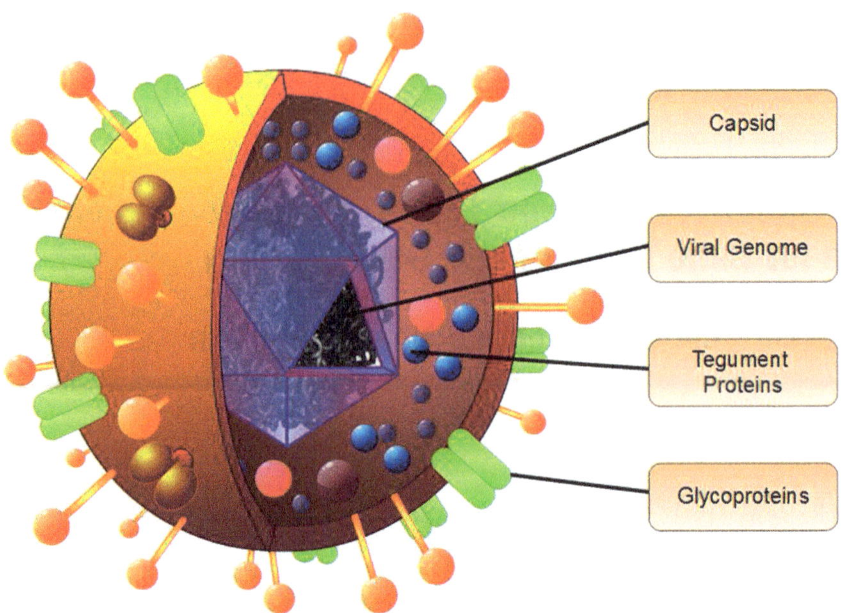

Fig. 1 The HSV-1 particle is composed of four layers: the lipid envelope carrying 12 virus encoded glycoproteins, the tegument layer, which is composed of some 20 different virus-encoded protein species, then the icosahedral capsid, which encloses the double-stranded DNA genome

centripetally into the central nervous system, to cause devastating encephalitis. For a comprehensive review on HSV-1 lytic and latent cycles, see reference [1].

The HSV-1 particle is made up of four layers. The viral genome is a 153-kbp double-stranded DNA molecule, enclosed in an icosahedral capsid that is surrounded by the tegument, a rather unstructured layer containing some 20 virus-encoded proteins. The tegument is delimited by the envelope, which is a lipid membrane of cellular origin, containing a dozen virus-encoded glycoproteins (*see* Fig. 1). HSV-1 enters epithelial cells and neurons by fusion of the virus envelope with the plasma or endosomal membranes, and the virus capsids are transported to the nuclear pores through association with microtubules, from where the viral DNA is released into the nucleus [2]. During lytic infection, the viral genome expresses more than 80 genes that are temporarily regulated in a cascade fashion, giving rise to three phases of gene expression. The expression cascade, which is regulated mainly at the transcriptional level, begins with the expression of the immediate-early (IE) genes. Four of these IE genes encode regulatory proteins (ICP0, ICP4, ICP22, and ICP27) that are responsible for controlling viral gene expression during subsequent early and late phases of the replication cycle and for inducing shutoff of cellular protein synthesis [3]. Transcription of IE genes occurs in the absence of de novo viral protein synthesis and is highly stimulated by a virion protein known as VP16, which is a powerful transcription factor that, in conjunction with cellular proteins, acts on DNA motifs present only in the IE regulatory regions to up-regulate expression [4]. The early gene products that are synthesized next include enzymes that, like thymidine kinase and ribonucleotide reductase, act to increase the pool of deoxynucleotides of the infected cells, and several replication proteins that are directly involved in viral DNA synthesis. The last temporal class of genes expressed are the late genes, which encode proteins involved in the packaging of virus DNA, as well as the structural proteins involved in the assembly of the virion particle, including the capsid, the tegument, and the glycoproteins [1]. Some of these structural proteins, such as the tegument protein VP16, play major regulatory roles in the next infectious cycle. Capsids are assembled in the nucleus but the tegument and the envelope are acquired in the cytoplasm, most probably by budding into endosomes, and are released by exocytosis at cell membranes [5].

During latency in sensory neurons, the viral genome remains as a circular chromatinized episome [6] within the cell nuclei, and undergoes dramatic changes resulting in an almost complete silencing of transcription. Only one region of the viral genome, known as the latency-associated transcript (LAT) locus, is actively transcribed during latency, due to the presence of a latency-associated promoter (LAP) that remains active during this phase of the infection, resulting in the synthesis of non-messenger RNA molecules, which accumulate in the nucleus of the latently infected neurons [7]. Recently, the LAT locus has been shown to express miRNA molecules that can down-regulate expression of lytic viral genes [8]. The latent virus genome can reactivate in response to a wide variety of stimuli that allow it to enter the lytic phase of the HSV-1 life cycle. For a more specific review on HSV-1 latency, see [9].

1.1.2 HSV-1-Derived Vectors

HSV-1-derived vectors have the capacity to deliver up to 150-kbp of foreign DNA to the nucleus of most proliferating and quiescent mammalian cells, making this family of vectors a very interesting tool for gene transfer and gene therapy. The uniqueness of HSV-1-based vectors stems from several properties of HSV-1: (1) the large capacity of the virion, which allows it to accommodate a very large genome, (2) the virus can infect many different cell types, both quiescent and proliferating mammalian cells, (3) the virus DNA will not integrate into host chromosomes, thus reducing the risk of insertional mutagenesis, (4) the complexity of the virus genome, which contains approximately 40 genes that are not essential for virus replication and can therefore be replaced by transgene sequences without disturbing virus production in cultured cells, (5) the efficiency to infect cells of the nervous system, (6) the ability to *trans*-synaptically spread from neuron to neuron in both anterograde and retrograde directions, and (7) the capacity to establish latent infections in neurons.

Three different types of vectors can be derived from HSV-1, which attempt to exploit one or more of the above properties, attenuated recombinant HSV-1 vectors, defective recombinant HSV-1 vectors, and HSV-1-derived amplicon vectors. For a recent comprehensive review article on HSV-1-based vectors, see [10] and references therein. This chapter will develop only the methods currently used to generate, produce, and titrate, both attenuated and defective recombinant HSV-1 vectors.

1.1.3 Replication-Defective Recombinant HSV-1 Vectors

Replication-defective recombinant HSV-1 vectors are created by either mutating or deleting essential genes for viral replication. Therefore, these mutants can grow only in specifically designed transformed cell lines, where they are complemented *in trans.* To date, several replication-defective HSV-1 vectors have been constructed in which the immediate-early (IE) genes, expressing regulatory-infected cell proteins (ICP) 0, 4, 22, 27, have been deleted in various combinations [11–14]. ICP4 and ICP27 are essential for replication and the deletion of one or both of these genes requires adequate complementing cell lines, such as the Vero-7b cell line [15], capable of providing *in trans* the proteins encoded by deleted viral genes. Defective HSV-1 recombinants are currently being used in gene therapy protocols to treat cancer-associated pain [16].

1.1.4 Attenuated Recombinant HSV-1 Vectors

These vectors carry deletions in one or more non-essential genes, resulting in viruses that retain the ability to replicate in vitro, but are compromised in vivo, in a context-dependent manner [17, 18]. Among the limitations to the use of HSV-1 is the fact that wild-type virus is highly pathogenic and entry in the brain can cause fatal encephalitis. Toxic and/or pathogenic properties of the virus must, therefore, be disabled prior its use as a gene delivery vector. Several genes involved in HSV-1 replication, virulence and immune evasion, non-essential for the viral life cycle in vivo, have been identified. These genes are usually involved in multiple interactions with cellular proteins, which optimize the ability of the virus to grow within cells. Understanding such interactions has permitted the deletion/modification of these genes, alone or in combination, to create virus mutants with a reduced ability to replicate in normal quiescent cells, but that can replicate in tumor or dividing cells [19] and which are currently used in clinical protocols to treat a variety of cancers, with encouraging results.

In spite of their fundamental biological differences, defective and attenuated recombinant HSV-1 vectors are constructed and prepared using very similar methodologies. The only significant practical difference is the need to use complementing cell lines to produce the defective recombinants. Classically, recombinant HSV-1 vectors were constructed by homologous recombination in eukaryotic cells, by co-transfecting the virus genome and a plasmid carrying the transgene of interest flanked by sequences homologous to the target locus on the HSV-1 genome to facilitate recombination. A more recent approach uses an HSV-1 genome cloned as a bacterial artificial chromosome. There, the transgene of interest is introduced into the virus genome by homologous recombination in bacteria. We will describe here both approaches.

1.2 Construction of Recombinant HSV-1 Vectors by Homologous Recombination in Eukaryotic Cells

Alterations of the HSV-1 genome in eukaryotic cells can be achieved in a number of ways. These usually require a two-step process in which portions of the virus genome, which have been cloned into plasmid vectors, are first altered in vitro; then the modified sequences are introduced into the virus genome and recombinant viruses are selected. Several methods have been described to insert DNA sequences into the viral DNA. Recombination into specific sites within the viral genome has been achieved in vitro using a site-specific recombination system derived from phage P1 [20–22]. It is possible, however, to significantly enhance the frequency of recombination using a two-step method through homologous recombination in cultured cells [23]. The first step is the insertion of a reporter gene cassette flanked by PacI or PmeI restriction enzymes sites not otherwise found in the viral genome. Green fluorescent protein (GFP) and LacZ are two convenient marker genes that allow easy selection of the mutated virus genome (Fig. 2). The second step is the substitution of the reporter gene with a second foreign DNA, carrying the transgene of interest, by digestion of the vector DNA with PacI or PmeI to remove the reporter gene and subsequent repair of the vector genome by homologous recombination with a transgene expression plasmid. Potential recombinants identified by a "clear plaque" phenotype (not expressing GFP or LacZ, Fig. 2) arise at high frequency (80–100 %). For details on the construction of recombinant vectors by homologous recombination in eukaryotic cells, refer to Subheadings 2.1 and 3.1.

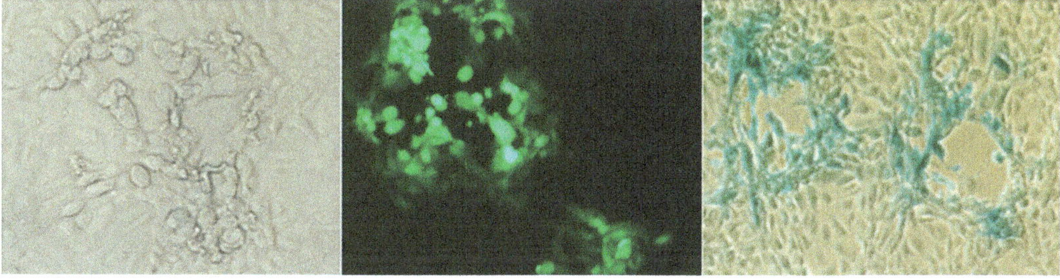

Fig. 2 Plaque phenotype of a recombinant HSV-1 expressing GFP (*middle panel*) and/or β-galactoside (*right panel*) reporter genes. Compare with the "clear plaque" phenotype (*left panel*)

1.3 Construction of Recombinant HSV-1 Vectors by Homologous Recombination in Bacteria: ET Recombination and GalK-Positive/Negative Selection

The cloning of large DNA virus genomes, such as that of HSV-1 [24, 25] as bacterial artificial chromosomes (BAC), has facilitated the easy construction of recombinant viruses by homologous recombination in *E. coli*. The protocol below describes the construction of recombinant HSV-1 by using the λ prophage homologous recombination system (red α and red β genes) and *galK* selection. The *galK* selection method is a two-step system: In the first step, a *galK* cassette, flanked by at least 50 nucleotides of homology to specified positions on the HSV-1 BAC DNA, is inserted via homologous recombination into the BAC (*galK-positive* selection). In the second step, the *galK* cassette is replaced by homologous recombination with an oligonucleotide or PCR product that contains appropriate homology arms and selection against *galK*. This method allows constructing a recombinant HSV-1 within 2–3 weeks. For details on the construction of recombinant HSV-1 vectors by homologous recombination in bacteria, refer to Subheadings 2.2 and 3.2.

2 Materials

2.1 Construction of Recombinant HSV-1 Vectors by Homologous Recombination in Eukaryotic Cells

1. T-175 cm^2 and T-75 cm^2 flasks, blue filter cup (Nunc A/s, Roskilde, Denmark).

2. 24-, 48-, 96-well cell culture cluster, polystyrene tissue culture treated (Corning Incorporated, Corning, NY).

3. 60 mm × 15 mm dish polystyrene (Corning Incorporated).

4. 10 mL tubes: PST test tubes with two-position closure cap, individually wrapped (Artiglass, Padova, Italy).

5. Pipette tips for p20, p200, p1000 with filter (Artiglass).

6. Tips for viral DNA with filter.

7. 2, 5, 10, 25 mL disposable serological pipettes (Corning Incorporated).

8. 15 and 50 mL centrifuge screw cap tube, polypropylene (Corning Incorporated).

9. 1.5 mL Eppendorf safe-lock tubes.

10. Cell scrapers 18 cm handle/1.8 cm blade and 25 cm handle/1.8 cm blade (BD Falcon, BD Biosciences, San Jose, CA).

11. 50 mL Reagent Reservoir, polystyrene (Costar®, Corning Incorporated Life Sciences, Tewksbury, MA).

12. Stericup, vacuum disposable filtration system, 0.22 μm and 0.45 μm membrane (Merk Millipore, Billerica, MA).

13. Phenol:Chloroform:Isoamyl alcohol (25:25:1) (Ambion®, Life Technologies™, Carlsbad, CA).

14. Chloroform:Isoamyl alcohol (25:1) (Ambion®, Life Technologies™).

15. Geneticin G418 (Roche, Basel, Switzerland).

16. Proteinase K (20 mg/mL) (Euroclone, Pero, Italy).

17. Trypsin: 0.25 % trypsin/0.02 % EDTA (Life Technologies™).

18. Sterile Glycerol (Sigma, St. Louis, MO).

19. Tris-EDTA buffer (TE): 10 mM Tris–HCl, pH 8.0, 1 mM EDTA.

20. Vero cells (African green monkey kidney) (American Type Culture Collection, ATCC, Teddington, UK). Complementing cell lines (such as VERO-7b cells) are required to propagate recombinant HSV-1 vectors with deletions in essential genes.

21. Cell culture medium: Dulbecco's Modified Eagle's Medium (DMEM) high glucose (Life Technologies™) supplemented with 10 % fetal bovine serum (FBS, Thermo Scientific™ HyClone™, Rockford, IL) and 2 mM glutamine, 100 units/mL penicillin, 100 μg/mL streptomycin.

22. 0.5 M EDTA Stock Solution: Dissolve 16.81 g of EDTA in 90 mL distilled water. Adjust pH to 8.0 with NaOH. Adjust volume to 100 mL with H_2O. Store at room temperature.

23. 1× PBS (Phosphate Buffered Saline): Dissolve 8 g of NaCl, 0.2 g of KCl, 1.44 g of Na_2HPO_4, 0.24 g of KH_2PO_4 in 800 mL distilled H_2O. Adjust pH to 7.4 and bring the volume to 1 L with additional distilled H_2O. Sterilize by autoclaving. Store at room temperature.

24. 1× TBS (Tris Buffered Saline): Dissolve 8 g of NaCl, 0.2 g of KCl, 3 g of Tris base in 800 mL of distilled H_2O. Adjust the pH to 7.4 with HCl and bring the volume to 1 litter with additional distilled H_2O. Sterilize by autoclaving. Store at room temperature.

25. Lysis buffer: 10 mM Tris–HCl, pH 8.0, 10 mM EDTA, 0.6 % SDS. The proteinase K 0.25 mg/mL is added at the moment of the use.

26. 10 % SDS (Sodium Dodecyl Sulfate) solution: 10 % (w/v) in distilled H_2O. Sterilize by passage through a 0.22 μm Stericup filter. Store at room temperature. Note: precipitation of SDS is not unusual, warm gently to re-dissolve.

27. 2× HBS (HEPES Buffered Saline): Dissolve in 800 mL distilled H_2O 5 g HEPES (0.021 M), 0.125 g $Na_2HPO_4 \cdot 2H_2O$* (0.702 mM), 8 g NaCl (0.137 M), 0.37 g KCl (5 mM), 1 g glucose** (5.6 mM). Adjust pH to 7.05 with NaOH 5 M and bring the volume to 1 L with additional distilled H_2O. Accurate pH of this solution is critical. Sterilize by passage through a 0.22-μm Stericup filter. Store at 4 °C.

   *(0.188 g if is Na_2HPO_4 7 hydrate, 0.251 g if is Na_2HPO_4 12 hydrate, 0.09966 g if is Na_2HPO_4 anhydrous; **1.11 g if is glucose monohydrate).

28. 2 M CaCl$_2$: Dissolve 29.4 g CaCl$_2$ in 70 mL distilled H$_2$O. Adjust the volume to 100 mL with additional distilled H$_2$O. Sterilize by passage through a 0.2-µm filter. Store in aliquots at –20 °C.

29. 1.5 % Methylcellulose overlay: add 1.5 g of methylcellulose to 100 mL PBS, pH 7.5 in a sterile bottle containing a stir bar. Autoclave the bottle on liquid cycle for 45 min. After the solution cools, add 350 mL of DMEM high glucose supplemented with 2 mM glutamine, 100-units/mL penicillin, 100-µg/mL streptomycin. Mix well, place the bottle on a stir plate at 4 °C overnight or until the methylcellulose is completely dissolved. Once the methylcellulose has been solubilized, add 50 mL of fetal bovine serum.

30. 1 % Crystal violet in solution (50:50 methanol: distilled H$_2$O v/v).

2.1.1 Materials and Solutions for Viral Stock Preparation and OptiPrep Gradient

Materials

1. T150–175 cm^2 tissue culture flasks (Nunc A/s).

2. 15 and 50 mL centrifuge screw cap tube (Corning Incorporated).

3. 50-mL tubes copolymer (Nalge Nunc International Corporation, Rochester, NY).

4. Polyallomer centrifuge tubes and plugs 5/8 × 2¾ in. 11.2 mL capacity (Opti Seal™, Beckman Coulter, Inc, Brea, CA).

5. Needles: 18G 1½ in.

6. Syringes: 10 cc.

7. Sonicator: e.g. Misonix Sonicator Ultrasonic Processor (Cole-Parmer International, Vernon Hills, IL).

8. 2½″ Cup Horn for sonicator (Misonix).

9. Centrifuge: e.g. Beckman Avanti J25 (Beckman Coulter).

10. Ultracentrifuge: e.g. Optima LE-80K (Beckman Coulter).

11. Rotor JA-20 Beckman Coulter.

12. Rotor Vti65.1 Beckman Coulter.

13. Autoclave.

14. Spectrophotometer.

15. Shaker.

16. Two water baths.

17. Hemocytometer.

18. Vortex.

Solutions

1. Iodixanol (OptiPrep ™, Axis- Shield PoC, Oslo, Norway) (Solution A).

2. To 2.8 mL 5 M NaCl add 6 mL HEPES 1 M, pH 7.3, and 1.2 mL EDTA 0.5 M, pH 8.0. Add H$_2$O to a final volume of 100 mL and sterile filtrate. Store at 4 °C (Solution B).

3. To 2.8 mL 5 M NaCl add 1 mL HEPES 1 M, pH 7.3, and 200 µL EDTA 0.5 M, pH 8.0. Add H_2O to a final volume of 100 mL and sterile filtrate. Store at 4 °C (Solution C).

4. Mix 5 volumes of OptiPrep™ and 1 volume of solution B (5:1) (Solution D). Prepare fresh before use, e.g. 10 mL OptiPrep™ and 2 mL B is sufficient for one Beckman, Opti Seal™ tube and one balancing tube.

5. Mix virus (0.5 mL) and solution C (4.3 mL) to obtain a total of 4.8 mL/Beckman Opti Seal™ tube. Sonicate the virus to break up clumps before adding it to solution C. Use solution C for balancing tubes. Use no more than virus supernatant obtained from three T175 tissue culture flasks for each Opti Seal™ polyallomer centrifuge tube. Prepare just before use (Solution E).

6. Top-up solution: Mix 1.27 mL solution C and 1 mL solution D. This equals a final concentration of 22 % OptiPrep™ in the gradient. Prepare fresh before use (Solution F).

2.2 Construction of Recombinant HSV-1 Vectors by Homologous Recombination in Bacteria

2.2.1 Generation of a Targeting DNA Fragment by PCR Amplification

1. Plasmid pgalK [26].

2. Oligonucleotide primers (Microsynth AG, Balgach, Switzerland).

3. QIA Quick PCR purification kit (Qiagen, Hilden, Germany).

4. Restriction endonuclease *Dpn*I (New England Biolabs, Ipswich, MA).

5. Electrophoresis grade agarose (e.g. Lonza, Rockland, USA).

6. TAE buffer 25×: Dissolve 121 g Tris base and 16.8 g EDTA in 970 mL H_2O, add 30 mL glacial acetic acid.

7. MinElute Gel Extraction Kit (Qiagen).

2.2.2 Preparation of Electrocompetent E. coli, Electroporation, and Selection of GalK-Positive and GalK-Negative Bacteria

1. *E. coli* SW102. *See* **Note 1**.

2. HSV-1 BAC; e.g. YE102 BAC [25].

3. Gene pulser cuvettes, 0.1 cm (BioRad, Hercules, CA).

4. M9 salts 1×: Dissolve 6 g Na_2HPO_4, 3 g KH_2PO_4, 1 g NH_4Cl, 0.5 g NaCl in 1 L H_2O; autoclave at 121 °C for 15 min, store at room temperature.

5. M63 minimal plates containing galactose, biotin, leucine, and chloramphenicol. Dissolve 7.5 g agar in 400 mL H_2O; autoclave (121 °C, 1 bar, 20 min). Cool down to 50 °C and add 100 mL 5× M63 medium, 0.5 mL 1 M $MgSO_4$ solution, 5 mL 20 % galactose (autoclaved stock solution), 2.5 mL biotin (0.2 mg/mL stock solution, autoclaved), 2.25 mL leucine (10 mg/mL stock solution, autoclaved), and 500 µL chloramphenicol (12.5 mg/mL).

6. M63 medium 5×: Dissolve 10 g $(NH_4)_2SO_4$, 68 g KH_2PO_4, 2.5 mg $FeSO_4·7H_2O$ in 800 mL H_2O, adjust pH to 7.0 with KOH, add H_2O to 1 L, autoclave.

7. McConkey indicator plates containing chloramphenicol: Dissolve 25 g McConkey agar in 500 mL H_2O, autoclave. Cool down to 50 °C and add 5 mL of 20 % galactose (autoclaved stock solution in H_2O) and 500-µL chloramphenicol (12.5 mg/mL).

8. M63 minimal plates containing glycerol, 2-deoxy-galactose (DOG), biotin, leucine, and chloramphenicol. Dissolve 1.5 g McConkey agar in 70 mL H_2O and autoclave. Cool down to 50 ° C and add 10 mL 2 % DOG (0.2 g in 10 mL H_2O) and autoclave. Cool down to 50 ° C and add 20 mL 5× M63 medium, 0.1 mL 1 M $MgSO_4$ solution, 1 mL 20 % galactose (autoclaved stock solution), 0.5 mL biotin (0.2 mg/mL stock solution, autoclaved), 0.45 mL leucine (10 mg/mL stock solution, autoclaved), and 100 µL chloramphenicol (12.5 mg/mL).

2.2.3 Isolation and Analysis of BAC DNA from E. coli (Miniprep Protocol)

1. Solution P1: 50 mM Tris–HCl pH 8.0, 10 mM EDTA, 00 µg/mL RNAseA. Filter sterilize, store at 4 ° C.

2. Solution P2: 0.2 N NaOH. Filter sterilize, store at room temperature.

3. Solution P3: 3 M potassium acetate, pH 5.5. Autoclave, store at 4 ° C.

4. Isopropanol.

5. Selected restriction endonucleases (New England Biolabs).

2.2.4 Transfection of Mammalian Cells with BAC DNA and Reconstitution of Recombinant HSV-1

1. Vero cells (African green monkey cells) (ATCC).

2. Dulbecco's modified Eagle medium (DMEM; Life Technologies™) with 10 % FBS.

3. 0.25 % trypsin/0.02 % EDTA (Life Technologies™).

4. Opti-MEM I reduced-serum medium (Life Technologies™).

5. Plasmid p116 (Dr. K. Tobler, University of Zurich, Zurich, Switzerland). *See* **Note 2**.

6. LipofectAMINE reagent (Life Technologies™).

7. Phusion™ DNA Polymerase (Finnzymes, Espoo, Finland).

8. DMSO (Finnzymes).

9. 5× GC buffer (Finnzymes).

10. Solution P1: 50 mM Tris–HCl, pH 8.0, 10 mM EDTA, 100 µg/mL RNase A (filter sterilize, store at 4 °C).

11. Solution P2: 0.2 N NaOH, 1 % SDS (filter sterilize, store at room temperature).

12. Solution P3: 3 M potassium acetate, pH 5.5 (filter sterilize, store at 4 ° C).

3   Methods

3.1   Construction of Recombinant HSV-1 Vectors by Homologous Recombination in Eukaryotic Cells

3.1.1   Preparation of Viral DNA for Transfection

Engineering a new recombinant virus requires purified infectious viral DNA along with a plasmid containing the specific sequences of interest flanked by sufficient amounts of viral sequences (at least 500–1,000 bp of flanking HSV-1 sequences) homologous to the targeted gene locus within the HSV-1 genome. The quality of the above reagents will determine the frequency and the efficacy of generating the desired recombinant virus. The quality of the viral DNA can be evaluated by its capability to produce plaques following transfection of 1–5 µg of viral DNA, depending on the deletions present in the HSV-1 genome. The protocol for this is as follows:

1. Infect a T75 cm^2 (8×10^6 cells/flask) or a T175 cm^2 (24×10^6 cells/flask) sub-confluent monolayer of cells, either Vero cells or a *trans*-complementing cell line such as Vero-7b, at a multiplicity of infection (MOI) of 1–3 plaque-forming units per cell (PFU/cell) with the parental HSV-1 virus strain that you wish to genetically modify. *See* **Note 3**.

2. Allow infection to proceed for ~18–24 h depending on the cells and the virus strain used.

3. Wait until the infection is completed, all the cells should be rounded-up but still adhere to the flask.

4. Remove the cells by tapping the flask or use a cell scraper to dislodge them.

5. Pellet the cells for 5–10 min at $1,204 \times g$ in a 15 mL or 50 mL centrifuge screw cap tube.

6. Wash the cells once with 5 mL PBS pH 7.5.

7. Lyse the cells with 2–3 mL of lysis buffer.

8. Incubate the tube at 37 °C overnight (ON) in an orbital shaker.

9. Extract DNA twice with Phenol:Chloroform:Isoamyl alcohol (25:25:1). It is important to invert the tube well to achieve proper mixing of the phases but not to be too vigorous. When transferring the aqueous phase into a new tube, go as close to the interface as possible. The DNA present at the interface is extremely viscous.

10. Extract twice with Chloroform:Isoamylic alcohol (24:1).

11. Transfer the aqueous phase into a new 15 mL centrifuge screw cap tube, going again as close to the interface as possible.

12. Add 2 volumes of cold isopropanol or cold ethanol. Mix well.

13. Spool the DNA on a heat-sealed glass Pasteur pipette, or store the mixture overnight at –20 °C. If spooling the DNA, remove the isopropanol or ethanol by capillary force and transfer the glass pipette with the DNA into a new tube containing H$_2$O or TE buffer. *See* **Note 4**.

Once the plasmid has been constructed (*see* **Note 5**) and the viral DNA has been prepared, it is possible to transfer the transgene from the plasmid to the HSV-1 genome, using the active recombination machinery of the virus, by co-transfecting the linearized plasmid DNA and the purified viral DNA into permissive cells such as Vero cells, or complementing cell lines if the viral genome is deleted in essential genes (*see* **Note 6**), using the calcium phosphate method (*see* **Note 7**).

1. The day before transfection, seed $8–9 \times 10^5$ Vero cells (in DMEM+10%FBS+pen/strep) for replication competent viral DNA, or the pertinent complementing cell line for replication-defective viral DNA, into 60 mm plates. Adjust the final volume to 3 mL of growth medium. The next morning, cells should have reached 70–80 % confluence. If so, it is possible to proceed with transfection.

2. For each transfection event, add 500 μL of HBS (pH 7.05) in a sterile 10 mL polystyrene tube.

3. To each tube add viral DNA (about 1–5 μg) using tips with wide ends, and then the linearized plasmid DNA (the amount of plasmid DNA should be equal to about 10× to 50× equivalents of viral DNA). Mix contents of each tube.

4. Add 30 μL of 2 M $CaCl_2$ to each tube, drop by drop, mix each tube immediately by blowing air into the tubes (use 2 mL pipettes) to facilitate the precipitate formation avoiding large clumps.

5. Incubate for 10–20 min at room temperature to allow the precipitate to form.

6. In the meantime, aspirate the medium from the plates and rinse the cells 3 times with 3 mL HBS.

7. Pipet the transfection mixture up and down, add it carefully drop by drop onto the cell monolayer, gently mix by moving the dish, and place the plates at 37 °C in the CO_2 incubator for 15–20 min.

8. Carefully overlay 3 mL DMEM+10 % FBS completed medium and incubate the transfected cells at 37 °C for 4–6 h in 5 % CO_2 incubator.

9. 4–6 h later, aspirate the medium from the plates and wash once with 3 mL DMEM +10%FBS completed medium.

10. Gently add 3 mL 20 % glycerol (in DMEM+10%FBS completed medium). Leave exactly for 3 min on the monolayer.

11. Carefully remove all the glycerol solution by aspiration and wash monolayer 4 times with DMEM+10%FBS completed medium.

12. Overlay monolayer with 3 mL DMEM+10%FBS completed medium and incubate at 37 °C in 5 % CO_2.

13. Observe the plates under the microscope for the production of HSV-1 cytopathic effect (CPE) indicating the presence of infectious foci. This usually takes 3–5 days depending on the virus and the cell type used to propagate the recombinants.

14. When the virus plaques become visible, harvest the monolayer into the medium, isolate the virus from the cell pellet by three cycles of freezing and thawing, sonicate, and centrifuge at $771 \times g$ for 5–10 min to eliminate the cellular debris. Store the unpurified virus stock at –80 °C for later use (isolation by limiting dilution).

3.1.3 Recombinant Virus Isolation by Limiting Dilution

The advantage of the limiting dilution approach to isolate individual recombinant viruses is that it avoids contamination of potential recombinants with parental viruses, which often occurs when using the standard plaque isolation technique, in which the single plaques are picked following methylcellulose or agarose overlay. The virus stock obtained after co-transfection (Subheading 3.1.2, **step 14**) should be used to isolate recombinants. If recombination led to a deletion of an essential gene, the screen of the recombinant can be confirmed on both the complementing cells (e.g. 7b Vero-derived cell line that expresses the HSV-1 IE genes ICP4 and ICP27, which are essential for virus replication) and non-complementing cells (e.g. Vero cells). Selection of a non-essential gene cannot be easily accomplished. For that situation, usually a marker gene is introduced in a first step, which then is substituted with the transgene of interest in a second step. To isolate the recombinant virus, it is necessary to go through at least three rounds of limiting dilution.

1. Titer the virus stock obtained from the co-transfection event, as described in Subheading 3.1.2.

2. Detach the cell monolayer with trypsin, count cells, and transfer $2–3 \times 10^6$ cells in a final volume of 2.0 mL in DMEM+10%FBS completed medium to a 15 mL centrifuge screw cap tube.

3. Add 20–30 PFU of virus stock to the cells. Rock at 37 °C for 1 h to adsorb the virus. *See* **Note 8**.

4. Add 8.0 mL DMEM+10 % FBS completed medium to the 2.0 mL of the infected cells to reach a final volume of 10 mL. Using a 50 mL reagent reservoir and a multichannel pipette, dispense 0.1 mL into each well of a 96-well plate. Incubate at 37 °C and wait for appearance of plaques.

5. Identify the wells containing single plaques and mark them. Carefully inspect the edges of the wells under high power to ensure that no additional plaques are present.

6. Once the plaques become visible (normally 2–3 days after infection, depending on the recombinants and the cells), freeze the plate at –80 °C, and thaw at 37 °C in an incubator. *See* **Note 9**.

7. Using a p200 Pipetman, scrape the cells from the bottom of each well identified to contain a single plaque and pipet the entire content of the well into an Eppendorf tube and freeze at −80 °C.

8. Repeat **steps 1–7** two more times (second and third limiting dilutions).

9. Plate out 2–3×10^6 cells into a 96-well plate again (cells in a final volume of 10 mL complete medium, 0.1 mL per well).

10. The next day, inoculate each well with 1/10 of the preliminary stock for each single virus in ~50 μL of medium. Adsorb virus as usual (at 37 °C for 1 h, rocking every 15 min) and then add another ~50 μL of medium to each well. It is useful at this point to infect positive and negative control viruses into some of the wells. Incubate at 37 °C and wait for the appearance of the plaques. If the recombinant virus is replication-defective, it is recommended to infect 24–48 well plates with at least half of the volume of each virus isolate to obtain sufficient amounts of viral DNA to perform a Southern blot.

11. Wait for all the cells in each well to round up. Aspirate the medium and add 200 μL of lysis buffer to the cells in preparation for Southern-blot analysis of viral DNA.

12. At this point, it is useful to characterize the recombinant viral DNA by Southern blot hybridization analyses using probes specific for the inserted sequences. *See* **Notes 10** and **11**.

3.1.4 Preparation of High-Titer Replication-Defective Recombinant Viral Stock

After the recombinant has been isolated, following three rounds of limiting dilution, prepare a midi-stock of the desired virus from a monolayer of cells in two to three 150–175 cm² tissue-culture flasks and determine the titer of the stock for preparation of the large stock (*see* Subheading 3.1.5). The following procedure is used to prepare a stock of replication-defective recombinant HSV-1. It can be scaled up or down depending on specific needs.

1. Seed 24×150—175 cm² flasks of complementing cells with 10×10^6 cells for each flask, to get confluent monolayers on the next day, in 20 mL of DMEM+10%FBS completed medium. Incubate at 37 °C in 5 % CO_2.

2. The next day wash the cells with TBS twice, add enough trypsin to cover the monolayer and detach the cells. Collect the cells in 50 mL Corning screw cap tubes and pellet at $1{,}204 \times g$ for 10 min. Discard supernatants and resuspend cell pellets in a small volume, combine all cell pellets in one 50 mL Corning screw cap tube, re-pellet and resuspend in a final volume of no more than 20 mL. Infect the cells in suspension with the recombinant virus at MOI of 0.05–0.08 PFU/cell. Gently rock the tube for 1 h at 37 °C. *See* **Note 12**.

3. Plate back infected cells into the 150–175 cm² flasks and add growth medium to a final volume of 20 mL.

4. Incubate until all the cells are infected (24–36 h post infection).

5. Scrape infected cells into the medium and pipet the suspension into 50 mL Corning screw cap tubes.

6. Pellet at $1,204 \times g$ for 15 min at 4 °C. Decant supernatant and store on ice.

7. Resuspend cell pellets in a small volume of supernatant, combine and re-pellet.

8. Resuspend the final pellet in 2–3 mL of medium in a 15 mL Corning screw cap tube. If you stop at this point put everything at −80 °C.

9. Freeze–thaw the cell pellet 3 times; vortex each time after thawing. After the final thawing, sonicate 3 times for 10–15 s with 10 s of pause in ice after each sonication. The virus suspension you obtain should be homogeneous.

10. Pellet the cell debris at $1,734 \times g$ for 15 min at 4 °C.

11. Transfer the supernatant derived from the pellet into 50-mL Oak Ridge polypropylene tube along with original viral supernatant. Spin down at $48,384 \times g$ for 30 min × 4 °C in JA-20 rotor.

12. Decant and discard supernatant. Carefully remove remaining supernatant and resuspend pellet by vortexing or pipetting in less than 1 mL growth medium. You can spin the tube to remove bubbles.

13. Do not freeze the virus but directly purify it in gradients (*see* Subheading 3.1.6). After purification, resuspend the virus in PBS, aliquot it in small volumes and freeze at −80 °C.

3.1.5 Titration of Virus Stock

The following procedure can be used to obtain the titer of virus stocks of any size.

1. One day prior to titration, prepare six-well tissue culture plates with 0.5×10^6 cells (e.g. Vero cells if no essential viral gene has been deleted, or corresponding complementing cells if the recombinant virus contains deletions in any essential gene) to titer the virus. The critical point is that on the day of titration the cell monolayer should be confluent.

2. Prepare a series of tenfold dilution (10^{-2}–10^{-10}) of the virus stock in Eppendorf tubes with 1 mL DMEM + 10 % FBS completed medium.

3. Add 100 µL of each dilution to confluent cells in a single well of a six-well cell culture plate.

4. Allow the virus to infect the cells for a period of 1 h at 37 °C in a CO_2 incubator. Rock the plates every 15 min to distribute the inoculum to all cells in the monolayer.

5. Aspirate off the virus inoculum, overlay the monolayer with 3 mL of 1.5 % methylcellulose in DMEM+10 % FBS completed medium.

6. Reincubate the plates for 3–5 days until well-defined plaques appear.

7. Aspirate off the methylcellulose and stain with 2 mL of 1 % crystal violet solution for 10–20 min. The stain fixes the cells and the virus.

8. Count the number of plaques per well, determine the average for each dilution (if it is in duplicate), and multiply by a factor of 10 to get the number of plaque forming units/mL (PFU/mL) for each dilution. Multiply this number by 10 to the power of the dilution to achieve the titer in PFU/mL. *See* **Note 13**.

3.1.6 Purification of Recombinant HSV-1 Stock

In order to purify the virus from cell debris or proteins and to prepare sufficient virus stock at high titer for use in preclinical experiments, the virus pellet can be resuspended in PBS and purified through sucrose, dextrane, or iodixanol gradients. The following protocol is based on the purification of the virus with iodixanol gradient (OptiPrep™ Axis-Shield, product No. 1030061). The iodixanol gradient is self-forming. Solutions B-F (*see* Subheading 2.1.1.2) must be kept on ice and the ultracentrifuge rotor precooled.

1. Prepare Opti Seal polyallomer centrifuge tubes for run in a rotor Vti65.1 Pipet 4.4 mL solution D into each tube.

2. Sonicate the virus stock to break up clumps before adding it to solution C to obtain the solution E.

3. Slowly add 4.8 mL solution E into each tube prepared at **step 1**. Be careful to avoid clogging of neck and bubble formation. To remove bubbles, use a syringe.

4. Fill up the tubes with solution F (about 1.5 mL for each tube).

5. Leave just a small air bubble in the neck of the tube and close it with the cap.

6. Balance tubes using a scale. If necessary, balance by adding solution F.

7. Dry the outside of the tubes if necessary and place them into the rotor. Place plugs and caps and close by using 120 in-lbs torque value.

8. Place rotor in centrifuge, close door, turn on vacuum, enter run specifications in ultracentrifuge: speed 296,516×*g*, time

from 4 to 15 h (depending if you want to collect the virus the same day of the following day), at 4 °C, maximum acceleration rate, no brake during deceleration.

9. During the run check if centrifuge attained full speed.

10. At the end of run turn off vacuum, remove rotor and carefully put the tubes on ice. A band will be visible in the middle of the gradient.

3.1.7 Collection of Virus Particles

1. Collect the band by puncturing the side of the tube 2–3 mm under the band with a needle and syringe. Be careful not to aspirate too much volume to avoid the collection of debris.

2. Place the collected virus into 30 mL Nalgene Centrifuge Oak Ridge tubes and add PBS to fill up the tube and to dilute the residual iodixanol solution coming with the collected virus (otherwise it will not be completely spin down due to the gradient solution).

3. Centrifuge the tubes in Beckman centrifuge, $48,384 \times g$ for 30 min at 4 °C in JA-20 rotor, to concentrate the virus. If you have performed your gradient in more than three ultracentrifuge tubes, place the collected virus into two JA-20 oak ridge tubes, dilute with PBS and centrifuge them.

4. Discard the supernatant and resuspend the pellet in about 1–2 mL of PBS. If the pellet is too clumped, leave it to disaggregate overnight on ice in a cold room.

5. Carefully transfer the virus in a 10 mL tube, sonicate it to break up clumps (2–3 times, 5–10 s each time, with 10 s pause in ice between sonications). The viral suspension should be homogeneous.

6. Aliquot the virus in small volumes in Eppendorf tubes.

7. Store the aliquots at –80 °C. *See* **Notes 14** and **15**.

3.2 Construction of Recombinant HSV-1 Vectors by Homologous Recombination in Bacteria

3.2.1 Generation of a GalK+ Targeting DNA Fragment by PCR Amplification

1. Design primers with 50-bp homology to either side of the targeting sequence on the HSV-1 genome, followed by sequences (underlined below) that bind to the *galK* cassette in plasmid pgalK, which serves as the template for the PCR reaction. Forward primer: 5′ 50-nucleotide homology arm-CGTGT TGACAATTAATCATCGGCA3′. Reverse primer: 5′ 50-nucleotide homology arm-TCAGCACTGTCCTGCTCCTT 3′.

2. Perform PCR amplification as follows: 10 μM of each primer, 10 μL 5× GC buffer, 3 μL DMSO, 10 μM dNTPs, 10 ng pgalK template DNA, 0.5 μL Phusion DNA Polymerase, H_2O to 50 μL; 94 °C for 15 s, 60 °C for 30 s, and 72 °C for 1 min, for 30 cycles.

3. Purify the PCR reaction by using the QIA Quick PCR purification kit.

4. Add 10 U *Dpn*I and incubate for 2 h at 37 °C. *See* **Note 16**.

5. Separate the fragments by agarose gel electrophoresis (1 % agarose in 1× TAE), and excise the band containing the *galK+* targeting DNA fragment. Purify the DNA by using the MinElute Gel Extraction Kit, precipitate and wash with ethanol, resuspend in 30 µL H_2O.

6. To determine the DNA concentration measure the absorbance at 260 nm (A_{260}) and 280 nm (A_{280}) using a UV spectrophotometer. *See* **Note 17**.

3.2.2 Preparation of Electrocompetent E. coli SW102

1. Inoculate 5 mL of LB medium with *E. coli* SW102. Incubate overnight at 32 °C on a shaker.

2. The next day, prepare the following: two water baths, one at 32 °C, the other at 42 °C; an ice/water slurry; 50 mL ice cold H_2O; a pre-cooled centrifuge (0 °C).

3. Inoculate 25 mL of LB medium in a 50 mL Erlenmeyer flask with 500 µL of the overnight culture. Incubate at 32 ° C in a shaking water bath until the OD_{600} reaches approximately 0.6.

4. Transfer 10 mL of the culture to another 50 mL Erlenmeyer flask and incubate for exactly 15 min at 42 °C.

5. Cool the culture in an ice/water slurry, transfer to two 15 mL Falcon tubes, and centrifuge for 5 min at 5,500×*g* and 0 °C.

6. Remove all supernatants and resuspend the pellet in 1 mL of ice cold H_2O by gently swirling the tube in the ice/water slurry.

7. Add 9 mL of ice cold H_2O and pellet again (5 min, 5,500×*g*, 0 °C). Repeat **step 6**.

8. Remove all supernatants by inverting the tubes on a paper towel. Resuspend the pellet in the remaining liquid (approximately 50 µL) and keep on ice until used for electroporation.

3.2.3 Electroporation of HSV-1 BAC DNA into E. coli SW102

1. Mix 25 µL of electrocompetent bacteria with 2 µL of HSV-1 BAC DNA (e.g. YE102bac) [25]. Transfer the suspension into a 0.1 cm cuvette and electroporate at 25 µF, 1.75 kV, and 200 Ω.

2. Recover the bacteria with 1 mL LB medium, and incubate the culture for 1 h at 32 ° C on a shaker. Plate 1:10, 1:100, and 1:1,000 dilutions (in LB medium) onto LB agar plates containing the appropriate antibiotic (e.g. 12.5 µg/mL of chloramphenicol for YE102bac).

<table>
<tr><td>

3.2.4 Electroporation of the GalK Targeting DNA into E. coli SW102 Containing the HSV-1 BAC and Screening for GalK-Positive Clones (galK-Positive Selection)

</td><td>

1. Prepare electrocompetent *E. coli* SW102 containing the HSV-1 BAC as described in Subheading 3.2.2, except that in steps 1 and 3, the LB medium should contain the appropriate antibiotic (e.g. 12.5 µg/mL of chloramphenicol for YE102bac).

2. Mix 25 µL of electrocompetent bacteria with 2 µL of the PCR product from Subheading 3.2.1, **step 5** (approx. 30 ng). Transfer suspension into a 0.1 cm cuvette and electroporate at 25 µF, 1.75 kV, and 200 Ω.

3. Recover the bacteria with 1 mL LB medium and transfer into a 15 mL Falcon tube. Add another 9 mL of LB medium and incubate culture for 1 h at 32 °C on a shaker.

4. Wash the bacteria twice in 1×M9 salts as follows: Transfer 1 mL of the culture into an Eppendorf tube and pellet for 15 s at 17,900×*g*. Remove the supernatant with a pipette. Resuspend the pellet in 1 mL 1× M9 salts and centrifuge again. Repeat this washing step once more (*see* **Note 18**). Then resuspend the pellet in 400 µL of 1× M9 salts and plate serial dilutions in 1× M9 salts (1:10, 1:100, 1:1,000) onto M63 minimal medium plates containing galactose and the appropriate antibiotic (e.g. 12.5 µg/mL of chloramphenicol for YE102bac). Incubate for 3 days at 32 ° C.

5. Streak several colonies onto McConkey indicator plates containing galactose and the appropriate antibiotic (e.g. 12.5 µg/mL of chloramphenicol for YE102bac). Red colonies will indicate *galK+* bacteria. The *galK*-positive selection step is normally very efficient and it is not necessary to further analyze the clones. However, to confirm the correct insertion of the *galK* cassette, HSV-1 BAC DNA can be prepared and analyzed as described in Subheading 3.2.6.

</td></tr>
<tr><td>

3.2.5 Electroporation of the Targeting DNA into Galk-Positive E. coli SW102 Containing the HSV-1 BAC and Screening for Galk-Negative Clones (Galk-Negative Selection)

</td><td>

1. Pick single bright red colonies (*galK+*) from Subheading 3.2.4, **step 5**, and inoculate 5 mL LB medium containing the appropriate antibiotic (e.g. 12.5 µg/mL of chloramphenicol for YE102bac).

2. Prepare electrocompetent bacteria as described in Subheading 3.2.2, except that in **steps 1** and **3**, the LB medium should contain the appropriate antibiotic (e.g. 12.5 µg/mL of chloramphenicol for YE102bac).

3. For insertions of foreign DNA, prepare a linear targeting DNA by PCR amplification as described in Subheading 3.2.1. The forward and reverse primers contain the same 50 nucleotides of targeting sequence at the 5′ end as the primers designed in Subheading 3.2.1 **step 1**, followed by sequences that bind to the 3′ and 5′ ends of the DNA fragment to be inserted. For HSV-1 gene deletions, design an oligonucleotide that contains the 50 nucleotides of 5′ targeting sequence

</td></tr>
</table>

followed by the 50 nucleotides of 3′ targeting sequence of Subheading 3.2.1, **step 1**. For point mutations, the altered nucleotide(s) can be inserted into an oligonucleotide as described above, in the center of the 50 nucleotides 5′ and 3′ targeting sequences. It is not necessary to use double-stranded oligonucleotides, although the efficiency of a double-stranded DNA is somewhat higher.

4. Mix 25 µL of electrocompetent bacteria from **step 2** with 2 µL of the targeting DNA from Subheading 3.2.5, **step 3** (100–200 ng). Transfer suspension into a 0.1 cm cuvette and electroporate at 25 µF, 1.75 kV, and 200 Ω.

5. Recover the bacteria with 10 mL LB medium and transfer into a 50 mL Erlenmeyer flask; incubate culture for 4.5 h at 32 ° C on a shaker.

6. Wash and dilute the bacteria as in Subheading 3.2.4, **step 4**. Then plate bacteria onto M63 minimal medium plates containing glycerol, leucine, biotin, 2-deoxy-galactose (DOG), and the appropriate antibiotic (e.g. 12.5 µg/mL of chloramphenicol for YE102bac). Incubate for 3 days at 32 °C. *See* **Note 19**.

3.2.6 Isolation and Characterization of HSV-BAC DNA from Small Bacterial Cultures

1. Inoculate single bacterial colonies into 5 mL LB medium containing the appropriate antibiotic in 15 mL Falcon tubes. Incubate overnight at 32 °C on a shaker.

2. Centrifuge tubes for 10 min at $2,000 \times g$ and 4 °C. Discard supernatant, resuspend pellet in 300 µL of solution P1, and transfer suspension into an Eppendorf tube.

3. Add 300 µL of solution P2, mix gently, and incubate for 5 min at room temperature.

4. Slowly add 300 µL of solution P3, mix gently, and incubate on ice for at least 5 min.

5. Centrifuge for 10 min at $10,600 \times g$ and 4 °C. Transfer supernatant into a new Eppendorf tube that contains 800 µL of isopropanol. Mix by inverting the tube several times and incubate on ice for at least 5 min.

6. Centrifuge for 15 min at $10,600 \times g$ and 4 °C. Remove supernatant and wash the pellet with 500 µL of 70 % ethanol. Invert the tube several times and then centrifuge again for 5 min at $10,600 \times g$ at 4 °C.

7. Aspirate supernatant and air-dry pellet at room temperature. Then, resuspend the DNA in 40 µL of TE buffer in a 37 ° C water bath. Use 10 µL of the DNA for restriction endonuclease analysis and agarose gel electrophoresis (0.5 % agarose in TAE) to confirm the mutation in the HSV-1 DNA.

3.2.7 Transfection
of Mammalian Cells
with BAC DNA
and Reconstitution
of Recombinant HSV-1

1. Maintain Vero cells in DMEM + 10 % FBS. Propagate the culture twice a week by splitting 1/5 in fresh medium (20 mL) into a new 75-cm² tissue culture flask.

2. On the day before transfection, remove culture medium, wash twice with PBS, add a thin layer of trypsin/EDTA, and incubate for 10 min at 37 °C to allow cells to detach from the plate. Count cells using a hemocytometer, and plate 1.2×10^6 cells per 60-mm-diameter tissue culture dish in 3 mL DMEM + 10 % FBS.

3. For each 60-mm dish, place 100 µL Opti-MEM I reduced-serum medium into each of two 15-mL conical tubes. To one tube, add 2 µg of HSV-BAC DNA and 0.2 µg of plasmid p116, which expresses Cre-recombinase with an NLS (kindly provided by K. Tobler, University of Zurich, Switzerland) (*see* **Note 20**). To the other tube, add 12 µL LipofectAMINE.

4. Combine the contents of the two tubes, mix well, and incubate for 45 min at room temperature.

5. Wash the cultures prepared the day before (**step 2**) once with 2 mL Opti-MEM I. Add 1.1 mL Opti-MEM I to the tube from **step 4** containing the DNA-LipofectAMINE transfection mixture (1.3 mL total volume). Aspirate medium from the culture, add the transfection mixture, and incubate for 5.5 h.

6. Aspirate the transfection mixture and wash the cells 3 times with 2 mL Opti-MEM I. After aspirating the last wash, add 3.5 mL DMEM + 6 % FBS and incubate 2–3 days.

7. Scrape cells into the medium using a rubber policeman. Transfer the suspension to a 15-mL conical centrifuge tube and place the tube containing the cells into a beaker of ice water. Submerge the tip of the sonicator probe ~0.5 cm into the cell suspension and sonicate 20 s with 20 % output energy. This disrupts cell membranes and liberates cell-associated virus particles.

8. Remove cell debris by centrifugation for 10 min at $960 \times g$, 4 °C and inoculate fresh Vero cells or appropriate complementing cells for plaque purification.

9. Characterize plaque-purified virus as follows: confirm the absence of the BAC sequences (e.g. PCR), determine growth properties and titers, analyze the genotype of the recombinant virus.

4 Notes

1. The *E. coli* SW102 strain is derived from *E. coli* DH10B and contains the λ prophage recombination system and a deletion in the galactokinase gene (*galK*). The *galK* function can be added *in trans,* which restores the ability of the bacteria to grow on galactose as carbon source.

2. Any plasmid that expresses Cre-recombinase with a nuclear localization signal can be used.

3. The 7b Vero-derived cell line [15] expresses the HSV-1 IE genes ICP4 and ICP27 required for replication of a recombinant HSV-1 virus deleted in both IE genes. Vero-7b cells are subjected after several passages to 2-week long selection with 1 mg/mL G418 (Sigma Chemicals).

4. Viral DNA is very long and fragile. It is thus critical to take extreme care when handling viral DNA by using wide-end pipette tips (tips for genomic DNA). Store the viral DNA at 4 ° C; do not freeze to avoid breaking the large DNA.

5. Example of a plasmid used to insert new sequences into the viral genome. pTZgJHE plasmid: The 2036-bp SalI-HindIII fragment from the HSV-1 genome (nucleotides 136308–138345) containing gene *Us5* (corresponding to HSV-1 glycoprotein J promoter and coding sequence) was cloned into SalI-HindIII of pTZ18U plasmid (Invitrogen™) using T4 DNA ligase (NEB). The resulting plasmid, pTZgJ, was used to generate plasmid pTZgJHE by insertion of GFP coding sequence driven by the cytomegalovirus (HCMV) promoter. The GFP cassette was inserted as a NruI-SphI fragment isolated from pcDNA3.1HygroGFP between the SphI (137626) and NruI (137729) sites of the virus genome, thereby introducing a deletion in *Us5* locus between the TATA box and the gJ coding sequence. pTZgJ can be used as a shuttle plasmid, in which the GFP cassette can be replaced with a transgene of interest.

6. Examples of recombinant HSV-1 viruses. A) Replication-defective viruses that require complementing cell lines: S0ZgJHE is a recombinant virus deleted in the ICP4 IE gene, with the GFP reporter gene driven by the HCMV promoter placed in the *Us5* locus (glycoprotein J) and the lacZ reporter gene, under ICP0 promoter, placed in the *UL41* locus (vhs, virion host shutoff). TOZ-GFP is a recombinant virus deleted in the ICP4, ICP27 and ICP22 IE genes, with the GFP reporter gene, under the control of HCMV promoter placed in the ICP22 locus and the lacZ reporter gene, under the control of the ICP0 promoter, in the *UL41* locus. In these cases, the viral DNAs and the recombinant plasmids containing the transgenes of interest can be co-transfected into a complementing cell line (Vero-7b), which provides, *in trans*, the essential viral genes ICP4 and ICP27. B) Replication-competent viruses that can be grown in Vero cells or other permissive cell lines: Viruses deleted in non-essential genes such as UL41, γ34.5, or TK.

7. The use of wide-boar Pipetman tips will help to prevent shearing of the viral DNA, increasing on this way the infectivity of the viral DNA preparation. The quality of the plasmid is crucial for the recombination efficiency. The size of the flanking HSV-1 sequences is crucial; longer sequences increase the efficacy of homologous recombination. The size of the insert can affect stability if it is too large; part of the insert can be lost over time or it will not be possible to obtain a purified isolate of the desired recombinant. It is important to linearize the plasmid to increase the recombination efficiency and to exclude that recombinants are isolated that have incorporated the entire transfer plasmid through a single crossover. The pH of the HBS is crucial for the transfection efficiency.

8. Prior to do the limiting dilution; it is recommended to sonicate the virus stock for a few seconds before infection in order to resuspend the virus particles. Clumps might cause a single plaque arising from two or more viruses.

9. As soon as the plaques become visible they should be picked.

10. The specific HSV-1 gene locus targeted for deletion/insertion can affect the recombination event or affect the stability of the desired recombinant, and sometimes it will not be possible to obtain a purified isolate of the desired recombinant.

11. Recombination into the repeat sequences can yield a mixture of viruses containing insertion into one copy of the gene, leading to rescue by the not deleted copy, and also in this case it will not be possible to obtain a purified isolate of the desired recombinant.

12. To prepare attenuated viral vectors, the amount of cells that should be infected can be lower since usually these recombinants grow well in vitro. These recombinants can infect attached cells. The cells can be prepared the day before and on the next day decant medium from the flasks and add the virus in an amount of serum-free medium, sufficient to cover the monolayer. The infection can be performed at an MOI of 0.01 PFU/cell.

13. The titer of a large viral stock should be determined at least in duplicate.

14. Following purification of the virus, it is necessary to add glycerol to a final concentration of 10 % to the virus stock in order to cryo-preserve it. If the virus is prepared to be used in animal experiments it should be aliquoted without glycerol but in small volumes to avoid thawing the vials twice.

15. Virus stocks should be maintained at a low passage number. Use one vial of a newly prepared stock as a stock for preparing all future stocks used in a series of experiments. In order to reduce

the chance of rescuing wild-type virus during the propagation of viruses carrying deletions of essential gene(s), stocks should be routinely prepared from single plaque isolates.

16. DpnI does not cleave non-methylated DNA amplified by PCR but cleaves the methylated template plasmid DNA isolated from *E. coli*.

17. A value of 1.0 for A_{260} is equivalent to 50 μg/mL of double-stranded DNA. Additionally, the ratio between A_{260} and A_{280} provides information about DNA purity. Typically, pure DNA preparations have an A_{260}/A_{280} value of 1.8. Do not use DNA preparations with a ratio below this value.

18. Washing with M9 salts is important to remove all residual-rich medium from the bacteria before plating on minimal medium plates.

19. The efficiency of the *galK*-negative selection step is low, and the majorities of the colonies that form under DOG selection are not correct (but may have point mutations in the *galK* gene or large deletions). To overcome the problem, longer homology arms could be designed to increase the frequency of recombination.

20. The HSV-1 sequences in BACs are normally flanked by loxP1 sites. This allows removing the bacterial sequences by Cre-recombinase during reconstitution of virus following transfection of BAC DNA in mammalian cells.

References

1. Roizman B, Knipe DM (2001) Herpes simplex viruses and their replication. In: Knipe DM, Howley PM (eds) Fields virology. Lippincot, Williams and Wilkins, Philadelphia, PA, pp 2399–2460

2. Marozin S, Prank U, Sodeik B (2004) Herpes simplex virus type 1 infection of polarized epithelial cells requires microtubules and access to receptors present at cell–cell contact sites. J Gen Virol 85:775–786

3. Honess RW, Roizman B (1975) Regulation of herpesvirus macromolecular synthesis: sequential transition of polypeptide synthesis requires functional viral polypeptides. Proc Natl Acad Sci U S A 72:1276–1280

4. Batterson W, Roizman B (1983) Characterization of the herpes simplex virion-associated factor responsible for the induction of alpha genes. J Virol 46:371–377

5. Skepper JN, Whiteley A, Browne H et al (2001) Herpes simplex virus nucleocapsids mature to progeny virions by an envelopment → deenvelopment → reenvelopment pathway. J Virol 75:5697–5702

6. Deshmane SL, Fraser NW (1989) During latency, herpes simplex virus type 1 DNA is associated with nucleosomes in a chromatin structure. Virol 63:943–947

7. Farrell MJ, Dobson AT, Feldman LT (1991) Herpes simplex virus latency-associated transcript is a stable intron. Proc Natl Acad Sci U S A 88:790–794

8. Umbach JL, Kramer MF, Jurak I et al (2008) MicroRNAs expressed by herpes simplex virus 1 during latent infection regulate viral mRNAs. Nature 454:780–783

9. Preston CM (2000) Repression of viral transcription during herpes simplex virus latency. J Gen Virol 8:1–19

10. Manservigi R, Argnani R, Marconi P et al (2007) Herpesvirus-based vectors for gene transfer, gene therapy, and the development of novel vaccines. In: Hefferon KL (ed) Virus expression vectors. Transworld Research Network, Kerala, India, pp 205–246

11. Krisky DM, Marconi PC, Oligino TJ et al (1998) Development of herpes simplex virus

replication-defective multigene vectors for combination gene therapy applications. Gene Ther 5:1517–1530

12. Wu N, Watkins SC, Schaffer PA et al (1996) Prolonged gene expression and cell survival after infection by a herpes simplex virus mutant defective in the immediate-early genes encoding ICP4, ICP27, and ICP22. J Virol 70: 6358–6369

13. Samaniego LA, Neiderhiser L, DeLuca NA (1998) Persistence and expression of the herpes simplex virus genome in the absence of immediate-early proteins. J Virol 72: 3307–3320

14. Berto E, Bozac A, Marconi P (2005) Development and application of replication-incompetent HSV-1-based vectors. Gene Ther 12(Suppl 1):S98–S102

15. Krisky DM, Wolfe D, Goins WF et al (1998) Deletion of multiple immediate-early genes from herpes simplex virus reduces cytotoxicity and permits long-term gene expression in neurons. Gene Ther 5:1593–1603

16. Wolfe D, Mata M, Fink DJ (2009) A human trial of HSV-mediated gene transfer for the treatment of chronic pain. Gene Ther 16:455–460

17. Advani SJ, Weischelbaum RR, Whitley RJ et al (2002) Friendly fire: redirecting herpes simplex virus-1 for therapeutic applications. Clin Microbiol Infect 8:551–563

18. Argnani R, Lufino M, Manservigi M et al (2005) Replication-competent herpes simplex vectors: design and applications. Gene Ther 12(Suppl 1):S170–S177

19. Nawa A, Luo C, Zhang L et al (2008) Non-engineered, naturally oncolytic herpes simplex virus HSV1 HF-10: applications for cancer gene therapy. Curr Gene Ther 8:208–221

20. Gage PJ, Sauer B, Levine M et al (1992) A cell-free recombination system for site-specific integration of multigenic shuttle plasmids into the herpes simplex virus type 1 genome. J Virol 66:5509–5515

21. Rinaldi A, Marshall KR, Preston CM (1999) A non-cytotoxic herpes simplex virus vector which expresses Cre recombinase directs efficient site-specific recombination. Virus Res 65: 11–20

22. Stricklett PK, Nelson RD, Kohan DE (1998) Site-specific recombination using an epitope tagged bacteriophage P1 Cre recombinase. Gene 215:415–423

23. Krisky DM, Marconi PC, Oligino T et al (1997) Rapid method for construction of recombinant HSV gene transfer vectors. Gene Ther 4:1120–1125

24. Saeki Y, Ichikawa T, Saeki A et al (1998) Herpes simplex virus type 1 DNA amplified as bacterial artificial chromosome in Escherichia coli: rescue of replication-competent virus progeny and packaging of amplicon vectors. Human Gene Ther 9:2787–2794

25. Tanaka M, Kagawa H, Yamanashi Y et al (2003) Construction of an excisable bacterial artificial chromosome containing a full-length infectious clone of herpes simplex virus type 1: viruses reconstituted from the clone exhibit wild-type properties in vitro and in vivo. J Virol 77:1382–1391

26. Warming S, Costantino N, Court DL et al (2005) Simple and highly efficient BAC recombineering using *galK* selection. Nucleic Acids Res 33:e36

Herpes Simplex Virus Type 1 (HSV-1)-Derived Amplicon Vectors for Gene Transfer and Gene Therapy

Cornel Fraefel, Peggy Marconi, and Alberto L. Epstein

Abstract

Amplicons are defective, helper-dependent, herpes simplex virus type 1 (HSV-1)-derived vectors. The main interest of these vectors as gene transfer tools stems from the fact that the amplicon vector genomes do not carry protein-encoding viral sequences. Consequently, they are completely safe for the host and non-toxic for the infected cells. Moreover, the complete absence of virus genes provides space to accommodate very large foreign DNA sequences, up to almost 150-kbp, the size of the virus genome. This large transgene capacity can be used to deliver complete gene loci, including introns and exons, as well as long regulatory sequences, conferring tissue-specific expression, or stable maintenance of the transgene in proliferating cells. During many years the development of these vectors and their application in gene transfer experiments was hindered by the presence of contaminating toxic helper virus particles in the vector stocks. In recent years however, two different methodologies have been developed that allow generating amplicon stocks either completely free of helper particles or only faintly contaminated with fully defective helper particles. This chapter summarizes these two methodologies.

Key words HSV-1, Amplicon vectors, Gene transfer, Gene therapy, Bacterial artificial chromosomes (BACs), CRE/loxP1 site-specific recombination

Abbreviations

A/H   Amplicon/helper
HC   Helper-contaminated stocks
HF   Helper-free stocks
PFU   Plaque forming units
TU   Transducing units

1   Introduction

In the accompanying chapter that focuses on herpes simplex virus type 1 (HSV-1)-based recombinant vectors, we have briefly described the structure of the virus particle and introduced some

Laura Lossi and Adalberto Merighi (eds.), *Neuronal Cell Death: Methods and Protocols*, Methods in Molecular Biology, vol. 1254, DOI 10.1007/978-1-4939-2152-2_21, © Springer Science+Business Media New York 2015

important conceptual elements of the molecular biology of HSV-1 which are required for understanding the generation and production of recombinant vectors. In the first part of this chapter we will introduce critical aspects of the virus genome and virus replication, which are conceptually important for the understanding of HSV-1 amplicon vector biology.

The linear double-stranded HSV-1 genome has a size of 153-kbp and can be divided into two segments, designated as L (long) and S (short). Each segment consists of unique sequences (UL and US) bracketed by inverted repeats. The repeats surrounding the L component are designated as *ab* and *b'a'*, while those surrounding the S component are designated *a'c'* and *ca* [1, 2]. The number of *a* sequence repeats at the L terminus and at the L/S junction is variable while at the S terminus there is a single *a* sequence. The HSV-1 genome can then be represented as:

$$a_n b\text{---------------}UL\text{----------------}b'a'_m c'\text{----------}US\text{--------}ca$$

where a_n and a_m are *a* sequences that can vary from 1 to more than 10 copies [3]. The UL sequence contains at least 58 genes whereas US contains at least 13 genes. The inverted repeats flanking UL (*b* and *b'*) each contains 3 genes whereas the inverted repeats flanking the US sequence (*c* and *c'*) each contains a single gene. Thus, the HSV-1 genome contains at least 79 canonical genes from which four are diploid [4]. It is however accepted today that the total number of genes probably exceeds 85 and that the virus genome encodes also several miRNA sequences. In addition, the virus genome carries two types of *cis*-acting sequences that are essential for virus multiplication: the origins of viral DNA synthesis and the cleavage/packaging signals.

The HSV-1 genome contains three origins of replication (ori). One origin, designated as ori_S, has a size of approximately a 130-bp, is located within the repeated sequences surrounding US, and hence is diploid [5]. The third origin, designated ori_L, is present in one copy near the center of the UL region of the genome [6] and is approximately 180-bp long. Mutant viruses lacking either ori_L or both copies of ori_S are replication competent, suggesting that all origins are functionally equivalent [7, 8]. The essential cleavage/packaging signals are located within the *a* sequences found at both ends of the virus genome as well as at the L/S junction [9]. These sequences contain both unique and repeated elements, and their size can vary in different HSV-1 strains from 250 to 500 nucleotides [10, 11].

The prevailing model for HSV-1 replication proposes that (a) the linear viral genome circularizes immediately after infection, probably in the cell nucleus, (b) the genome then replicates as a head-to-tail concatemer, probably as a result of a complex rolling circle mechanism that includes recombination events between

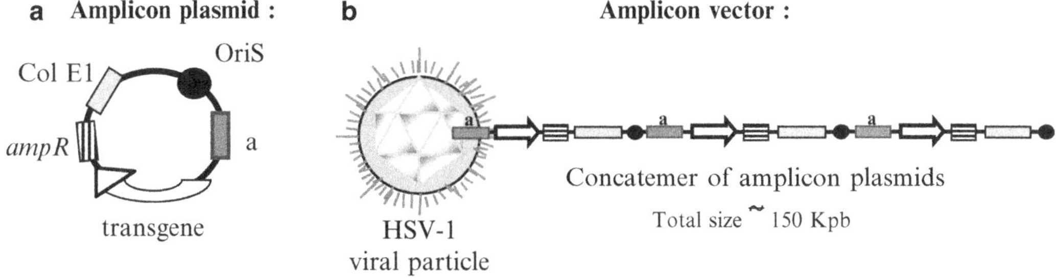

Fig. 1 Structure of the amplicon plasmid and amplicon vector. (**a**) An amplicon plasmid is a standard *Escherichia coli* plasmid, containing one bacterial origin of replication (ColE1) and one gene conferring resistance to an antibiotic (generally *ampR*), carrying in addition one HSV-1 origin of replication (oriS), one HSV-1 packaging signal (*a*) and the transgenic sequence of interest (represented as an *arrow*). (**b**) An amplicon vector is an HSV-1 viral particle containing a concatemer of the amplicon plasmid DNA, up to around 150-kbp

repeated sequences, and (c) concatemeric DNA is cleaved at the *a* sequences into unit-length genomes during packaging in preformed capsids (for a review on HSV-1 replication, see [12]).

1.1 HSV-1-Based Amplicon Vectors

Amplicon vectors, or amplicons [13], are HSV-1 particles identical to wild type HSV-1 from the structural, immunological and host-range points of view, but which carry a concatemeric form of a DNA plasmid, named the amplicon plasmid, instead of the viral genome. An amplicon plasmid (Fig. 1a) is a standard *E. coli* plasmid carrying one origin of virus DNA replication (generally oriS) and one packaging signal (*a*) from HSV-1, in addition to the transgene sequences of interest [14, 15]. One major advantage of amplicons as gene transfer tools is the fact that they carry no virus genes and consequently do not induce synthesis of virus proteins. Therefore, these vectors are fully non-toxic for the infected cells and non-pathogenic for the inoculated organisms. Furthermore, the absence of virus genes in the amplicon genome strongly reduces the risk of reactivation, complementation or recombination with latent or resident HSV-1 genomes.

The versatility of amplicons stems from the fact that during their production the amplicon genome will replicate, like HSV-1, via a rolling circle-like mechanism, generating long concatemers composed of tandem repeats of the amplicon plasmid ([16] and Fig. 1b). Since HSV-1 particles will always package around 150-kbp DNA, the size of the virus genome, the number of repeats that a particular amplicon vector will carry and deliver, will depend on the size of the original amplicon plasmid [17]. Therefore, an amplicon plasmid of around 5-kbp will be repeated some 30 times in the amplicon vector, while a very large amplicon plasmid, carrying a 150-kbp genomic locus, will generate amplicon vectors carrying a single repeat of this sequence. A second major benefit that arises from the lack of virus genes in the amplicon plasmid is that most of the 150-kbp capacity of the HSV-1 particle can be used to accommodate very large pieces of foreign DNA. This is undoubtedly the most outstanding property of amplicons, as there is no other viral vector system available displaying the capacity to deliver such a large amount of foreign DNA to the nuclear environment of mammalian cells.

1.2 Preparation of Non-toxic Amplicon Vector Stocks

Since HSV-1-amplicon vectors carry no viral genes they are replication-defective and depend on helper functions for production. The helper genome should provide all virus functions required to replicate and package the amplicon genome (including replication, structural, and DNA packaging proteins), but should lack packaging signals to avoid packaging of the helper genome itself. It is critical that amplicon stocks that will be used for gene transfer and gene therapy do not contain contaminating helper particles, which can be toxic and induce immune responses. While during many years it was not possible to generate such helper-free amplicon vectors, two different methodologies have more recently been developed that allow generating amplicon stocks either completely free of helper particles or only faintly contaminated with fully defective helper particles, as described in the following paragraphs.

1.2.1 Production of Amplicon Vectors by Co-transfecting Amplicon Plasmids DNA and Helper Genomes

Helper functions can be provided by replication-competent, but packaging-defective HSV-1 genomes cloned as set of cosmids [18] or bacterial artificial chromosome (BAC) [19]. Following transfection into mammalian cells, sets of cosmids that overlap and represent the entire HSV-1 genome can form circular replication-competent viral genomes via homologous recombination. These reconstituted viral genomes give rise to infectious virus progeny. Similarly, BACs that contain the entire HSV-1 genome also produce infectious virus progeny in transfected cells. If the viral DNA packaging/cleavage (*a*) signals are deleted from the HSV-1 cosmids or HSV-1 BACs, reconstituted virus genomes are packaging-defective; however, in the absence of the *a* signals, these genomes can still provide all helper functions required for the replication and packaging of co-transfected amplicon DNA. The resulting amplicon vector stocks are essentially free of helper virus contamination. To improve safety, in the latest version of this strategy [19] the helper genome carried by the BAC lacks a gene encoding one essential virus function (generally ICP27) and its length is oversized, thus further avoiding packaging. Amplicon plasmids are therefore replicated and packaged in a cell line complementing the lacking virus function, and often co-transfected with a plasmid expressing this function, as illustrated in Fig. 2. For details on the preparation of amplicon vectors following this approach, refer to Subheadings 2.1 and 3.1.

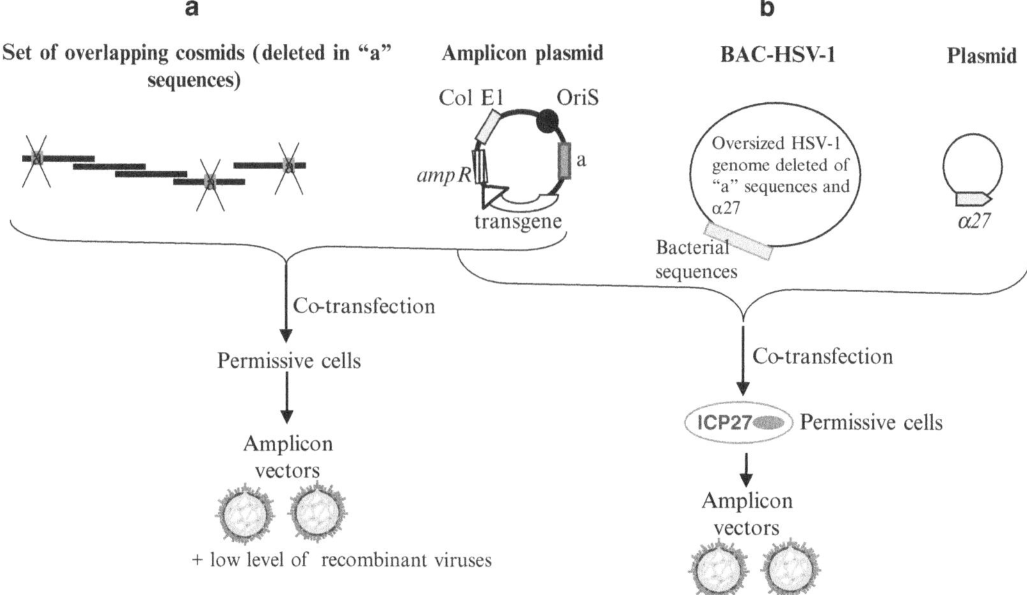

Fig. 2 DNA-based methods for amplicon vector production. Transfection of cells with either a set of cosmids carrying overlapping HSV-1 fragments or a bacterial artificial chromosome (BAC) carrying a modified HSV-1 genome. (**a**) A set of 5 overlapping cosmids, deleted for the packaging signals (*a*), is co-transfected with the amplicon plasmid into permissive cells. This allows the generation by homologous recombination of a virus genome, able to express all transacting functions but that cannot be packaged into HSV-1 particles, as well as a stock of amplicon vectors. (**b**) A BAC-HSV-1 genome deleted for both the packaging *a* signals and the α27 gene is co-transfected with both the amplicon plasmid and a plasmid encoding ICP27 protein. As the BAC-HSV-1 genome is oversized and deleted for the packaging signals it cannot therefore be packaged into HSV-1 particles, allowing the production of helper virus-free HSV-1 amplicon vector stocks

1.2.2 Production of Amplicon Vectors Using the Cre/Loxp1 Site-Specific Recombination System

Alternatively, amplicon vector stocks containing only very low amounts of otherwise defective helper virus can be prepared using a system based on the deletion of the *a* signals of the helper virus genome by Cre/loxP1-based site-specific recombination, to inhibit packaging of the helper genome in the cells that are producing the amplicons [20]. This helper virus, named HSV-1-LaLΔJ, carries a unique and ectopic *a* signal, flanked by two loxP1 sites in parallel orientation. This is therefore a Cre-sensitive virus that cannot be packaged in Cre-expressing cells due to deletion of the floxed packaging signal. Nevertheless, some helper genomes can escape action of the Cre-recombinase, allowing the production of some contaminating helper virus particles. For this reason, the two genes surrounding the cleavage/packaging signal, which respectively encode a virulence factor known as ICP34.5 and the essential protein ICP4, were deleted from the HSV-1-LaLΔJ helper genome [20]. Although the amplicon stocks prepared with this helper virus (in a complementing cell line expressing both Cre and ICP4 proteins) still can contain a small amount of contaminating helper

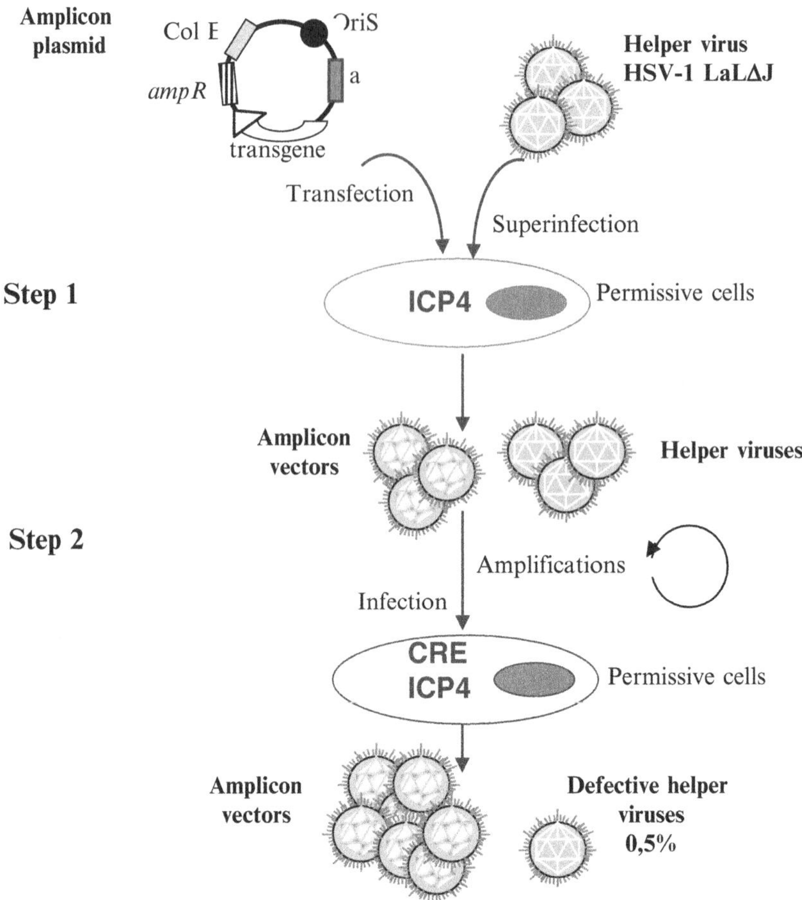

Fig. 3 Helper virus-based methods for amplicon vector production. Step 1: After transfection of ICP4 *trans-complementing* cells with the amplicon plasmid, the viral proteins required for the replication and packaging of this plasmid are supplied by super-infection with helper virus (HSV-1-LaLΔJ). Serial passages of the resulting amplicon stock can be done on the same cells, resulting in large-scale vector production. At this step, the ratio of amplicon to helper particles is generally 1:1. Step 2: Cells expressing both ICP4 and Cre-recombinase are infected with the contaminated amplicon stock produced in Step 1. The Cre-recombinase will induce deletion of the packaging sequences (*a*) of the HSV-1-LaLΔJ genome, therefore avoiding its packaging. The resulting amplicon vector stock can contain a very small amount of HSV-1-LaLΔJ helper virus (0.05–0.5 %) but this helper virus is completely replication-defective and non-pathogenic

particles, these particles are replication incompetent and cannot spread upon infection of target cells or tissues. The amplicon packaging process using HSV-1-LaLΔJ as the helper virus includes two steps: a first one, in ICP4-complementing cells, allows generating large amounts of helper-contaminated amplicon vectors, while the second one, in ICP4 and Cre-recombinase expressing cells, allows eliminating by Cre-mediated deletion of the packaging signal, most of the contaminating helper viruses (*see* Fig. 3).

Use of the HSV-1-LaLΔJ helper virus system generally results in the production of large stocks of amplicon vectors only barely contaminated (0.05–0.5 %) with defective, non-pathogenic helper virus particles. For details on the preparation of amplicon vectors following this strategy, refer to Subheadings 2.2 and 3.2.

2   Materials

2.1   Packaging of HSV-1 Amplicon Vectors Using a Replication-Competent, Packaging-Defective HSV-1 Genome Cloned as a BAC

2.1.1   Preparation of HSV-1 BAC DNA

The following paragraphs describe the materials required for preparation of: (1) large amounts of HSV-BAC DNA; (2) co-transfection of amplicon DNA and HSV-BAC DNA into VERO 2-2 cells by cationic liposome-mediated transfection; and (3) conc entration/purification and titration of packaged amplicon vector stocks.

1. *E. coli* clones of HSV-1 BAC fHSVΔpacΔ27ΔKn and plasmid pEBHICP27 [19].

2. LB medium containing 12.5-µg/mL chloramphenicol.

3. Plasmid Maxi Kit (Qiagen, Hilden, Germany), which includes Qiagen-tip 500 columns and buffers P1, P2, P3, QBT, QC, QGT, and QF.

4. TE buffer pH 7.4.

5. Restriction endonucleases *Hin*dIII and *Kpn*I (New England Biolabs, Ipswich, MA).

6. TAE electrophoresis buffer (10×): 24.2 g Tris base, 5.71 mL glacial acetic acid, 3.72 g $Na_2EDTA \cdot 2H_2O$, H_2O to 1 L. Store at room temperature.

7. Graduated snap-cap tubes 17×100 mm (e.g. BD Falcon, BD Biosciences, San Jose, CA), sterile.

8. Sorvall™ GSA and SS-34 rotors (Thermo Scientific™, Rockford, IL)

9. 120 mm-diameter folded filters (Schleicher & Schüll, Dassel, Germany).

10. Ultra Clear Centrifuge tubes 13×51 mm (Beckman, Munich, Germany).

11. TV 865-ultracentrifuge rotor (Sorvall™).

12. 1 mL disposable syringes.

13. 21-gauge and 36-gauge hypodermic needles.

14. UV-lamp (366 nm).

15. Dialysis cassettes, Slide-A-Lizer 10K (10,000 MWCO; Pierce, Rockford, USA).

16. UV spectrophotometer (e.g. Ultrospec 3000, Pharmacia and Upjohn Company, Kalamazoo, MI).

17. Bench centrifuge.

18. Vortex.

19. Slide-A-Lyzer Dialysis Cassettes, 10K MWCO (Thermo Scientific™).

2.1.2 Preparation of HSV-1 Amplicon Vector Stocks

1. Vero 2-2 cells [21].

2. Any amplicon plasmid. *See* **Note 1**.

3. Dulbecco's modified Eagle medium (DMEM, Life Technologies™, Carlsbad, CA) with 10 and 6 % fetal bovine serum.

4. G418 (Geneticin, Life Technologies™).

5. 0.25 % trypsin/0.02 % EDTA (Life Technologies™).

6. Opti-MEM I reduced-serum medium (Life Technologies™).

7. HSV-1 BAC fHSVΔpacΔ27ΔKn and pEBHICP27 plasmid DNA [19] (Dr. Y. Saeki, Ohio State University, Columbus, OH, USA: saeki.6@osu.edu).

8. HSV-1 amplicon DNA (maxiprep DNA isolated from *E. coli*).

9. LipofectAMINE Plus reagent (Life Technologies™).

10. Phosphate-buffered saline (PBS).

11. 10, 30, and 60 % (w/v) sucrose in PBS.

12. 75-cm² tissue culture flasks.

13. Humidified 37 °C, 5 % CO_2 incubator.

14. 60-mm-diameter tissue culture dishes.

15. 15-mL conical centrifuge tubes.

16. Dry ice/ethanol bath.

17. Probe sonicator.

18. 0.45 μm syringe-tip filters (Sarstedt polyethersulfone membrane filters).

19. 20 mL disposable syringes.

20. 30 mL centrifuge tubes (e.g. Beckman Ultra-Clear 25 × 89 mm and 14 × 95 mm, Beckman Coulter Inc., Brea, CA).

21. Sorvall™ SS-34 rotor (Thermo Scientific™, Rockford, IL).

22. Fiber-optic illuminator.

23. Ultracentrifuge (Sorvall™) with Beckman SW28 and SW40 rotors (Beckman Coulter).

2.1.3 Harvesting, Purification, and Titration of HSV-1 Amplicon Vectors

1. Vero cells (clone 76; ECACC, Salisbury, UK); BHK cells (clone 21; ECACC); 293 cells (ATCC, LGC Standards, Teddington, UK).

2. DMEM (e.g. Life Technologies™) supplemented with 10 and 2 % FBS.

3. 4 % (w/v) paraformaldehyde solution.

4. 5-Bromo-4-chloro-3-indolyl-b-d-galactopyranoside (X-gal) staining solution: 20 mM $K_3Fe(CN)_6$, 20 mM $K_4Fe(CN)_6 \cdot 3H_2O$, 2 mM $MgCl_2$ in PBS pH 7.5. Filter sterilize and store up to 1 year at 4 °C. Before use, equilibrate solution to 37 °C and add 20 µL/mL of 50 mg/mL X-gal in DMSO. Store X-gal solution in 1-mL aliquots up to several years at –20 °C in the dark.

5. GST solution: 2 % (v/v) goat serum and 0.2 % (v/v) Triton X-100 in PBS. Store up to 1 month at 4 °C.

6. Primary and secondary antibodies specific for detection of the transgene product.

7. 24-Well tissue culture plates.

8. Humidified 37 °C, 5 % CO_2 incubator.

9. Inverted fluorescence microscope.

10. Inverted light microscope.

11. Hemocytometer.

12. Rubber policeman.

13. Fiber-optic illuminator.

2.2 Packaging of Amplicon Vectors Using a Replication-Incompetent, Cre/loxP1 Sensitive Helper Virus

2.2.1 Preparation of the Defective Helper Virus

The following paragraph describes the materials required to produce large amounts of defective HSV-1-LaLΔJ helper virus.

1. Vero (African green monkey cells, ATCC).

2. Vero-7b [22] cell line.

3. Gli36 cell line [23].

4. TE-Cre-Grina cell line [20].

5. Six-well tissue culture plates (Corning Incorporated, Corning, NY).

6. Growth medium: DMEM (Invitrogen™ Life Technologies™) supplemented with 10 % fetal bovine serum (FBS, Invitrogen™), 100 units/mL penicillin and 100 µg/mL streptomycin, both from Invitrogen™. All cell lines are maintained at 37 °C in humidified incubators containing 5 % CO_2.

7. Maintenance medium: medium 199 (Lonza, Basel, Switzerland) supplemented with 1 % FBS.

8. PBS.

9. G418 (Life Technologies™).

10. Opti-MEM (Ultra-MEM, Lonza).

11. LipofectAMINE Plus reagent (Invitrogen™).

12. Polystyrene roller bottles (VWR International LLC, Radnor, PA).

3 Methods

***3.1 Packaging
of HSV-1 Amplicon
Vectors Using
a Replication-
Competent,
Packaging-Defective
HSV-1 Genome
Cloned as a BAC***

*3.1.1 Preparation
of HSV-1 BAC DNA*

1. Prepare a 17×100-mm sterile snap-cap tube containing 5 mL LB/chloramphenicol medium. Inoculate with frozen long-term culture of the HSV-1 BAC clone (fHSVΔpacΔ27ΔKn). Incubate for 8 h at 37 °C in a shaker.

2. Transfer 1 mL of the culture into each of four 2-L flasks containing 1,000 mL sterile LB/chloramphenicol medium, and incubate for 16 h at 37 °C, with shaking.

3. Distribute the 4 L overnight culture into six 250 mL polypropylene centrifuge tubes and pellet by centrifugation for 10 min at $4,000 \times g$, 4 °C. Decant medium, fill polypropylene centrifuge tubes again with bacterial culture, and repeat centrifugation.

4. After the last centrifugation, invert each tube on a paper towel for 2 min to drain all liquid. Resuspend each of the pellets in 5 mL buffer P1 and combine the six aliquots. Add 130 mL buffer P1 and distribute to four fresh 250-mL polypropylene centrifuge tubes (40 mL per tube).

5. Add 40 mL buffer 2 to each centrifuge tube, mix by inverting the tubes four to six times, and incubate for 5 min at room temperature.

6. Add 40 mL buffer P3 and mix immediately by inverting the tubes six times. Incubate the tubes for 20 min on ice. Invert the tube once more and centrifuge for 30 min at $16,000 \times g$, 4 °C.

7. Filter the supernatants through a folded filter (120 mm diameter) into four fresh 250-mL polypropylene centrifuge tubes.

8. Precipitate the DNA with 0.7 volumes (84 mL per tube) isopropanol, mix gently, and centrifuge immediately for 30 min at $17,000 \times g$, 4 °C.

9. Remove the supernatants and mark the locations of the pellet. Wash the DNA pellet by adding 20 mL cold 70 % ethanol to each and centrifuge for 30 min at $16,000 \times g$, 4 °C.

10. Carefully remove the supernatants and resuspend each of the four pellets in 2 mL TE buffer, pH 7.4. Pool the four solutions (total volume 8 mL) and add 52 mL QGT buffer (final volume 60 mL).

11. Equilibrate two Qiagen-tip 500 columns with 10 mL buffer QBT, and allow the columns to empty by gravity flow.

12. Transfer the solution through a folded filter (120 mm diameter) into Qiagen-tip 500 columns (30 mL per column), and allow the liquid to enter the resin by gravity flow.

13. Wash each column twice with 30 mL buffer QC, and then elute DNA from each column with 15 mL prewarmed (65 °C) buffer QF into a 30 mL centrifuge tube.

14. Precipitate the DNA with 0.7 volumes (10.5 mL) isopropanol, mix, and immediately centrifuge for 30 min at $20,000 \times g$, 4 °C.

15. Carefully remove the supernatants from **step 14** and mark the locations of the pellets on the outside of the tubes. Wash the pellets with chilled 70 % ethanol and, if necessary, re-pellet at the same settings as in **step 14**.

16. Aspirate the supernatants completely. Resuspend each pellet in 3 mL TE buffer (pH 7.4) for several hours at 37 °C.

17. Prepare two Beckman Ultra Clear Centrifuge tubes (13×51 mm) with 3 g CsCl and add the DNA solution from **step 16** (3 mL per tube). Mix the solution gently until salt is dissolved. Add 300-µL ethidium bromide (10 mg/mL in H_2O) to the DNA/CsCl solution. Then overlay the solution with 300 µL paraffin oil and seal the tubes.

18. Centrifuge for 17 h at $218,500 \times g$, 20 °C.

19. Two bands of DNA, located in the center of the gradient, should be visible in normal light. The upper band consists of linear and nicked circular HSV-1 BAC DNA. The lower band consists of closed circular HSV-1 BAC DNA.

20. Harvest the lower band using a disposable 1 mL syringe fitted with a 21-gauge hypodermic needle under UV-light and transfer it into a microfuge tube.

21. Remove ethidium bromide from the DNA solution by adding an equal volume of n-butanol in TE/CsCl (3 g CsCl dissolved in 3 mL TE, pH 7.4).

22. Mix the two phases by vortexing and centrifuge at $210 \times g$ for 3 min at room temperature in a bench centrifuge.

23. Carefully transfer the lower, aqueous phase to a fresh microfuge tube. Repeat **steps 21–23** four to six times until the pink color disappears from both the aqueous phase and the organic phase.

24. Add an equal volume of isopropanol, mix and centrifuge at $210 \times g$ for 3 min at room temperature. Transfer the aqueous phase to a fresh microfuge tube.

25. To remove the CsCl from the DNA solution, dialyze 6 h against TE, pH 7.4 at 4 °C. Then, change the TE buffer and dialyze overnight. For dialysis, the DNA solution is injected into a dialysis cassette (Slide-A-Lyzer Dialysis Cassette, 10 K MWCO) using a 1 mL disposable syringe fitted with a 36-gauge hypodermic needle. After dialysis, the solution is recovered from the dialysis cassette by using a fresh 1 mL disposable

syringe fitted with a 36-gauge hypodermic needle. The DNA solution is then transferred to a clean microfuge tube and stored at 4 °C. After characterization of the DNA (concentration and restriction enzyme analysis), store DNA at 4 °C.

26. Determine the absorbance of the DNA solutions from **step 25** at 260 nm (A_{260}) and 280 nm (A_{280}) using an UV spectrophotometer. From 4 L of bacterial cultures, HSV-BAC DNA yields are typically in the range of 200–300 μg.

27. Verify the HSV-1 BAC DNA by restriction endonuclease analysis (e.g. *Hin*dIII, *Kpn*I). Separate the fragments overnight by electrophoresis on a 0.4 % agarose gel at 40 V in 1× TAE electrophoresis buffer (*see* **Note 2**), using high-molecular-weight DNA and 1-kb DNA ladder as size standards. Stain with ethidium bromide (1 mg/mL in H_2O) and compare restriction fragment patterns with the published HSV-1 sequence [24].

3.1.2 Preparation of Plasmid DNA (Maxiprep Protocol–Qiagen®)

1. Prepare a 17 × 100 mm sterile snap-cap tube containing 5 mL LB/chloramphenicol medium. Inoculate with frozen long-term culture of the *E. coli* harboring the plasmid. Incubate for 8 h at 37 °C in a shaker.

2. Transfer 1 mL of the culture into a 1 L flask containing 200 mL sterile LB medium with the appropriate antibiotic, and incubate for 16 h at 37 °C, with shaking.

3. Transfer the overnight culture into a 250 mL polypropylene centrifuge tube and pellet by centrifugation for 10 min at 4,000 × *g*, 4 °C. Decant medium and invert the tube on a paper towel for 2 min to drain all liquid. Resuspend the pellet in 10 mL buffer P1.

4. Add 10 mL buffer P2, mix by inverting the tube four to six times, and incubate for 5 min at room temperature.

5. Add 10 mL chilled buffer P3 and mix immediately by inverting the tube six times. Incubate the tube for 20 min on ice. Invert the tube once more and centrifuge for 30 min at 16,000 × *g*, 4 °C.

6. Filter the supernatants through a folded filter (120 mm diameter) into a 30 mL centrifuge tube.

7. Equilibrate a Qiagen-tip 500 column with 10 mL buffer QBT, and allow the column to empty by gravity flow.

8. Transfer the solution from **step 6** into the Qiagen-tip 500 column, and allow the liquid to enter the resin by gravity flow.

9. Wash the column twice with 30 mL buffer QC, and then elute DNA from the column with 15 mL pre-warmed (65 °C) buffer QF into a 30 mL centrifuge tube.

10. Precipitate the DNA with 0.7 volumes (10.5 mL) isopropanol, mix, and immediately centrifuge for 30 min at 20,000×*g*, 4 °C.

11. Carefully remove the supernatant from **step 10** and mark the location of the pellet on the outside of the tube. Wash the pellet with chilled 70 % ethanol and, if necessary, re-pellet at the same settings as in **step 10**.

12. Aspirate the supernatant completely. Resuspend the pellet in 200 µL TE buffer (pH 7.4), and determine the DNA concentration using a UV spectrophotometer.

3.1.3 Transfection of Vero 2-2 Cells and Harvesting, Concentration, and Purification of Packaged Amplicon Vectors

1. Maintain Vero 2-2 cells in DMEM/10 % FBS containing 500 µg/mL G418. Propagate the culture twice a week by splitting 1/5 in fresh medium (20 mL) into a new 75 cm² tissue culture flask. *See* **Note 3**.

2. On the day before transfection, remove culture medium, wash twice with PBS, add a thin layer of trypsin/EDTA, and incubate for 10 min at 37 °C to allow cells to detach from plate. Count cells using a hemocytometer and plate 1.2×10^6 cells per 60 mm-diameter tissue culture dish in 3 mL DMEM/10 % FBS.

3. For each 60 mm dish, place 250 µL Opti-MEM I reduced-serum medium into each of two 15-mL conical tubes. To one tube, add 0.6-µg amplicon DNA, 2 µg of the HSV-1 BAC DNA and 0.2-µg pEBHICP27 DNA. Mix the tube and slowly add 10 µL PLUS reagent. Incubate the tube for 5 min at room temperature, mix and incubate for another 5 min. To the other tube, add 15 µL LipofectAMINE.

4. Combine the contents of the two tubes, mix well, and incubate for 45 min at room temperature.

5. Wash the cultures prepared the day before (**step 2**) once with 2 mL Opti-MEM I. Add 1.1 mL Opti-MEM I to the tube from **step 4** containing the DNA-LipofectAMINE transfection mixture (1.3 mL total volume). Aspirate medium from the culture, add the transfection mixture, and incubate for 5.5 h.

6. Aspirate the transfection mixture and wash the cells three times with 2 mL Opti-MEM I. After aspirating the last wash, add 3.5 mL DMEM/6 % FBS and incubate for 2–3 days.

7. Scrape cells into the medium using a rubber policeman. Transfer the suspension to a 15 mL conical centrifuge tube and place the tube containing the cells into a beaker of ice water. Submerge the tip of the sonicator probe ~0.5 cm into the cell suspension and sonicate 20 s with 20 % output energy. This disrupts cell membranes and liberates cell-associated vector particles.

8. Remove cell debris by centrifugation for 10 min at 1,400×*g*, 4 °C and filter the supernatant through a 0.45-μm syringe-tip filter attached to a 20 mL disposable syringe into a new 15 mL conical tube. Remove a sample for titration, then divide the remaining stock into 1 mL aliquots, freeze them in a dry ice/ethanol bath, and store at –80 °C. Alternatively, concentrate (**steps 9a** and **10a**) or purify and concentrate (**steps 9b** and **10b**) the stock before storage.

Protocol a (pelleting):

9a. Transfer the vector solution from **step 8** to a 30 mL centrifuge tube and spin 2 h at 20,000×*g*, 4 °C.

10a. Resuspend the pellet in a small volume (e.g. 300 μL) of 10 % sucrose. Remove a sample of the stock for titration, then divide into aliquots (e.g. 30 μL) and freeze in a dry ice/ethanol bath. Store at –80 °C.

Protocol b (gradient centrifugation):

9b. Prepare a sucrose gradient in a Beckman Ultra-Clear 25×89-mm centrifuge tube by layering the following solutions in the tube: 7 mL 60 % sucrose; 7 mL 30 % sucrose; 3 mL 10 % sucrose. Carefully add the vector stock from **step 8** (up to 20 mL) on top of the gradient and centrifuge for 2 h at 100,000×*g*, 4 °C, using a Beckman SW28 rotor.

10b. The interface between the 30 and 60 % sucrose layers appears as a cloudy band when viewed with a fiber-optic illuminator. Aspirate the 10 and 30 % sucrose layers from the top and collect the virus band at the interface between the 30 and 60 % layers. Transfer to a Beckman Ultra-Clear 14×95-mm centrifuge tube, add ~15 mL PBS, and pellet virus particles for 1 h at 100,000×*g*, 4 °C, using a Beckman SW40 rotor. Resuspend the pellet in a small volume (e.g. 300 μL) of 10 % sucrose. Divide into aliquots (e.g. 30 μL) and freeze in a dry ice/ethanol bath. Store at –80 °C. Before freezing, retain a sample of the stock for titration.

3.1.4 Titration of HSV-1 Amplicon Vector Stocks

1. Plate cells (e.g. Vero 7b, BHK 21, or 293 cells) at a density of $1.0×10^5$ per well of a 24-well tissue culture plate in 0.5 mL DMEM/10 % FBS. Incubate overnight.

2. Aspirate the medium and wash each well once with PBS. Remove PBS and add 0.1 μL, 1 μL, or 5 μL samples collected from vector stocks, diluted to 250 μL each in DMEM/2 % FBS.

3. Incubate for 1–2 days. Remove the inoculum and fix cells for 20 min at room temperature with 250 μL of 4 % paraformaldehyde, pH 7.0. Wash the fixed cells three times with PBS,

then proceed (depending on the transgene) with a detection protocol such as green fluorescence (**step 4a**), X-gal staining (**steps 4b** and **5b**), or immunocytochemical staining (**steps 4c–6c**).

Protocol a (fluorescence detection):

4a. Detect cells expressing the gene for EGFP (*see* **Note 4**): Examine the culture from **step 3** (before or after fixation) using an inverted fluorescence microscope. Count green fluorescent cells and determine the vector titer in transducing units (TU)/mL by multiplying the number of transgene-positive cells by the dilution factor. *See* **Notes 5** and **6**.

Protocol b (X-gal detection):

4b. Detect cells expressing the *E. coli lacZ* gene: Add 250 µL X-gal staining solution per well of the 24-well tissue culture plate from **step 3**, and incubate for 4–12 h (depending on the cell type and the promoter regulating expression of the transgene) at 37 °C.

5b. Stop the staining reaction by washing the cells three times with PBS. Count blue cells using an inverted light microscope, and determine the vector titer in TU/mL by multiplying the number of transgene-positive cells by the dilution factor.

Protocol c (immunocytochemical staining):

4c. Detect transgene-expressing cells by immunocytochemical staining: Add 250 µL GST solution per well of the 24-well tissue culture plate from **step 3** (to block nonspecific binding sites and to permeabilize cell membranes) and let stand for 30 min at room temperature. Replace the blocking solution with the primary antibody (diluted in GST) and incubate overnight at 4 °C.

5c. Wash the cells three times with PBS, leaving the solution in the well for 10 min each time. Add secondary antibody (diluted in GST) and incubate for at least 4 h at room temperature.

6c. Wash the cells twice with PBS and develop according to the appropriate visualization protocol. Count transgene-positive cells using an inverted light microscope and determine the vector titer as TU/mL by multiplying the number of the transgene-positive cells by the dilution factor.

3.2 Packaging of Amplicon Vectors Using a Replication-Incompetent, Cre/ loxP1-Sensitive Helper Virus

This protocol describes methods (1) to prepare, purify, and titrate the HSV-1-LaLΔJ helper virus, and (2) to prepare, purify, and titrate amplicon stocks using HSV-1-LaLΔJ as the helper virus. HSV-1-LaLΔJ [20] is a defective virus lacking both copies of the gene encoding the essential ICP4 protein, as well as both copies of the neurovirulence factor ICP34.5. In addition, it lacks all native cleavage/packaging signals *a*, and contains, instead, a unique ectopic *a* signal bordered by two loxP1 sites in parallel orientation, which has been introduced into the gC (UL44) locus of the virus genome. Therefore, the virus also lacks the non-essential glycoprotein gC. HSV-1-LaLΔJ can grow only in cells complementing the essential ICP4 function, such as the Vero-derived 7b cell line [22]. In contrast, this virus cannot grow in cells expressing Cre-recombinase, such as the TE-Cre-Grina cell line, even if this cell line expresses ICP4 [20] due to the deletion of the *a* sequences.

3.2.1 Production, Purification, and Titration of HSV-1-LaLΔJ Helper Virus Stocks

1. As before quoted, HSV-1-LaLΔJ is a defective recombinant virus. Therefore, to prepare stocks of this virus, follow the instructions described in the accompanying chapter on HSV-1 recombinant vectors, protocol 3.1.4. However, since HSV-1-LaLΔJ lacks the gene encoding ICP4, it should be grown in ICP4-expressing cells, such as the 7b Vero-derived cell line [22]. These cells grow in DMEM medium supplemented with 10 % FBS, 2 mM l-glutamine, 100-units/mL penicillin and 100-μg/mL streptomycin. Geneticin (G418) should be added every four passages (1 mg/mL), to avoid losing the complementing ICP4 gene.

2. To titrate the helper virus stock, follow the instructions described in Subheading 3.1.5 of the accompanying Chapter 20. The virus should be titrated simultaneously in complementing cells, such as Vero-7b, and in non-complementing Vero cells, to allow detection of unwanted replication-competent particles that can sometimes be generated by recombination between the virus genome and the ICP4 gene located in the cellular genome; such revertant viruses should produce lysis plaques in Vero cells. If this is the case, start the production again, infecting the complementing cells at a very low MOI (lower than 0.05 PFU/cell), using plaque-purified defective virus.

3. To purify virus stock if required, follow the instruction described in Subheading 3.1.6 of the accompanying Chapter 20.

3.2.2 Production of Amplicon Vectors Using Cre/loxP1 Site-Specific Recombination

The production of amplicon vector stocks using HSV-1-LaLΔJ as the helper virus is a two-step protocol, described in detail in reference [20] and illustrated in Fig. 3. In the first step, stocks of amplicons contaminated with large amounts of helper virus particles are produced in ICP4-complementing cells (Vero-7b cells). In the second step, stocks of vectors containing only very small amounts of replication-defective helper viruses are prepared in cells expressing both ICP4 and Cre-recombinase (TE-Cre-Grina cells) as follows:

Generation of P0 Stock (Helper-Contaminated, HC)

1. The day before transfection, plate 5×10^6 Vero-7b cells in a 75-cm^2 tissue culture flask with growth medium (*see* Subheading 2.2.1).

2. Transfect the amplicon plasmid. Per 75 cm^2 cell culture flask: Mix 6 μg amplicon plasmid DNA + 750 μL Opti-MEM + 30 μL plus reagent. Incubate for 15 min at room temperature, and

add a mix of 45 μL LipofectAmine + 750 μL Opti-MEM. After 15 min at room temperature, add the transfection mix to the Vero-7b cells in 10 mL Opti-MEM medium and incubate at 37 °C in 5 % CO_2.

3. After 3 h add 10 mL Opti-MEM medium to the cells and incubate the cultures overnight.

4. Infect the transfected cells with helper virus as follows:

5. One day after transfection, discard medium from the flask. *See* **Note 7**.

6. Rinse cells once with maintenance medium (*see* Subheading 2.2.1) and discard medium.

7. Add 3 mL maintenance medium containing the helper virus diluted to a MOI of 0.5 PFU/cell.

8. Place the flask on a shaker for 1½ h, if possible under CO_2 atmosphere.

9. Discard medium, rinse twice with maintenance medium.

10. Add 20 mL maintenance medium.

11. Incubate for 48 h at 37 °C in a 5 % CO_2 incubator.

12. Collect helper-contaminated amplicon vector stocks: At 48 h post-infection, most cells should have rounded and show cytopathic effect (CPE) typical of HSV-1. Scrape the cells into the medium and transfer the suspension into 50 mL Falcon tubes.

13. Spin down at $771 \times g$ for 10 min at 4 °C.

14. Transfer the supernatant (SN) to a 35 mL oak ridge tube (this is SN1).

15. Resuspend the cell pellet in 1 mL PBS and disrupt the cells either by 3 cycles of freezing–thawing or using a water sonicator (3 times 30 s in cold water). Then, spin down at $771 \times g$ for 10 min at 4 °C.

16. Discard the pellet containing cell debris and store the supernatant containing the viral particles (this is SN2) at −80 °C.

17. Spin down SN1 at $18,000 \times g$ at 4 °C for 1 ½ h. Discard the supernatant and resuspend the pellet containing virus particles in 1 mL PBS.

18. Add to SN2 and keep this final P0 vector stock at −80 °C until titration.

Titration of Amplicon Vectors and Helper Virus in P0 Stocks

1. One day prior to titration of the P0 virus stock prepare 6-well tissue culture plates with 1×10^6 Gli36 cells, Vero-7b cells or Vero cells per well. These three cell lines are propagated in growth medium as described in Subheading 2.2.1. *See* **Note 8**.

2. Prepare a series of tenfold dilutions (10^{-2}–10^{-10}) of the vector stock in Eppendorf tubes with 1 mL growth medium without serum.

Table 1
Titers, ratios, and amounts of amplicon vectors and helper particles (*see* Note 14)

	P0 (HC)	P1 (HC)	P2 (HC)	P3 (HF)
Titer amplicon (TU/mL)	10^7	10^8	10^9	10^8
Titer helper (PFU/mL)	3×10^7	5×10^7	10^8	5×10^5
Ratio A/H	1:3	2:1	10:1	200:1
Amount (mL)	0.5	1	5–10	5–10

3. Infect cells as described in Subheading 3.1.5 of the accompanying Chapter 20 on HSV-1 recombinant vectors.

  (a) Determination of the titer of amplicon particles: One day after infection, if the amplicon vectors express GFP, count green fluorescent Gli36 cells and determine the average for each dilution (at least in duplicate). Then multiply this number by the dilution factor to determine the titer of the vector stock in transduction units (TU)/mL. If the amplicon expresses LacZ, use the X-gal staining protocol described in Subheading 3.1.4.

  (b) Determination of the helper virus titer: Three days after infection, fix, stain, and count the number of plaques per well in the Vero 7b monolayers. Determine the average for each dilution (at least in duplicate), and multiply by the dilution factor to calculate the number of plaque forming units/mL (PFU/mL).

  (c) Determination of the titer of replication-competent revertant virus: Proceed exactly as in 3b but infecting non-trans-complementing Vero cells (*see* **Note 9**). *See* Table 1.

For more details on the titration of amplicon vectors and helper virus, *see* Subheadings 3.1.4 (this chapter, for amplicons) and 3.1.5 (accompanying Chapter 20 on HSV-1 recombinant vectors, for helper particles).

Amplification from P0 to P1 and Titration of P1 Stocks (Helper-Contaminated)

The stock of helper-contaminated amplicon vectors should be expanded by amplifying the P0 vector stock in Vero-7b cells.

1. The day before infection, plate 1.3×10^7 Vero-7b cells per 175-cm^2 tissue culture flask.

2. Infect cells using the P0 vector stock:

3. Add 5 mL maintenance medium containing the vector stock diluted to a MOI of 0.3 PFU/cell.

4. Place the flask on a shaker for 1 h 30, if possible under CO_2 atmosphere.

5. Discard medium, rinse cells twice with maintenance medium.

6. Add 30 mL maintenance medium.

7. Then proceed as in Subheading 3.2.2.1, **step 11** to generate the P1 vector stock.

8. Titrate P1 vector stock as in Subheading 3.2.2.2 (*see* **Note 10**). *See* Table 1.

Amplification from P1 to P2 and Titration of P2 Stocks (Helper-Contaminated)

1. It is often convenient to further amplify the vector stock. To this end, expand the amount of Vero-7B cells to be infected as much as desired. If you prefer to use 175 cm^2 tissue culture flasks, proceed as in Subheading 3.2.2.3 but expanding the number of tissue culture flasks and scaling up the procedure. However, instead of using 175 cm^2 flasks, at this step we prefer to infect cells in roller bottles, which are easier to manipulate and require less medium. The protocol is as follows:

2. Seed 2×10^7 Vero-7b cells/roller bottle in 100 mL of growth medium. Since cells in roller bottles are not incubated in a CO_2 atmosphere, CO_2 should be added to the growth medium using a pipette connected to a CO_2 tube, until CO_2 bubbles fill up the bottle.

3. Turn the roller bottles at a speed of 0.4 rounds per min. Cells generally become confluent (10^8 cells/bottle) in 4–5 days at 37 °C.

4. Infect cells with the P1 vector stock. Discard medium from each roller bottle and add 20 mL of maintenance medium containing the vector stock diluted to a MOI of 0.3 PFU/cell.

5. Two hours later add maintenance medium to a final volume of 100 mL per roller bottle.

6. Incubate for 48 h at 37 °C constantly turning the bottles at a speed of 0.4 rounds per min.

7. When CPE is maximum, which generally occurs at 48 h post-infection, collect the vectors as in Subheading 3.2.2.1, **step 12**, but scale up the number of tubes. This is the P2 vector stock.

8. Titrate P2 vector stock as described in Subheading 3.2.2.2 (*see* **Note 11**). See also Table 1.

Production and Titration of P3 Amplicon Vector Stocks (Helper-Free, HF)

1. Plate 1.3×10^7 TE-Cre-Grina cells per 175 cm^2 tissue culture flask.

2. The following day, infect cells with the P2 vector stock at an MOI of 3 TU/cell. At this dilution of the amplicon vector, the concentration of the helper virus in the stock is approximately 0.3–0.5 PFU/cell. If the concentration of helper virus in the stock is too low, add more helper virus. *See* **Note 12**.

3. Place the flask on a shaker for 1½ h, if possible under CO_2 atmosphere.

4. Discard medium, rinse cells twice with maintenance medium.

5. Add 30 mL maintenance medium per flask.

6. Incubate for 48 h at 37 °C in 5 % CO_2 incubator.

7. To collect the virus, proceed as in protocol Subheading 3.2.2.1, **step 12**. This will be the P3 vector stock ("helper-free" (HF) vector stock).

8. Titrate the amplicon vectors and the helper virus as in Subheading 3.2.2.2 (*see* **Note 13**). *See* Table 1.

9. If you wish to purify your amplicon stocks for *in vivo* applications, either helper-contaminated or helper-free stocks, follow the protocol described in Subheading 3.1.6 of the accompanying Chapter 20 on recombinant HSV-1 vectors.

4   Notes

1. An HSV-1 amplicon plasmid is an *E. coli* plasmid containing one origin of DNA replication and one cleavage/packaging signal (*a*) from HSV-1. It usually carries also a reporter gene expressing GFP, LacZ, or luciferase, which allows to easily titrate the vector stock and to identify the infected cells. It contains, in addition, a multiple cloning site where the desired transgene sequences can be introduced. It is produced and purified like any standard bacterial plasmid.

2. Treat gel with care; 0.4 % gels are very delicate.

3. Cells are incubated in a humidified 37 °C, 5 % CO_2 incubator throughout the protocol. All solutions and equipment coming into contact with cells must be sterile.

4. Use of an amplicon vector that expresses an easily detectable reporter (e.g. EGFP) is strongly recommended when establishing the packaging protocol in the laboratory. Although the quality of the DNA and condition of the cells are of prime concern, other components, e.g. lipids, DNA concentration, and incubation times may also influence transfection and packaging efficiency.

5. The titers expressed as transducing units per milliliter (TU/mL) are relative. Factors influencing relative transduction efficiencies include the cells used for titration, the promoter regulating the expression of the transgene, the transgene, and the sensitivity of the detection method.

6. The vector titers realized with amplicons that contain the standard ~1-kb *ori* should be in the range of 10^6–10^7 TU/mL

before concentration. The recovery of transducing units after concentration/purification is around ~50 %.

7. Before infecting the transfected cells, confirm that transfection was efficient, resulting in at least 30 % of cells expressing the reporter transgene (generally GFP). If this is not the case, it is better to transfect again, using fresh cells and/or optimizing the transfection procedure.

8. While the number of physical particles is an intrinsic property of the virus stock, independent of the cell types to be infected, the number of infectious particles, hence the titer of a virus or of a vector stock strongly depends on the susceptibility of the cells. In the case of helper virus-free amplicon vectors, some cell types, such as Gli36 cells (a human glioblastoma cell line), give very high vector titers, while Vero-derived cell lines give much lower vector titers. In contrast, Vero or Vero-derived cells give very good titers of the helper virus.

9. In a typical P0 situation we obtain an amplicon to helper ratio of about 1:3. We usually do not observe the generation of replication-competent virus particles.

10. At this step the ratio of amplicon to helper particles generally inverts in favor of amplicon particles (from 2:1 to 5:1). The titers of P1 are generally one order of magnitude higher than in P0.

11. At this step, the ratio of amplicon to helper particles increases in favor of amplicon particles (from 5:1 to 10:1) while the titers of the stock can be substantially increased, depending on the number of tissue culture flasks infected.

12. The critical point here is that each cell should receive at least one amplicon particle. The infected cells will become round but without displaying an open CPE, as the helper particles cannot spread in these cells.

13. We usually observe less than 1 % contamination of the vector stock with defective helper particles (ratio of amplicon to helper particles ranges between 100:1 and 500:1). However, the titer of the amplicon vectors is generally one order of magnitude lower than that of the P2 stock used to infect TE-Cre-Grina cells.

14. Table 1 presents results obtained in a typical vector preparation. Values can be somewhat different depending on the nature and size of the amplicon plasmid, on the passage number of cell lines, and on the efficiency of transfection in P0. Note that "helper-free" stocks obtained using this strategy can be contaminated to a very low extent with replication-defective helper viruses.

References

1. Sheldrick P, Berthelot N (1975) Inverted repetitions in the chromosome of herpes simplex virus. Cold Spring Harb Symp Quant Biol 39(Pt 2):667–678

2. Wadsworth S, Jacob RJ, Roizman B (1975) Anatomy of herpes simplex virus DNA. II. Size, composition, and arrangement of inverted terminal repetitions. J Virol 15:1487–1497

3. Roizman B, Sears AM (1990) Herpes simplex viruses and their replication. In: Fields BN, Knipe DM (eds) Virology, 2nd edn. Raven Press Ltd., New York, pp 1795–1841

4. Roizman B, Ward PL (1994) Herpes simplex genes: the blueprint of a successful human pathogen. Trends Genet 10:267–274

5. Stow ND, McMonagle EC (1983) Characterization of the TRS/IRS origin of DNA replication of herpes simplex virus type 1. Virology 130:427–438

6. Weller SK, Spadaro A, Schaffer JE et al (1985) Cloning, sequencing and functional analysis of oriL, a herpes simplex virus type 1 origin of DNA synthesis. Mol Cell Biol 5:930–942

7. Igarashi K, Fawl R, Roller RJ et al (1983) Construction and properties of a recombinant herpes simplex virus 1 lacking both S-component origins of DNA synthesis. J Virol 67:2123–2132

8. Polvino-Bodnar M, Orberg PK, Schaffer PA (1987) Herpes simplex virus type 1 oriLis not required for virus replication or for the establishment and reactivation of latent infection in mice. J Virol 61:3528–3535

9. Deiss LP, Chou J, Frenkel N (1986) Functional domains within the *a* sequence involved in the cleavage-packaging of herpes simplex virus DNA. J Virol 59:605–618

10. Mocarski ES, Deiss LP, Frenkel N (1985) The nucleotide sequence and structural features of a novel US-*a* junction present in a defective herpes simplex virus genome. J Virol 55:140–146

11. Mocarski ES, Roizman B (1981) Site-specific inversion sequence of the herpes simplex virus genome: domain and structural features. Proc Natl Acad Sci U S A 78:7047–7051

12. Bataille D, Epstein AL (1995) Herpes simplex virus type 1 replication and recombination. Biochimie 77:787–795

13. Spaete RR, Frenkel N (1982) The herpes simplex virus amplicon: a new eucaryotic defective-virus cloning-amplifying vector. Cell 30:295–304

14. Spaete RR, Frenkel N (1985) The herpes simplex virus amplicon: analyses of cis-acting replication functions. Proc Natl Acad Sci U S A 82:694–698

15. Vlazny DA, Frenkel N (1981) Replication of herpes simplex virus DNA: localization of replication recognition signals within defective virus genomes. Proc Natl Acad Sci U S A 78:742–746

16. Boehmer PE, Lehman IR (1997) Herpes simplex virus DNA replication. Annu Rev Biochem 66:347–384

17. Kwong AD, Frenkel N (1984) Herpes simplex virus amplicon: effect of size on replication of constructed defective genomes containing eucaryotic DNA sequences. J Virol 51:595–603

18. Fraefel C, Song S, Lim F et al (1996) Helper virus-free transfer of herpes simplex virus type 1 plasmid vectors into neural cells. J Virol 70:7190–7197

19. Saeki Y, Fraefel C, Ichikawa T et al (2001) Improved helper virus-free packaging system for HSV amplicon vectors using an ICP27-deleted, oversized HSV-1 DNA in a bacterial artificial chromosome. Mol Ther 3:591–601

20. Zaupa C, Revol-Guyot V, Epstein AL (2003) Improved packaging system for generation of high levels non-cytotoxic HSV-1 amplicon vectors using Cre-loxP1 site-specific recombination to delete the packaging signals of defective helper genomes. Human Gene Ther 14:1049–1063

21. Smith IL, Hardwicke MA, Sandri-Goldin RM (1992) Evidence that the herpes simplex virus immediate early protein ICP27 acts post-transcriptionally during infection to regulate gene expression. Virology 186:74–86

22. Krisky DM, Wolfe D, Goins WF et al (1998) Deletion of multiple immediate-early genes from herpes simplex virus reduces cytotoxicity and permits long-term gene expression in neurons. Gene Ther 5:1593–1603

23. Kashima T, Vinters HV, Campagnoni AT (1995) Unexpected expression of intermediate filament protein genes in human oligodendroglioma cell lines. J Neuropathol Exp Neurol 54:23–31

24. McGeoch DJ, Dalrymple MA, Davison AJ et al (1988) The complete DNA sequence of the long unique region in the genome of herpes simplex virus type 1. J Gen Virol 69:1531–15374

Chapter 22

Bone Marrow Transplantation for Research and Regenerative Therapies in the Central Nervous System

David Díaz, José Ramón Alonso, and Eduardo Weruaga

Abstract

Bone marrow stem cells are probably the best known stem cell type and have been employed for more than 50 years, especially in pathologies related to the hematopoietic and immune systems. However, their potential for therapeutic application is much broader (because these cells can differentiate into hepatocytes, myocytes, cardiomyocytes, pneumocytes or neural cells, among others), and they can also presumably be employed to palliate neural diseases. Current research addressing the integration of bone marrow-derived cells in the neural circuits of the central nervous system together with their features and applications are *hotspots* in current Neurobiology. Nevertheless, as in other leading research lines the efficacy and possibilities of their therapeutic application depend on the technical procedures employed, which are still far from being standardized. In this chapter we shall explain one of these procedures in depth, namely the transplantation of whole bone marrow from harvested bone marrow stem cells for subsequent integration into the encephalon.

Key words Bone marrow ablation, Bone marrow harvesting, Bone marrow stem cells, Cell therapy, Cell transplantation

1 Introduction

Bone marrow stem cells (SC) constitute the best known population of the adult stem cell group. They were identified 50 years ago as the cells responsible for the formation of blood cell populations [1]. Besides their main function, it was demonstrated that bone marrow SC can differentiate in vivo into elements of a wide variety of tissues [2]. More precisely, in the late 1990s it was also reported that bone marrow-derived cells (BMDC) could differentiate into cells of the central nervous system (CNS) in vivo [3]. Three years later, two parallel works also demonstrated that BMDC could not only become glial cells but also neurons [4, 5].

Traditionally, bone marrow SC have been employed in clinical practice since 1956, when E. Donnall Thomas performed the first bone marrow transplantation in a patient with leukemia [6, 7].

Laura Lossi and Adalberto Merighi (eds.), *Neuronal Cell Death: Methods and Protocols*, Methods in Molecular Biology, vol. 1254, DOI 10.1007/978-1-4939-2152-2_22, © Springer Science+Business Media New York 2015

Since then, the methodologies for therapeutic treatments employing bone marrow cells have been refined. In addition, and taking into account their unexpected plasticity, bone marrow SC are currently being applied as a therapeutic tool in different models of disease (heart and coronary diseases), infarctions, stroke, graft versus host disease, bone diseases, cancers and neural diseases), and not only those related to the hematopoietic system [8].

Regarding the use of bone marrow SC in research and possible therapies in the CNS, current methodologies offer a wide range of possibilities. One of the most widely employed techniques for these purposes involves the ablation of recipient bone marrow and its subsequent replacement by BMDC so that these can reach the recipient's encephalon. In this chapter, we shall explain this methodology, dealing with the procedures for bone marrow ablation and bone marrow SC harvesting and administration.

2 Materials

Note that the suppliers are only for orientation and they may change, depending on the location of researchers.

2.1 Bone Marrow Ablation

1. Source of ^{137}Cs gamma radiation: Gammacell 1000 Elite (MDS Nordion, Ottawa, Canada).

2. Busulfan, $C_6H_{14}O_6S_7$ (Sigma-Aldrich, Steinheim, Germany).

3. Dimethylsulfoxide (DMSO).

4. Phosphate buffered Saline (PBS) pH 7.4 (25 °C): For 1 L mix 8 g NaCl, 0.2 g KCl, 1.44 g Na_2HPO_4 and 0.21 g KH_2PO_4; adjust pH with 1 M HCl.

2.2 Bone Marrow Stem Cell Extraction

1. Iscove's Modified Dulbecco's Medium (IMDM; Invitrogen; Carlsbad, USA).

2. Lysis buffer for erythrocytes; 140 mM NH_4Cl, 17 mM Tris-base; adjust pH to 7.4 with 1 M HCl ADD.

3. 70-μm pore size filter (BD Falcon; Bedford, MA, USA).

4. Falcon tubes (BD Falcon).

2.3 Cell Transplantation

1. 30G syringes (BD Falcon).

2. LE5016 clamps for mice (PanLab; Barcelona, Spain).

3. InfraCare HP3621 heating lamp (Philips Ibérica, Madrid, Spain).

3 Methods

The ablation of the recipient's bone marrow is performed before systemic injections—whether in blood vessels, intraperitoneal or intrahepatic—that are usually aimed at the replacement of such bone marrow. It is not possible to detect the own BMDC within the encephalon of a given research animal. Therefore, regardless of whether the aim is to study the neural integration of BMDC or the therapeutic properties of such integration, it is necessary to label these cells, generally employing fluorescent proteins (the donors are usually transgenic animals that express them constitutively). Accordingly, these systemic transplantations should ensure the replacement of the recipient's bone marrow by a new one whose derivatives are easily distinguishable. Ablation of the recipients' own bone marrow is necessary to avoid problems of rejection and to achieve an empty niche for the new labeled cells [9, 10]. There are different possibilities of bone marrow ablation, and their pros and cons are summarized in Table 1.

3.1 Bone Marrow Ablation

3.1.1 Physical Ablation

Physical ablation involves the use of ionizing irradiation to destroy the bone marrow of the recipient and is the most common ablative methodology [3, 11–16].

1. Place the animal under a ^{137}Cs gamma radiation source. Different devices are useful for this purpose, ranging from special devices for mice to re-calibrated devices for human irradiation. ^{60}Co sources, although older, are also effective.

2. Irradiate the animal with a minimum dose of 7.5 Gy [13]. Higher doses can be also applied, but their side effects are stronger (*see* **Notes 1** and **2**). If doses lower than 7.5 Gy are employed, it is necessary to irradiate the animal twice or three times to achieve suitable bone marrow ablation (Table 2). Note that the irradiation time depends on the dose chosen and the progressive exhaustion of the radiation source.

Table 1
Pros and cons of ablation procedures

Methodology	Pros	Cons
Physical ablation	Widely employed Opens the blood–brain barrier Accelerates the reconstitution of the bone marrow	Severe secondary effects on cell proliferation
Chemical ablation	No major secondary effects Allows transplants in newborns (busulfan)	Does not open the blood–brain barrier Lower integration rates of BMDC into the brain?
Genetic ablation	No secondary effects Allows more physiological-like experiments Allows transplants in newborns	Needs specific mouse strains

Table 2
Doses of radiation employed for bone marrow ablation

4.5–5.5 Gy (multiple sessions)	[3, 11, 23, 26]
7.5 Gy	[12–16]
8 Gy	[21, 22]
9.5 Gy	[19, 29, 31]
10–11 Gy	[2, 25]

3.1.2 Chemical Ablation

The second possibility for ablating the bone marrow of recipients is the use of chemicals. The most widely employed compounds are also used in human chemotherapy against cancer, busulfan being the most widely employed [17, 18].

1. Weigh 25 mg of busulfan.

2. Dissolve busulfan in 500 µL DMSO.

3. Warm 24.5 mL PBS to 60 °C and maintain this temperature.

4. Add the mixture of DMSO and busulfan to the PBS and mix thoroughly. Keep it at 60 °C again.

5. Inject 20 µL/g of body weight intraperitoneally (which is equivalent to a dose of 20 g /kg). *See* **Note 3**.

Busulfan can also be injected in pregnant females in order to obtain bone marrow-ablated offspring. To achieve this, the previous procedures should be followed taking into account two variations:

1. A dose of 15.5 µL/g of the final mixture (15.5 g/kg).

2. The busulfan injections should be performed twice (on embryonic days 18.5 and 19.5) and between the caudal-most pair of nipples, to ensure a suitable effect in the embryos.

3.2 Bone Marrow Stem Cell Extraction

In order to obtain bone marrow cells, when working with small animals it is necessary to sacrifice the donors. The specific procedures to obtain such cells are as follows:

1. Sacrifice mice by cervical dislocation. When employing rats, it is also possible to decapitate them. Anesthetics are not recommended since they may interact with the mobility of bone marrow SC.

2. Remove hind legs and dissect femurs and tibias separately (*see* **Note 4**). It is important to handle the fragile bones of mice with great care in the extraction process to avoid contamination of the bone marrow. The humerus can be also dissected to obtain a higher number of cells.

3. Carefully remove muscles and other tissues surrounding the bones. Once the bones have been harvested, the following procedures should be performed under a laminar-flow hood to guarantee sterile conditions.

4. Perform injections of Iscove's Modified Dulbecco's Medium (IMDM) medium at both epiphyses of the bones to wash the bone marrow. It is recommended to perform a hole on each epiphysis prior to injecting the medium with the needle of the syringe.

5. Perform three or four injections of 1 mL of medium at each end of the bone to properly wash the bone marrow: the bone metaphysis should remain empty, becoming transparent or white instead of dark red. Other standard cell culture mediums (e.g. DMEM, Dulbecco's Modified Eagle's medium) can be also employed.

6. Collect the bone marrow wash on a 70-μm pore size filter placed on the top of a 50 mL Falcon collection tube.

7. Add more IMDM medium to the filter and gently disaggregate the bone marrow pellets with the aid of a sterile blunt object (e.g. the protective cap of the syringe needle). Stir this cell suspension gently with the blunt object to filter it into the collection tube.

8. Repeat the previous step until the collection tube contains 50 mL of medium (with the filtered cells in suspension).

9. Centrifuge the tube at $364 \times g$ for 5 min at 4 °C (*see* **Note 5**).

10. Remove the supernatant.

11. Re-suspend the pellet in 5 mL of lysis buffer for 5 min to break up erythrocytes.

12. Stop the reaction by adding 45 mL of 0.1 M PBS (achieving a final volume of 50 mL).

13. Take an aliquot (e.g. 100 μL) to estimate the number of cells collected in the final volume of 50 mL. Thoma or Neubauer chambers and also automatic cell counters can be used for this purpose.

14. Centrifuge the tube at $364 \times g$ for 5 min at 4 °C (*see* **Note 5**).

15. Remove the supernatant.

16. Re-suspend the pellet in PBS and keep the final suspension on ice to ensure cell survival. It is important to note that there is a maximum volume that can be transplanted in each animal.

    - For intravenous administration in adult mice, a volume of 150–200 μL should be employed; for younger mice, a slightly lower volume, about 100–150 μL, should be used.

    - For intraperitoneal injections, higher volumes can be employed.

    - Newborns are difficult to inject intravenously and hence intrahepatic injections are usually employed [18, 19]. Taking into account the tiny size of pups, no more than 50 μL should be injected.

Table 3
Number of cells for systemic transplants

≤1 million	[24, 29]
2–3 million	[3]
5–8 million	[11, 13–16, 20, 23, 25]
10 million	[18, 21, 16, 31]
10–20 million	[12]
30 million	[22]

- Finally, the desired number of cells to be transplanted in each receptor is also a factor to take into account (Table 3). Thus, the volume of PBS employed to obtain the final cell suspension depends on the desired concentration of cells (according to the volume and number of cells to be transplanted).

3.3 Cell Transplantation

There are several possibilities for administering bone marrow SC, but here we shall describe the three methods most commonly used:

- Intraperitoneal administration is the easiest method and this needs no further explanation. It is employed frequently [5, 12, 20–22]. It should be emphasized that if the procedure is not performed properly, the cells may be injected into the gut or bladder. It is therefore important to hold the animal with its head directed downwards to ensure the displacement of the organs to the thoracic cavity and enable adequate injection into the peritoneal cavity.

- Intrahepatic injections are especially suggested for newborns [18] since their skin allows the visualization of the liver (a dark red color). Injections into this structure have no additional complications (*see* **Note 6**).

- Intravenous injections require more expertise. These injections are often made in the tail vein [3, 4, 11, 13–16, 20, 21, 23–27], but they can also be performed in the retro-orbital plexus [19, 28, 29] or in the maxillo-facial vein (for pups; [17]).

There are several steps to be taken into account when performing these transplants (*see* **Note 7**):

1. The animal must be immobilized properly. For injections in the tail vein, there are some clamps for mice and rats that allow the immobilization of the animal, keeping its tail outside the device to access the blood vessels easily. For other types of injections, the animal can be anesthetized.

2. It is recommended to induce vasodilatation (especially for tail vein injections). For this purpose warm water or infrared lamps can be used. The blood vessel to be injected must be perfectly localized.

3. The needles should be thin, but their inner diameter wide enough to allow the injection of the cell suspension without massive cell lysis. 30G needles for mice and 26G needles for rats are recommended.

4. During injection, the bevel of the needle should be oriented upwards.

5. All these blood vessels are very superficial, so the needle should be inserted as parallel to the vessel as possible, and never injected deeply.

6. Once the needle is in the vein, when injecting the cells it should be possible to depress the plunger of the syringe easily, without resistance. No edema should appear and, also, it is easy to appreciate the displacement of the blood inside the blood vessel due to the liquid injected. A common signal of well performed injections (inside the blood vessel) is a small bleeding of the wound caused by the needle.

4   Notes

1. Radiation (for physical ablation) has side effects: proliferating cells are susceptible to such radiation and indeed this is the basis of radiation therapy to treat tumors or neoplastic disorders. Thus, not only bone marrow SC but also other proliferating cells are killed by irradiation, including those whose niche lies inside the encephalon. In addition radiation has a dose-dependent effect [30] and any dose necessary to ablate bone marrow –7.5 Gy or higher impairs neurogenesis [13]. Moreover, radiation also affects encephalic regions other than neurogenic zones [18].

2. Conversely, some of the side effects of radiation can be beneficial for the cells arriving at the brain (even for cell therapies); e.g. it has been reported that radiation accelerates blood reconstitution in bone marrow-transplanted animals [20], and radiation also facilitates BMDC incorporation [31], presumably due to an opening of the blood–brain barrier [32, 33].

3. Temperature of the busulfan solution should not fall below 50 °C. Therefore, it must be injected quickly to avoid excessive cooling from 60 °C.

4. To facilitate the removal of the hind legs of the donors, it is suggested that a cut be made in the fur of the lumbar region, then pulling the skin backwards to expose the underlying mus-

culature, in particular that of the legs. Then, the legs are removed and dissected by cutting them at the level of the hip, avoiding any harm to the femurs.

5. The first pellet of bone marrow cells (obtained after the first centrifugation) is red because of the presence of erythrocytes. After the lysis of these red cells, the second pellet should have a much paler color.

6. Owing to the small size of newborns, only small volumes of cell suspensions should be injected in these animals. Accordingly, the concentration of cells should be relatively high in order to transplant a suitable number. Therefore, any loss of cell suspension also involves the loss of many cells. Transplants in pups often flow back slightly because of their small size. Thus, to avoid an excessive loss of cells, care should be taken to keep the needle inside the body of the animal for a few seconds after the injection, after which it must be removed slowly.

7. The procedures explained above are aimed at achieving a bone marrow replacement for studying subsequent cell integration and its effects in the CNS. However, these procedures for bone marrow ablation and transplantation can also be employed in other research lines.

Acknowledgements

This work was supported by the Ministerio de Investigación y Ciencia (BFU2010-18284), Junta de Castilla y León and Centre for Regenerative Medicine and Cell Therapy of Castilla y León. The authors also express their gratitude to N. Skinner for revising the English version of the manuscript.

References

1. Becker AJ, McCulloch EA, Till JE (1963) Cytological demonstration of the clonal nature of spleen colonies derived from transplanted mouse marrow cells. Nature 197:452–454

2. Krause DS, Theise ND, Collector MI et al (2001) Multi-organ, multi-lineage engraftment by a single bone marrow-derived stem cell. Cell 105:369–377

3. Eglitis MA, Mezey E (1997) Hematopoietic cells differentiate into both microglia and macroglia in the brains of adult mice. Proc Natl Acad Sci U S A 94:4080–4085

4. Brazelton TR, Rossi FM, Keshet GI et al (2000) From marrow to brain: expression of neuronal phenotypes in adult mice. Science 290:1775–1779

5. Mezey E, Chandross KJ, Harta G et al (2000) Turning blood into brain: cells bearing neuronal antigens generated in vivo from bone marrow. Science 290:1779–1782

6. Thomas ED (1983) Bone marrow transplantation in leukemia. Haematol Blood Transfus 28:11–15

7. Thomas ED (1983) Bone marrow transplantation. A lifesaving applied art. An interview with E. Donnall Thomas, MD. JAMA 249:2528–2536

8. Tögel F, Westenfelder C (2007) Adult bone marrow-derived stem cells for organ regeneration and repair. Dev Dyn 236:3321–3331

9. Fuchs E, Tumbar T, Guasch G (2004) Socializing with the neighbors: stem cells and their niche. Cell 116:769–778

10. Li L, Xie T (2005) Stem cell niche: structure and function. Annu Rev Cell Dev Biol 21:605–631

11. Johansson CB, Youssef S, Koleckar K et al (2008) Extensive fusion of haematopoietic cells with Purkinje neurons in response to chronic inflammation. Nat Cell Biol 10:575–583

12. Álvarez-Dolado M, Pardal R, García-Verdugo JM et al (2003) Fusion of bone-marrow-derived cells with Purkinje neurons, cardiomyocytes and hepatocytes. Nature 425:968–973

13. Díaz D, Recio JS, Baltanás FC et al (2011) Long-lasting changes in the anatomy of the olfactory bulb after ionizing irradiation and bone marrow transplantation. Neuroscience 173:190–205

14. Recio JS, Álvarez-Dolado M, Díaz D et al (2011) Bone marrow contributes simultaneously to different neural types in the central nervous system through different mechanisms of plasticity. Cell Transplant 20:1179–1192

15. Díaz D, Lepousez G, Gheusi G et al (2012) Bone marrow cell transplantation restores olfaction in the degenerated olfactory bulb. J Neurosci 32:9053–9058

16. Díaz D, Recio JS, Weruaga E et al (2012) Mild cerebellar neurodegeneration of aged heterozygous PCD mice increases cell fusion of Purkinje and bone marrow-derived cells. Cell Transplant 21:1595–1602

17. Piquer-Gil M, García-Verdugo JM, Zipancic I et al (2009) Cell fusion contributes to pericyte formation after stroke. J Cereb Blood Flow Metab 29:480–485

18. Espejel S, Romero R, Álvarez-Buylla A (2009) Radiation damage increases Purkinje neuron heterokaryons in neonatal cerebellum. Ann Neurol 66:100–109

19. Massengale M, Wagers AJ, Vogel H et al (2005) Hematopoietic cells maintain hematopoietic fates upon entering the brain. J Exp Med 201:1579–1589

20. Nygren JM, Liuba K, Breitbach M et al (2008) Myeloid and lymphoid contribution to non-haematopoietic lineages through irradiation-induced heterotypic cell fusion. Nat Cell Biol 10:584–592

21. Corti S, Locatelli F, Strazzer S et al (2002) Modulated generation of neuronal cells from bone marrow by expansion and mobilization of circulating stem cells with in vivo cytokine treatment. Exp Neurol 177:443–452

22. Corti S, Locatelli F, Donadoni C et al (2004) Wild-type bone marrow cells ameliorate the phenotype of SOD1-G93A ALS mice and contribute to CNS, heart and skeletal muscle tissues. Brain 127:2518–2532

23. Priller J, Persons DA, Klett FF et al (2001) Neogenesis of cerebellar Purkinje neurons from gene-marked bone marrow cells in vivo. J Cell Biol 155:733–738

24. Kemp K, Gordon D, Wraith DC et al (2011) Fusion between human mesenchymal stem cells and rodent cerebellar Purkinje cells. Neuropathol Appl Neurobiol 37:166–178

25. Vallieres L, Sawchenko PE (2003) Bone marrow-derived cells that populate the adult mouse brain preserve their hematopoietic identity. J Neurosci 23:5197–5207

26. Weimann JM, Johansson CB, Trejo A et al (2003) Stable reprogrammed heterokaryons form spontaneously in Purkinje neurons after bone marrow transplant. Nat Cell Biol 5:959–966

27. Magrassi L, Grimaldi P, Ibatici A et al (2007) Induction and survival of binucleated Purkinje neurons by selective damage and aging. J Neurosci 27:9885–9892

28. Nern C, Wolff I, Macas J et al (2009) Fusion of hematopoietic cells with Purkinje neurons does not lead to stable heterokaryon formation under noninvasive conditions. J Neurosci 29:3799–3807

29. Chen KA, Cruz PE, Lanuto DJ et al (2011) Cellular fusion for gene delivery to SCA1 affected Purkinje neurons. Mol Cell Neurosci 47:61–70

30. Tada E, Yang C, Gobbel GT et al (1999) Long-term impairment of subependymal repopulation following damage by ionizing irradiation. Exp Neurol 160:66–77

31. Wiersema A, Dijk F, Dontje B et al (2007) Cerebellar heterokaryon formation increases with age and after irradiation. Stem Cell Res 1:150–154

32. Pachter JS, de Vries HE, Fabry Z (2003) The blood-brain barrier and its role in immune privilege in the central nervous system. J Neuropathol Exp Neurol 62:593–604

33. Yuan H, Gaber MW, Boyd K et al (2006) Effects of fractionated radiation on the brain vasculature in a murine model: blood-brain barrier permeability, astrocyte proliferation, and ultrastructural changes. Int J Radiat Oncol Biol Phys 66:860–866

Part IV

Neuronal Death in Nonmammalian Models

Detection of Activated Caspase-8 in Injured Spinal Axons by Using Fluorochrome-Labeled Inhibitors of Caspases (FLICA)

Antón Barreiro-Iglesias and Michael I. Shifman

Abstract

Here, we present a detailed protocol for the detection of activated caspase-8 in axotomized axons of the whole-mounted lamprey spinal cord. This method is based on the use of fluorochrome-labeled inhibitors of caspases (FLICA) in ex vivo tissue. We offer a very convenient vertebrate model to study the retrograde degeneration of descending pathways after spinal cord injury.

Key words FLICA, Caspases, Apoptosis, Spinal cord injury, Cell death, Caspase-8, Extrinsic apoptotic pathway, Lamprey, Axotomy

1 Introduction

In mammals, spinal cord injury (SCI) leads to permanent disability because axons do not regenerate across the trauma zone and are unable to reconnect to their appropriate targets. An important goal of current SCI research is to promote the regeneration of interrupted axons [1]. A prerequisite for this would be the prevention of retrograde degenerative responses that may induce retrograde neuronal death or atrophy. It is generally accepted that several types of central nervous system (CNS) neurons die after axotomy. Examples include retinal ganglion cells [2–4] and motor neurons [5, 6]. Most of the studies have also reported death of at least some spinal-projecting neurons after SCI in opossum [7], rats [8–12], and humans [13–15]. The death of spinal-projecting neurons after SCI appears to involve an apoptotic process, as suggested by the appearance of TUNEL staining and activated caspase-3 immunoreactivity in pontine reticular neurons [9] and corticospinal neurons [10, 11]. However, the specific molecular mechanisms of axotomy-induced degeneration of spinal-projecting neurons have not yet been elucidated.

Laura Lossi and Adalberto Merighi (eds.), *Neuronal Cell Death: Methods and Protocols*, Methods in Molecular Biology, vol. 1254, DOI 10.1007/978-1-4939-2152-2_23, © Springer Science+Business Media New York 2015

Compared with mammalian models, lampreys have several unique advantages for the study of retrograde degeneration of lesioned spinal-projecting pathways. After a complete high spinal cord transection, lampreys recover normal appearing locomotion and many of their spinal axons regenerate selectively in their correct paths [16, 17]. The lamprey brainstem contains several uniquely identifiable reticulospinal neurons. These include the Mauthner neurons, which have crossed descending axons, and several pairs of Müller cells, which have ipsilateral descending axons. Because the axons of these neurons project almost the entire length of the spinal cord, they are all axotomized by a high spinal cord transection. Among these identifiable reticulospinal neurons, some are consistently "bad regenerators" and others are consistently "good regenerators" [18, 19]. Therefore, in the lamprey, the long-term fates of individual neurons after SCI can be correlated with their intrinsic regenerative abilities.

In previous studies, spinal cord transection induced the delayed death of lamprey reticulospinal neurons that, at earlier times post-injury, had been identified as bad regenerators [20, 21]. Evidence for cell death included the disappearance of Nissl staining, the loss of neurofilament expression, the absence of labeling when using retrograde tracers [20], and the earlier staining of these neurons with Fluoro-Jade C [21], a marker for degenerating neurons. The appearance of TUNEL staining [20] and activated caspases [22] in the axotomized perikarya suggested that the death and degeneration of these bad-regenerating neurons in lamprey was apoptotic. Thus the lamprey is a simple and convenient vertebrate model for the in vivo study of molecular mechanisms underlying the death or degeneration of spinal-projecting neurons after SCI.

In apoptosis, caspases are responsible for proteolytic cleavages that lead to cell disassembly (effector caspases) and are involved in upstream regulatory events (initiator caspases) [23]. In vertebrates, caspase-dependent apoptosis occurs through two main pathways—the intrinsic and extrinsic pathways [24]. The intrinsic or mitochondrial pathway is activated by genomic or metabolic stress, the presence of unfolded proteins, and other factors that lead to permeabilization of the outer membrane of mitochondria and the release of apoptotic proteins, mainly cytochrome c, into the cytosol. Progression through the intrinsic pathway usually leads to the formation of the apoptosome complex and the activation of the initiator caspase-9 [23]. The extrinsic or death receptor pathway involves activation of transmembrane death receptors by their ligands. This process usually leads to activation of initiator caspase-8 or -10 [23]. Therefore, activation of caspases is a hallmark of apoptosis. We have reported the development of a new method to detect caspase activation in identifiable reticulospinal neurons of the ex vivo whole-mounted lamprey brain, which was previously only used for in vitro assays [22]. This method is based on the use of fluorochrome-labeled inhibitors of caspases (FLICA). More recently, a study by the Selzer's group [25] further confirmed, by combining FLICA and TUNEL staining, that the FLICA method is a reliable way of detecting caspase activation and cell death in lamprey whole-mount preparations of the brain. Using this method, we observed that activated caspase-8 is present in the cell bodies of "poor survivor/bad regenerator" neurons (i.e. the M1, M2, M3, I1, I2, Mth, B1, B3, and B4 neurons) 2 weeks after a complete spinal cord transection [22]. This indicated that the extrinsic pathway of apoptosis is activated in these neurons after axotomy. However, it is still not known whether caspase-8 activation occurs first in the axotomized axon at the site of injury or directly in the cell body through the activation of retrograde signaling.

Here, we are presenting the detailed and adapted protocol for detection of the initial activation of caspase-8 in axotomized axons of descending neurons using the specific activated caspase-8 FLICA reagent. We were able to show that it is possible to detect the initial activation of caspase-8 in axotomized axons of the whole-mounted lamprey spinal cord after a spinal transection.

2 Materials

2.1 Reagents

1. Tricaine methanesulfonate (MS-222; Sigma Chemicals, St. Louis, MO).

2. Sodium chloride.

3. Potassium chloride.

4. Calcium chloride-2 hydrate crystalline ($CaCl_2$).

5. HEPES (e.g. Sigma Chemicals). Caution: Irritating to eyes, respiratory system, and skin. Use gloves, safety glasses, and mask.

6. Hydrochloric acid (HCl). Caution: Can cause skin burns, eye damage, and respiratory irritation. Avoid breathing dust/fume/gas/mist/vapors/spray. Use protective gloves, safety glasses, and mask.

7. Sodium hydroxide (NaOH). Caution: Causes severe skin burns and eye damage. Use protective gloves, safety glasses, and mask.

8. Sylgard 184 (Dow Corning, Midland, MI).

9. Image-iT LIVE Green Caspase-8 Detection Kit (Life Technologies™; La Jolla, CA). Kit components used in this study: 1 vial (component A of the kit) of the lyophilized FLICA reagent (FAM-LETD-FMK) for the specific detection of activated caspase-8, 1 vial, dimethylsulfoxide (DMSO; component D; may cause respiratory tract, skin or eye irritation) and 10× apoptosis wash buffer (component F).

10. Paraformaldehyde. Store at 4 °C. Caution: Toxic by inhalation and if swallowed. Use gloves, safety glasses, and mask.

11. Phosphate buffered saline (PBS) 10× concentrate (e.g. Sigma Chemicals).

12. Distilled water (dH$_2$O).

13. Prolong™ anti-fade reagent (Life Technologies™).

2.2 Equipment

1. Stereomicroscope (e.g. Nikon SMZ-U; Nikon, Tokyo, Japan).

2. Scalpel #3 and blades #11 (Fine Science Tools®, Foster City, CA).

3. Forceps #4 and #5 (Dumont, Montignez, Switzerland).

4. Castroviejo scissors #8 (Fine Science Tools®).

5. Superfrost® Plus glass slides (Menzel, Braunschweig, Germany).

6. Coverglass 24 mm × 60 mm (Menzel).

7. Fluorescence microscope (e.g. Nikon Eclipse 80i; Nikon).

8. Digital camera (e.g. CoolSNAP ES; Roper Scientific, Sarasota, FL).

2.3 Solutions and Fixatives

1. Lamprey Ringer solution: 137 mM NaCl, 2.9 mM KCl, 2.1 mM CaCl$_2$, 2 mM HEPES buffer, pH to 7.4 with HCl or NaOH. Store at 4 °C.

2. 0.1 % MS-222 in Ringer solution: To prepare 100 mL, add 0.1 g MS-222 to 100 mL Ringer solution. Prepare just before use.

3. 1× PBS solution: To prepare 100 mL, add 10 mL of the 10× concentrate to 90 mL of dH$_2$O. Store at room temperature.

4. 4 % paraformaldehyde solution in 1× PBS: To prepare 400 mL, add 16 g paraformaldehyde to 280 mL dH$_2$O, and heat it up to 60 °C with continuous stirring, add 8 drops of NaOH and wait until the solution clears, remove from the heat and add 10 mL of 10× PBS, make up to 400 mL with dH$_2$O and filter it. Prepare just before use and wait until it cools down.

3 Methods

3.1 Animals

Wild type larval sea lampreys (*Petromyzon marinus* L.), 10–14 cm in length (4–7 years old), were obtained from streams feeding Lake Michigan. In the laboratory, larvae were maintained in groups of 50–100 animals in 50 gallon freshwater tanks at 16 °C. The tanks are lined with an inch of gravel for the larvae to burrow in, and the water is charcoal filtered and conditioned with air stones for at least 48 h to remove chlorine and other impurities before placing the lampreys in them. Experiments were approved by the Institutional Animal Care and Use Committee at Temple University.

3.2 Spinal Cord Transection

1. Anesthetize animals by immersion in lamprey Ringer solution containing 0.1 % MS-222.

2. Place the anesthetized larvae in a Petri dish half-filled with Sylgard and covered with ice-cold Ringer solution. Keep the Petri dish with the animal on ice during the surgery under the stereomicroscope.

3. Expose the spinal cord from the dorsal midline at the level of the fifth gill by making a longitudinal incision with the scalpel (#11).

4. Use a forceps (#4) to hold the body walls open while performing the spinal cord transection.

5. Perform a complete transection of the spinal cord with Castroviejo scissors (#8).

6. Visualize the spinal cord cut ends and realign under the stereomicroscope by using a forceps (#4).

7. Close the wound in the dorsal body wall after surgery, and keep the animals on ice for 1 h allowing the wound to air dry.

8. Allow animals to recover in aerated fresh water tanks (1 larva per tank) at room temperature for the chosen time points (2 h, 1 day, and 1 week). Use non-injured animals and animals with a hemisection of the left spinal cord (incomplete transection) as controls.

9. Examine each transected animal 2 h after surgery to confirm that there was no movement caudal to the lesion site. *See* **Note 1**.

3.3 Detection of Activated Caspase-8 in Axons of Whole-Mounted Ex-Vivo Tissue

Detection of activated caspase-8 using the FLICA reagent was done in the ex vivo dissected brain and spinal cord rostral to the site of injury (*see* **Note 2**). Therefore, this allowed the analysis of caspase-8 activation in both the descending axons and the neuronal perikarya. The FLICA reagent (FAM-LETD-FMK; component A of the kit) is composed of a fluoromethyl ketone (FMK) moiety, which can react covalently with a cysteine, and a caspase-specific amino acid sequence. In the case of caspase-8, the recognition sequence is leucine-glutamic acid-threonine-aspartic acid (LETD). A fluorescein group is attached as a reporter. The FLICA reagent interacts with the enzymatic reactive center of the activated caspase-8 via the specific recognition sequence, and then attaches covalently through the FMK moiety. Since the FLICA reagent is cell permeant, unbound FLICA molecules diffuse out of the axons and cells during subsequent washes.

1. Prepare the 150× FLICA reagent stock solution: Add 50 µL of DMSO (component D of the kit) to the vial containing the lyophilized FLICA reagent and mix until the reagent is completely dissolved. *See* **Note 3**.

2. Prepare the 1× FLICA reagent solution just before starting the dissection of the brain/rostral spinal cord: Add 1 μL of the 150× FLICA reagent stock solution to 149 μL of 1× PBS.

3. At the chosen time points after spinal cord transection, re-anesthetize the larval sea lampreys with 0.1 % MS-222 and place them in the Petri dish with Sylgard filled with ice-cold Ringer solution.

4. Perform the brain/rostral spinal cord dissection on ice under the stereomicroscope. *See* **Note 4**.

5. To dissect the brain/rostral spinal cord first make a transverse incision between the nostril and the pineal organ with the scalpel (#11).

6. Remove the tissue surrounding the brain using a pair of Castroviejo scissors (#8) while holding it with a pair of forceps (#5).

7. Once the brain is exposed, gently remove the choroid plexus, which is located on top of the brain, by using a pair of forceps (#5).

8. Dissect out the brain/rostral spinal cord by holding the spinal cord with a pair of forceps (#5) while using a pair of Castroviejo scissors (#8) to cut the cranial nerves.

9. Immediately after dissection put the ex-vivo brain/rostral spinal cord in the 1× labeling solution containing the FLICA reagent (from **step 2**). Incubate protected from light (*see* **Note 5**) for 1 h at 37 °C. *See* **Note 6**.

10. Prepare the 1× wash buffer from 10× apoptosis wash buffer (component F of the kit) by adding 9 parts of distilled water to 1 part of 10× apoptosis wash buffer. *See* **Note 7**.

11. Wash the brain/rostral spinal cord 9 × 15 min in 1× wash buffer at room temperature (protected from light) on a nutator. *See* **Note 8**.

12. Cut the posterior and cerebrotectal commissures of the brain along the dorsal midline with the Castrovejo scissors, deflect laterally the alar plates and pin them flat to a small strip of Sylgard.

13. Fix brains/rostral spinal cords in 4 % paraformaldehyde in 1× PBS overnight at 4 °C.

14. Wash the preparations 4 × 15 min with 1× PBS, mount on Superfrost Plus glass slides, and coverslip with Prolong™ anti-fade reagent. *See* **Note 9**.

3.4 Image Acquisition

1. Photomicrographs were taken with a 20× objective using a Nikon Eclipse 80i microscope (*see* **Note 10**) equipped with a CoolSNAP ES camera.

2. Adjust brightness and contrast minimally using the Adobe Photoshop CS4 software.

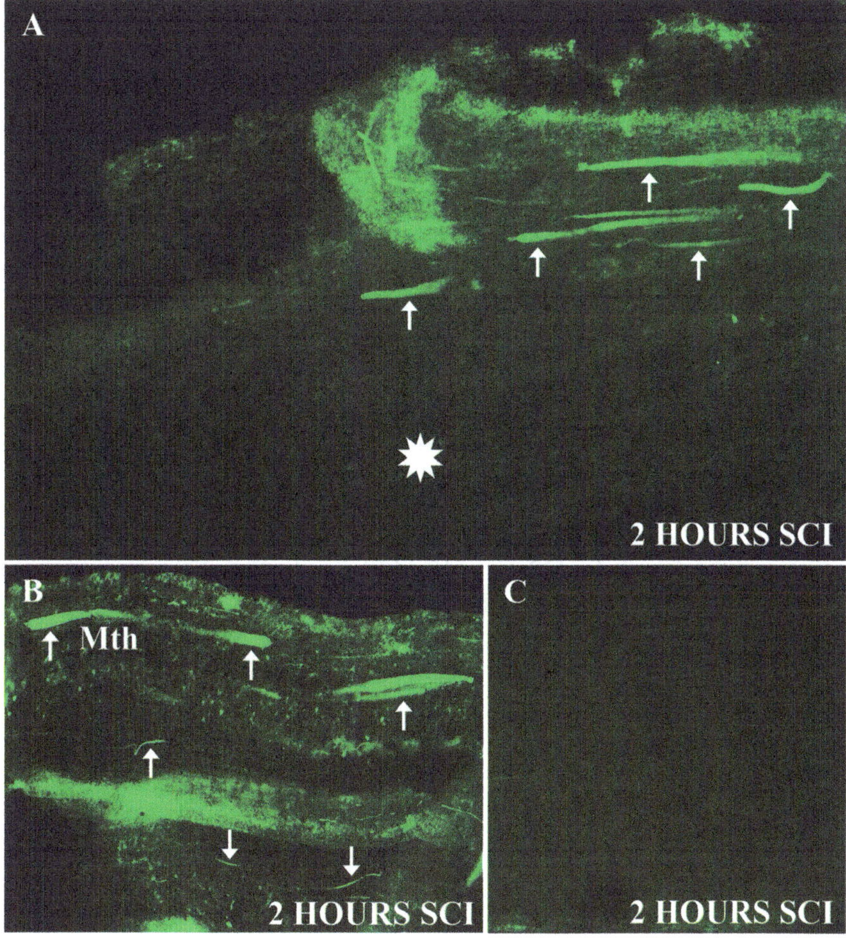

Fig. 1 Dorsal views of whole-mounted spinal cords of larval sea lampreys showing FAM-LETD-FMK labeling in axotomized axons (*arrows*) 2 h after a spinal cord hemisection (**a**) or a complete transection (**b** and **c**). (**a**) and (**b**) are photomicrographs of the spinal cord close to the site of injury and (**c**) is a photomicrograph at a midway level between the site of injury and the brain. Note the absence of labeled axons in the non-injured side of the SC in (**a**) (*asterisk*). Rostral is to the *right* in all panels. *Abbreviations*: *Mth* Mauthner neuron, *SCI* spinal cord injury

Figures 1 and 2 show some exemplificative results that can be obtained with the protocol described in this chapter. As a control to confirm that axotomy actually causes activation of caspase-8 in descending axons, spinal cord hemisections were performed and, 2 h later, the dissected brain/rostral spinal cords were processed for detection of activated caspase-8 with the FLICA reagent as explained. In these animals, activated caspase-8 was observed in descending axons on the injured side but not on the non-injured side of the spinal cord (Fig. 1a). Two hours after a complete spinal cord transection axotomized descending axons were labeled by the FLICA reagent near the site of injury (Fig. 1b), but not at midway levels between the site of injury and the brainstem-spinal cord transition (obex; Fig. 1c) or at the level of the obex (not shown).

At one-day after a complete spinal cord transection, FLICA staining filled the sealed tips of descending axotomized axons close to the site of injury (Fig. 2a, b). A simple oval shape and absence of filopodia/lamellipodia characterizes these axon tips, as previously reported for axons retrogradely labeled with fluorescent tracers [26]. Interestingly, at this post-injury time a few labeled descending axons were observed at midway levels between the site of injury and the obex (Fig. 2c), whereas no labeling was observed in the obex (Fig. 2d) or in the cell bodies of spinal-projecting neurons in the brainstem (Fig. 2e).

At 1-week post-transection, FAM-LETD-FMK-labeled descending axons with the typical axon tips were observed away and rostral to the site of injury (Fig. 2f, g), indicating that they had undergone retraction, as previously reported [26]. At this post-transection time, labeled descending axons were observed at the level of the obex (Fig. 2h), and labeled "poor survivor/bad regenerator" spinal-projecting neurons were observed in the brain (e.g. the M2 and M3 Müller cells; Fig. 2i).

Taken together these results show that the appearance of activated caspase-8 in the cell bodies of "poor survivor/bad regenerator" neurons is preceded by caspase-8 activation in the descending axons at the site of spinal cord transection.

4   Notes

1. A transection was tentatively considered complete if the animal could move only its head and body rostral to the lesion site.

2. The specific FLICA reagent binds active caspase-8, therefore the neurons have to be alive when applying it.

3. The unused portion of the 150× FLICA reagent can be stored in small aliquots protected from light at –20 °C, and then they will be stable for several months.

4. The cold temperature preserves the CNS alive for the subsequent labeling steps.

5. The FLICA reagents are coupled to a fluorochrome, so they are light sensitive. It is necessary to always handle them in the dark.

6. The incubation with the FLICA reagent can also be done at 4 °C [25]. This lower temperature is needed if the FLICA labeling is subsequently combined with TUNEL or in situ hybridization.

7. It is also possible to use 1× PBS for the washes.

8. Since the method was adapted from its use in cell culture assays to whole-mounted brain/spinal cord preparations, we increased the number of washes to get rid of the unbound FLICA molecules.

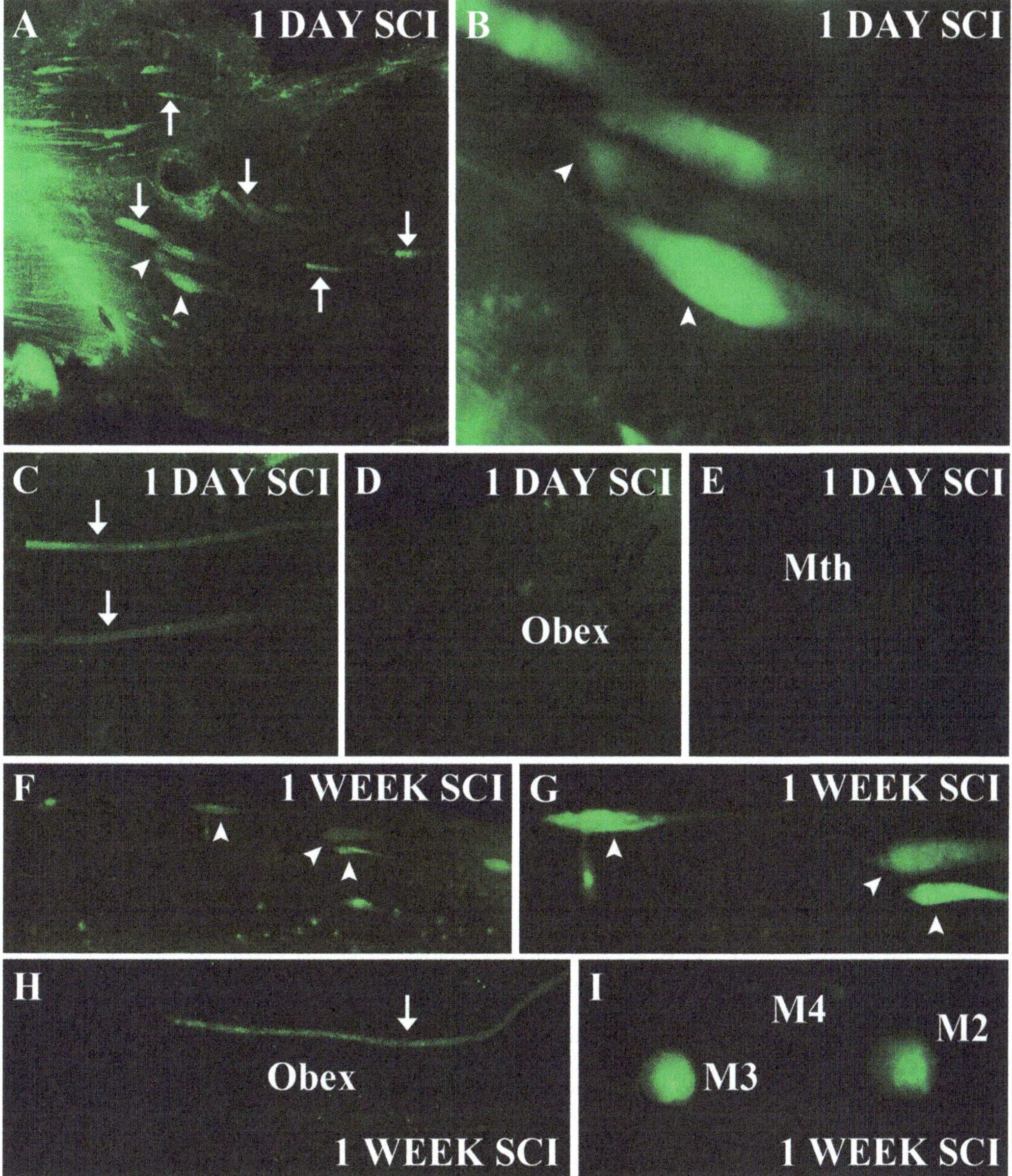

Fig. 2 Dorsal views of whole-mounted spinal cords and brains of larval sea lampreys showing FAM-LETD-FMK labeling in axotomized descending axons and neurons. (**a**) Whole-mounted spinal cord close to the site of transection showing labeled descending axons 1 day after a complete spinal cord injury (*arrows*). Note that some of these axons show the typical axon tip of axotomized axons (*arrowheads*). A detail of (**a**) is shown in (**b**). (**c**) Whole-mounted spinal cord at midway levels between the site of transection and the obex showing labeled descending axons 1 day after a complete SCI. (**d**) Whole-mounted spinal cord at the level of the obex showing the absence of labeling 1 day after injury in this region. (**e**) Whole-mounted brain showing the absence of labeling in a Mauthner neuron 1 day after injury. (**f**) Whole-mounted spinal cord at midway levels between the site of transection and the obex showing labeled descending axons 1 week after a complete SCI that have retracted from the site of injury. Note that some of these axons show the typical axon tip of axotomized axons (*arrowheads*). A detail of (**f**) is shown in (**g**). (**h**) Whole-mounted spinal cord at the level of the obex showing the presence of a labeled axon 1 week after injury. (**i**) Whole-mounted brain showing the soma of two "poor survivor/bad regenerator" labeled Müller neurons (M2 and M3) 1 week after injury. Rostral is to the *right* in all spinal cord figures and to the *top* in brain figures. Ventricle is to the *right* in brain figures. *Abbreviations: M2-M4* Müller neurons, *Mth* Mauthner neuron, *SCI* spinal cord injury

9. Avoid bubbles while mounting by pressing carefully the coverslip from one edge of the slide to the other with a lancet to push out the bubbles from the tissue.

10. Any model of fluorescence microscope can be used in these experiments as long as it has the suitable filter set.

Acknowledgments

A. Barreiro-Iglesias was supported by a Postdoctoral Fellowship from the Xunta de Galicia. M. I. Shifman was supported by the Shriners Research Foundation (grant number: SHC-85310).

References

1. Thuret S, Moon LD, Gage FH (2006) Therapeutic interventions after spinal cord injury. Nat Rev Neurosci 7(8):628–643

2. Berkelaar M, Clarke DB, Wang YC et al (1994) Axotomy results in delayed death and apoptosis of retinal ganglion cells in adult rats. J Neurosci 14:4368–4374

3. Kermer P, Klöcker N, Labes M et al (1999) Activation of caspase-3 in axotomized rat retinal ganglion cells in vivo. FEBS Lett 453: 361–364

4. Chaudhary P, Ahmed F, Quebada P et al (1999) Caspase inhibitors block the retinal ganglion cell death following optic nerve transection. Brain Res Mol Brain Res 67:36–45

5. Vanderluit JL, McPhail LT, Fernandes KJ et al (2000) Caspase-3 is activated following axotomy of neonatal facial motoneurons and caspase-3 gene deletion delays axotomy-induced cell death in rodents. Eur J Neurosci 12: 3469–3480

6. Chan YM, Yick LW, Yip HK et al (2003) Inhibition of caspases promotes long-term survival and reinnervation by axotomized spinal motoneurons of denervated muscle in newborn rats. Exp Neurol 181:190–203

7. Fry EJ, Stolp HB, Lane MA et al (2003) Regeneration of supraspinal axons after complete transection of the thoracic spinal cord in neonatal opossums (*Monodelphis domestica*). J Comp Neurol 466:422–444

8. Feringa ER, Vahlsing HL (1985) Labeled corticospinal neurons one year after spinal cord transection. Neurosci Lett 58:283–286

9. Wu KL, Chan SH, Chao YM et al (2003) Expression of pro-inflammatory cytokine and caspase genes promotes neuronal apoptosis in pontine reticular formation after spinal cord transection. Neurobiol Dis 14:19–31

10. Hains BC, Black JA, Waxman SG (2003) Primary cortical motor neurons undergo apoptosis after axotomizing spinal cord injury. J Comp Neurol 462:328–341

11. Lee BH, Lee KH, Kim UJ et al (2004) Injury in the spinal cord may produce cell death in the brain. Brain Res 1020:37–44

12. Klapka N, Hermanns S, Straten G et al (2005) Suppression of fibrous scarring in spinal cord injury of rat promotes long-distance regeneration of corticospinal tract axons, rescue of primary motoneurons in somatosensory cortex and significant functional recovery. Eur J Neurosci 22:3047–3058

13. Holmes G, May WP (1909) On the exact origin of the pyramidal tracts in man and other mammals. Proc R Soc Med 2:92–100

14. Nielson JL, Sears-Kraxberger I, Strong MK et al (2010) Unexpected survival of neurons of origin of the pyramidal tract after spinal cord injury. J Neurosci 30:11516–11528

15. Nielson JL, Strong MK, Steward O (2011) A reassessment of whether cortical motor neurons die following spinal cord injury. J Comp Neurol 519:2852–2869

16. Selzer ME (1978) Mechanisms of functional recovery and regeneration after spinal cord transection in larval sea lamprey. J Physiol 277: 395–408

17. Rodicio MC, Barreiro-Iglesias A (2012) Lampreys as an animal model in regeneration studies after spinal cord injury. Rev Neurol 55: 157–166

18. Davis GR Jr, McClellan AD (1994) Extent and time course of restoration of descending brainstem projections in spinal cord-transected lamprey. J Comp Neurol 344:65–82

19. Jacobs AJ, Swain GP, Snedeker JA (1997) Recovery of neurofilament expression selectively in regenerating reticulospinal neurons. J Neurosci 17:5206–5220

20. Shifman MI, Zhang G, Selzer ME (2008) Delayed death of identified reticulospinal neurons after spinal cord injury in lampreys. J Comp Neurol 510:269–282

21. Busch DJ, Morgan JR (2012) Synuclein accumulation is associated with cell-specific neuronal death after spinal cord injury. J Comp Neurol 520:1751–1771

22. Barreiro-Iglesias A, Shifman MI (2012) Use of fluorochrome-labeled inhibitors of caspases to detect neuronal apoptosis in the whole-mounted lamprey brain after spinal cord injury. Enzyme Res 2012:835731

23. Ola MS, Nawaz M, Ahsan H (2011) Role of Bcl-2 family proteins and caspases in the regulation of apoptosis. Mol Cell Biochem 351:41–58

24. Riedl SJ, Salvesen GS (2007) The apoptosome, signaling platform of cell death. Nat Rev Mol Cell Biol 8:405–413

25. Hu J, Zhang G, Selzer ME (2013) Activated caspase detection in living tissue combined with subsequent retrograde labeling, immunohistochemistry or in situ hybridization in whole-mounted lamprey brains. J Neurosci Methods 220:92–98

26. Jin LQ, Zhang G, Jamison C Jr, Takano H, Haydon PG, Selzer ME (2009) Axon regeneration in the absence of growth cones, acceleration by cyclic AMP. J Comp Neurol 515:295–312

Chapter 24

Generation of Zebrafish Models by CRISPR/Cas9 Genome Editing

Alexander Hruscha and Bettina Schmid

Abstract

The CRISPR/Cas system identified in archaea has been adopted and optimized for genome editing purposes in zebrafish. In vitro transcribed guide RNA and Cas9 mRNA are microinjected into fertilized zebrafish embryos to edit the zebrafish genome. Here, we describe how to design a gRNA, a fast method for in vitro transcription of gRNA from oligonucleotides, microinjection into fertilized zebrafish embryos, and a PCR-based restriction fragment length assay to identify mutations at the gRNA target site.

Key words Zebrafish, Genome editing, CRISPR, Cas9, gRNA, In vitro transcription, Microinjection, PCR, Mutation detection, RFLP

1   Introduction

Novel genome editing techniques including zinc finger nucleases (ZFN), TAL effector nucleases (TALEN), and the Clustered Regularly Interspaced Short Palindromic Repeats (CRISPR)/ CRISPR-associated system (Cas) allow genomic modifications at selected target sites. All techniques share the ability to identify and bind to a target site within the genome, and to recruit a nuclease to this target site to induce a double-strand break [1, 2]. For zebrafish genome editing, the CRISPR/Cas9 system is the fastest and most convenient approach to induce mutations since laborious cloning steps can be omitted [3–5]. Target sequences of 20 nucleotides (N20) adjacent to a PAM sequence (NGG), which is necessary for Cas9 nuclease to cleave DNA, can be easily identified by BLAST searches. The chosen target sequence is then transferred into a guide RNA (gRNA) design template and the gRNA template is ordered as a single-strand oligonucleotide (oligo) harboring a T7 polymerase binding site for in vitro transcription. Highly efficient gRNA are obtained by in vitro transcription (IVT) from the oligo directly rather than subcloning the oligo prior IVT.

Laura Lossi and Adalberto Merighi (eds.), *Neuronal Cell Death: Methods and Protocols,* Methods in Molecular Biology, vol. 1254, DOI 10.1007/978-1-4939-2152-2_24, © Springer Science+Business Media New York 2015

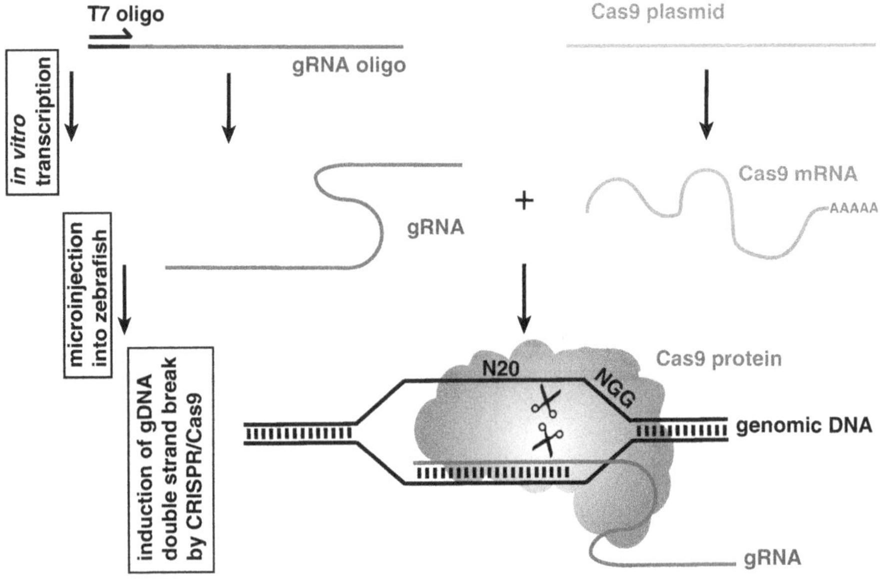

Fig. 1 Principle of the CRISPR/Cas9 system to induce mutations in zebrafish genome editing. The procedure consists of the following steps: (1) The chosen target sequence is transferred into a guide RNA (gRNA) design template ordered as a single-strand oligonucleotide (oligo) harboring a T7 polymerase binding site for in vitro transcription (*top left*); (2) Cas9 mRNA is generated by in vitro transcription from a linearized plasmid (*top right*); (3) The gRNA and Cas9 mRNA are coinjected into fertilized embryos where the translated Cas9 protein binds and cleaves the target DNA sequence (*bottom*)

This direct IVT approach is more time and cost-efficient. Cas9 mRNA is also generated by IVT from a linearized plasmid. gRNA and Cas9 mRNA are coinjected into fertilized embryos where translated Cas9 protein then binds and cleaves the target sequence (Fig. 1).

If the target site is mutated, efficiency can be analyzed by various methods, such as a classical restriction fragment length polymorphism (RFLP), high resolution melting analysis (HRMA), a CelI digest, or a T7 Endonuclease I assay. We describe the RFLP method, since it is the least expensive and most robust read out. The efficiency of the CRISPR/Cas9 is highly dependent on the gRNA and can reach up to 98 % mutagenesis rate [3].

2   Materials

1. MEGAshortscript T7 Transcription Kit (Ambion®, Technologies™, Gaithersburg, MD).

2. mMessage mMachine SP6 Transcription Kit (Ambion®).

3. RNAsecure (Ambion®).

4. NucleoSpin Gel and PCR Clean up (Macherey-Nagel, Düren, Germany).

5. Ethanol pa.

6. Nuclease-free water (Ambion®).

7. ApaI (New England Biolabs—NEB, Ipswich, MA).

8. CutSmart Buffer (NEB).

9. Annealing Buffer 10×: 1 M NaCl, 10 mM EDTA, 100 mM Tris, pH 7.5 with HCl.

10. TE buffer: 10 mM Tris, 1 mM EDTA, pH 8.0 with HCl.

11. Proteinase K stock: 17 mg/mL in destH$_2$O, Proteinase K (PCR grade, Roche, Basel, Switzerland).

12. PCR Buffer: 100 mM Tris, 500 mM KCl, 15 mM MgCl$_2$, 0.1 % gelatin, pH 8.3 with HCl.

13. PCR Mix: 60 μL dATP (100 mM) + 60 μL dTTP (100 mM) + 60 μL dGTP (100 mM) + 60 μL dCTP (100 mM) + 36.6 mL nuclease-free H$_2$O + 6 mL PCR Buffer.

14. GoTaq (Promega, Fitchburg, WI).

15. T7 oligo (5′-GCGTAATACGACTCACTATAG-3′).

16. gRNA design template: 5′-GCGTAATACGACTCACTATA GNNNNNNNNNNNNNNNNNNNNNGTTTTAGAG CTAGAAATAGCAAGTTAAAATAAGGCTAGTCC GTTATCAACTTGAAAAAGTGGCACCGAGTCGGTG C TTT-3′ (*see* Subheading 3.1).

17. SpCas9 plasmid (Addgene, http://www.addgene.org/ *see* also Subheading 3.4 and **Note 12**).

18. TURBO® DNase (Ambion®).

19. DNA analysis software.

20. Waterbath.

21. Thermocycler.

22. Microcentrifuge.

23. Spectrophotometer, e.g. NanoDrop (Wilmington, DE).

24. Gel electrophoresis apparatus.

25. Microinjection system.

26. Methanol.

27. 96-well plate.

3 Methods

3.1 Oligo Design for gRNA Transcription

1. Copy gRNA design template (Subheading 2, **item 15**) into DNA analysis software (for convenience shown as single strand, referred as "upper strand" or "non-template strand")

2. Identify genomic target sequence N20 with NGG PAM motif (Fig. 2a). *See* **Note 1**.

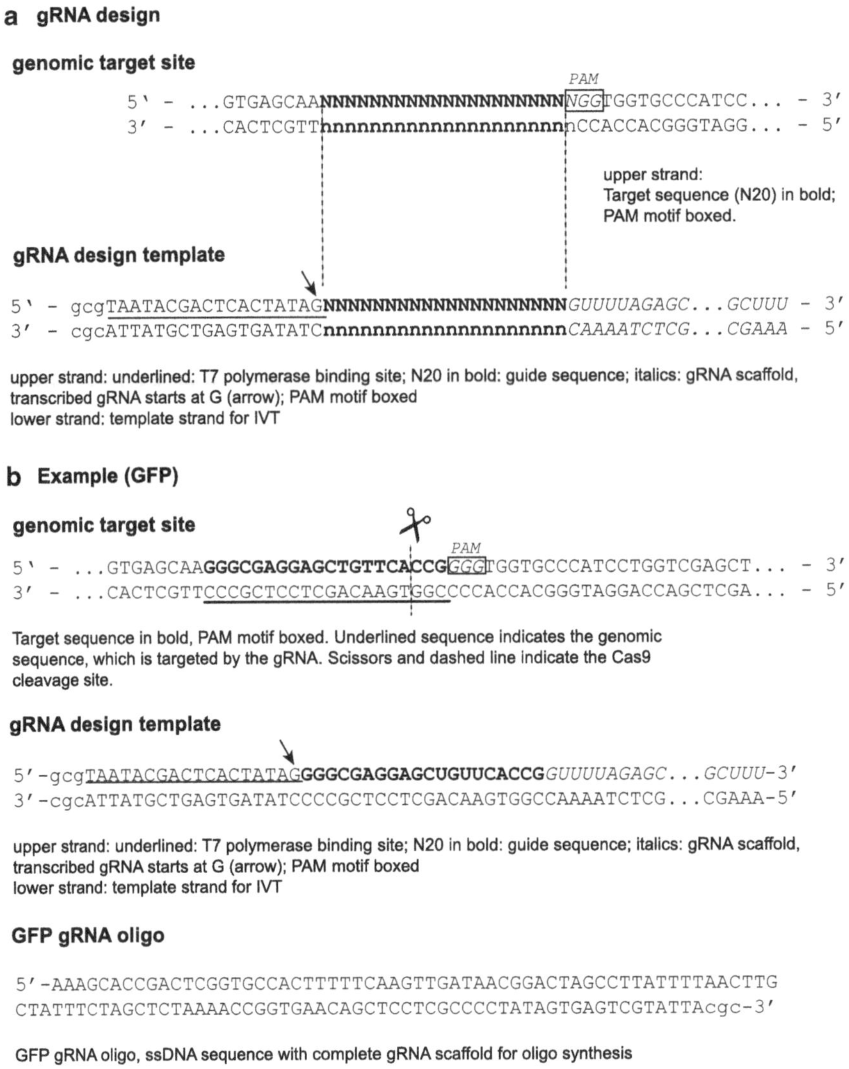

a   gRNA design

genomic target site

```
                                                                PAM
    5` - ...GTGAGCAANNNNNNNNNNNNNNNNNNNNNNGGTGGTGCCCATCC... - 3'
    3' - ...CACTCGTTnnnnnnnnnnnnnnnnnnnnnnnCCACCACGGGTAGG... - 5'
```

upper strand:
Target sequence (N20) in bold;
PAM motif boxed.

gRNA design template

```
    5` - gcgTAATACGACTCACTATAGNNNNNNNNNNNNNNNNNNNNGUUUUAGAGC...GCUUU - 3'
    3' - cgcATTATGCTGAGTGATATCnnnnnnnnnnnnnnnnnnnnCAAAATCTCG...CGAAA - 5'
```

upper strand: underlined: T7 polymerase binding site; N20 in bold: guide sequence; italics: gRNA scaffold,
transcribed gRNA starts at G (arrow); PAM motif boxed
lower strand: template strand for IVT

b   Example (GFP)

genomic target site

```
                                       PAM
    5` - ...GTGAGCAAGGGCGAGGAGCTGTTCACCGGGGTGGTGCCCATCCTGGTCGAGCT... - 3'
    3' - ...CACTCGTTCCCGCTCCTCGACAAGTGGCCCCACCACGGGTAGGACCAGCTCGA... - 5'
```

Target sequence in bold, PAM motif boxed. Underlined sequence indicates the genomic
sequence, which is targeted by the gRNA. Scissors and dashed line indicate the Cas9
cleavage site.

gRNA design template

```
    5'-gcgTAATACGACTCACTATAGGGGCGAGGAGCUGUUCACCGGUUUUAGAGC...GCUUU-3'
    3'-cgcATTATGCTGAGTGATATCCCCGCTCCTCGACAAGTGGCCAAAATCTCG...CGAAA-5'
```

upper strand: underlined: T7 polymerase binding site; N20 in bold: guide sequence; italics: gRNA scaffold,
transcribed gRNA starts at G (arrow); PAM motif boxed
lower strand: template strand for IVT

GFP gRNA oligo

```
5'-AAAGCACCGACTCGGTGCCACTTTTTCAAGTTGATAACGGACTAGCCTTATTTTAACTTG
CTATTTCTAGCTCTAAAACCGGTGAACAGCTCCTCGCCCCTATAGTGAGTCGTATTAcgc-3'
```

GFP gRNA oligo, ssDNA sequence with complete gRNA scaffold for oligo synthesis

Fig. 2 Principle for the design of a gRNA template (**a**) and exemplification of the design of a GFP gRNA (**b**)

3. Check genome for potential off-target site via BLAST search. *See* **Note 1**.

4. Replace guide sequence N20 of the upper strand in the gRNA design template by the target sequence of the genomic target site proximal of the PAM sequence (Fig. 2a).

5. Use the reverse strand of the gRNA design template for ssDNA gRNA oligo synthesis (shown for GFP gRNA in Fig. 2b). *See* **Note 2**.

6. Order PAGE purified gRNA oligo. *See* **Note 3**.

7. Prepare 100 µM oligo stocks in destH₂O, store at –20 °C.

3.2 RNAsecure Treatment (See Note 4) and Oligo Annealing

1. Prewarm waterbath to 60 °C.

2. Set up the following reaction in a PCR tube: 4.3 µL T7 oligo (100 µM in H_2O), 4.3 µL gRNA oligo (100 µM in H_2O), 1.0 µL Annealing Buffer 10×, 0.4 µL RNAsecure.

3. Seal tube (parafilm).

4. Incubate at 60 °C for 10 min to inactivate all RNases. *See* **Note 5**.

5. Remove reaction from waterbath.

6. Preheat waterbath to 95 °C.

7. Denature DNA for 5 min at 95 °C.

8. Slowly cool down waterbath containing the reaction tubes to room temperature for best annealing results (takes several hours).

3.3 gRNA In Vitro Transcription (MEGAshortscript T7 Transcription Kit)

1. Set up IVT reaction (*see* **Note 6**): 2 µL T7 10× Reaction Buffer (Kit), 4 µL H_2O, 2 µL ATP, 2 µL CTP, 2 µL GTP, 2 µL UTP, 4 µL template DNA (annealed oligos), 2 µL T7 Enzyme Mix.

2. Mix content.

3. Spin mixture down in a microcentrifuge at $2,000 \times g$ for 1 s.

4. Transcribe for 4 h at 37 °C.

5. Add 2 µL TURBO® DNase to reaction, mix and spin down.

6. Incubate reaction for 15 min at 37 °C.

7. Add 115 µL nuclease-free water to reaction and 15 µL Ammonium Acetate Stop Solution.

8. Mix reaction and precipitate RNA by adding 300 µL ethanol.

9. Chill for at least 30 min at –20 °C. *See* **Note 7**.

10. Centrifuge at $14,000 \times g$ at 4 °C for 30 min.

11. Remove supernatant.

12. Wash RNA pellet with 800 µL 70 % ethanol. *See* **Note 8**.

13. Centrifuge at $14,000 \times g$ at 4 °C for 30 min.

14. Remove supernatant. *See* **Note 9**.

15. Evaporate ethanol residues at room temperature (5–30 min). *See* **Note 10**.

16. Resuspend RNA in 10 µL nuclease-free water.

17. Measure RNA concentration (spectrophotometer, NanoDrop) and check RNA integrity on agarose gel. *See* **Note 11**.

18. Store 1.5 µL RNA aliquots at –80 °C.

3.4 Cas9 Plasmid Preparation

1. To linearize Cas9 plasmid, i.e. pCS2+/SpCas9 [3], mix 30 µg plasmid, 50 U ApaI, 5 µL CutSmart Buffer up to 50 µL with destH$_2$O. *See* **Note 12**.

2. Incubate for 2 h at 25 °C.

3. Load 100 ng undigested plasmid next to digested plasmid on an agarose gel to confirm complete digest.

4. Purify linearized plasmid using NucleoSpin Gel and PCR Clean up, elute DNA in RNase-free buffer. *See* **Note 13**.

5. Measure DNA concentration with spectrophotometer.

3.5 Cas9 In Vitro Transcription (mMessage mMachine SP6 Transcription Kit)

1. Use approximately 500 ng linearized plasmid as template for in vitro transcription. *See* **Note 14**.

   (a) Follow manufacturer's instruction considering following optimizations:

   (b) Transcribe for 2 h at 37 °C.

   (c) After transcription treat reaction with 1 μL TURBO® DNase for 15 min.

   (d) Follow the lithium chloride precipitation instructions, wash RNA pellet and evaporate ethanol residues.

   (e) Resuspend RNA in 10 μL nuclease-free water.

2. Measure concentration (spectrophotometer, NanoDrop) and check RNA integrity on agarose gel.

3. Store 1.5 μL aliquots at –80 °C. *See* **Note 15**.

3.6 Injection

1. Mix 1.5 μL gRNA plus 1.5 μL Cas9 mRNA. Working concentration should be higher than 600 ng/μL Cas9 mRNA and 2.0 μg/μL gRNA. *See* **Note 16**.

2. Inject directly into the cell during early 1-cell stage. *See* **Note 17**.

3.7 Embryo Lysis

1. Fix at 2 dpf 20 injected embryos and 4 uninjected control embryos per clutch in methanol (*see* **Note 18**), keep remaining injected siblings.

2. Transfer embryos in 96-well plate, remove methanol and evaporate remaining methanol at 55 °C.

3. Add 50 μL lysis buffer (45 μL TE buffer + 5 μL Proteinase K stock solution) per well.

4. Seal 96-well plate and lyse embryos at 55 °C for 2–4 h.

5. Deactivate Proteinase K for 5 min at 95 °C.

6. Use 1–2 μL for PCR.

3.8 PCR and Restriction Fragment Length Polymorphisms (RFLP) Analysis

1. Prepare the following Mastermix for PCR amplification: 15 μL PCR Mix, 0.02 μL forward primer (100 μM), 0.02 μL reverse primer (100 μM), 0.2 μL GoTaq Polymerase. *See* **Note 19**.

2. Provide 15 μL Mastermix per well.

3. Add 1.0–2.0 μL genomic DNA.

4. Run PCR.

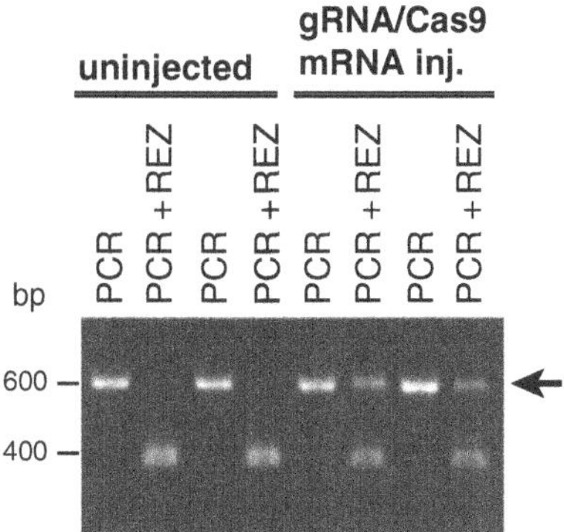

Fig. 3 Analytical digest of PCR fragments containing the target site generated from genomic DNA of embryos injected with gRNA and Cas9 mRNA. Restriction enzyme (REZ) digest leaves part of the PCR fragment undigested upon gRNA/Cas9 injection due to a mutated target site (*arrow*). Undigested band intensity correlates to mutagenesis efficiency

Use 8 µL PCR product for RFLP analysis (*see* **Note 20**). Buffer, amount of enzyme, incubation time and temperature according to the manufacturer's instructions: 8 µL PCR product, 1.3 µL 10× REZ buffer, 2–10 U REZ, fill up to 13 µL with nuclease-free H_2O.

5. Digest up to 4 h at optimal temperature. *See* **Note 21**.

6. Load PCR product next to the corresponding digest and estimate CRISPR efficiency (Fig. 3).

3.9 Screen for Inherited Genome Modifications

1. Raise siblings of positively tested injected embryos to adulthood (P0).

2. Cross the P0 fish to wildtype fish.

3. Check offspring for germ line transmission. *See* **Note 22**.

4. Raise siblings of positively tested P0 offspring (F1) and analyze fin biopsies for genome modification.

5. Sequence modified alleles and select for modifications of interest.

6. Cross positive F1 fish again to wildtype, raise offspring to establish stable fish strain.

4 Notes

1. If possible choose a target site, which is close to a REZ recognition site. Most CRISPR/Cas9-induced dsDNA breaks are repaired by non-homologous end joining leading to mutations (indels) at the Cas9 cut site and loss of REZ binding site. Genomic region will be amplified later and analyzed for loss

of this REZ site to determine CRISPR/Cas9 efficiency. Both genomic DNA strands are suitable for targeting, the only requirement being the PAM motif NGG. The PAM motif NAG was also shown to be sufficient for CRISPR/Cas9 cleavage, but efficiency was much lower [6]. Choice of target sequence and potential off-target cut offs always depend on the specific application. We therefore recommend to check specificity manually (i.e. BLAST search at www.ensembl.org) and to choose a proper target site upon the specific requirements. Alternatively, there are also some helpful tools online, i.e. http://crispr.mit.edu which also include first experimental results in cell culture models in their algorithm for off-target prediction [7]. Furthermore, it is essential to sequence the genomic target region of the wildtype fish strain (i.e. AB) you will use for injections. Most polymorphisms are not annotated and might differ between different labs and wildtype strains.

2. The shown gRNA design for GFP can be used as positive control for design, transcription, injection, and genotyping. Resulting gRNA is highly efficient and lead to almost 100 % mutagenesis rate (CRISPR efficiency).

3. For our previous study we used HPLC-purified oligos for gRNA transcription [3]. To increase success rate we would recommend PAGE-purified oligos, since the fraction of full-length oligos is supposed to be higher.

4. Since long oligos and the IVT reactions are rather expensive, we treat our oligos and dsDNA templates with RNAsecure, which is an affordable irreversible RNase inhibitor.

5. Take precautions when working with RNA: use only RNase-free consumables, filter-tips, DEPC or RNAsecure-treated buffers.

6. For several reactions at once, premix dNTPs and add 8 μL mixed dNTPs to each reaction.

7. Precipitated RNA can be chilled for up to 2 nights at –20 °C. However, 60 min at –20 °C is sufficient.

8. Carefully add ethanol to the RNA pellet. Do not resuspend the pellet.

9. Centrifuge at 14,000 × g at 4 °C for 30 min, remove ethanol using a 1 mL tip, centrifuge again to spin down remaining ethanol and remove all remaining ethanol using a 10 μL tip.

10. Do not overdry RNA pellet. 5 min is usually sufficient. Overdried RNA will appear as a high molecular smear during gel electrophoresis.

11. MEGAshortscript T7 Transcription Kit produces high amounts of RNA. You might have to dilute the RNA before quantification.

12. First experiments in the field were performed using the bacterial codon usage of SpCas9. Optimized codon usage for vertebrate

organisms like *Homo sapiens* increases RNA stability and translation efficiency. In this chapter, we refer to our first construct with the *S. pyogenes* codon usage [3]. New SpCas9 constructs with *Homo sapiens* codon usage are available upon request (manuscript in preparation) or from other labs provided by Addgene.

13. Linearized plasmid can be purified via column purification. It is in most cases sufficient for RNase removal. Additional RNAsecure treatment can inactivate remaining RNases.

14. Cas9 can be transcribed in several reactions in parallel. Check concentration and integrity of each reaction. Then pool reactions, which contain highest concentration and best RNA quality. Store 1.5 μL aliquots at –80 °C and test one aliquot of the pooled RNA for efficiency. To avoid time-consuming PCR readouts, use i.e. a GFP target site in a GFP reporter fish strain as a control for CRISPR/Cas9 activity. The gRNA described in Fig. 2 will give even in lower concentration close to 100 % CRISPR efficiency.

15. RNA aliquots can be stored at –80 °C for several months.

16. High concentration of gRNAs or Cas9 RNA itself seems not to be toxic. Occasionally, observed toxicity is in most cases due to low RNA purity or ethanol residues of the RNA preparation.

17. It is essential for early and high translation level of injected RNA to inject into the cell. Injections into the yolk lead to decreased activity due to lower or delayed translation of the RNA.

18. Analysis of single embryos allows an accurate estimation of the CRISPR efficiency and indicates variation due to injection. Raise only siblings of injection sessions with low variability and high efficiency. The amount of embryos to be raised should be estimated based on this result.

19. The provided PCR protocol is based on a protocol used for the Tübingen2000 screen [8]. It is very robust when working with unpurified DNA like digested larvae and most of all tested primer pairs give specific PCR products. However, a gradient-PCR for establishment of each primer pair is recommended.

20. There are several ways to determine cutting efficiencies. Very common are the T7 Endonuclease I and the CelI assay (SURVEYOR). Those assays rely on digesting mismatches in a heteroduplex formed of a mixture of mutated (indels or mismatches) and wildtype amplicons. Both approaches have some disadvantages: enzymes are quite expensive, PCR-cleanup is required and direct differentiation of homozygous mutant and wildtype is not possible. Furthermore, proofreading polymerases are recommended and possible SNPs in the whole amplicon region have to be determined first [9]. HRMA on the other side is not feasible for efficiency tests, since indels in injected embryos lead to an unpredictable change of the

melting curve, which allows no estimation of mutagenesis rate. But HRMA is a reliable fast system for SNP detection, i.e. for stably transmitted point mutations [10]. In our hands RFLP is the most robust readout and depending on the enzyme quite affordable.

21. So far we tested more than 35 REZ (NEB and Fermentas) [9, 10]. All of them showed sufficient activity in unpurified PCR products. Use a PCR-cycler with heated lid, if temperature above 37 °C is required for optimal REZ activity.

22. Depending on the CRISPR/Cas9 efficiency it might be advantageous to perform a "pre-screen." For low efficient CRISPR and rare knock ins, raise injected fish and analyze fin biopsies. Only test fish which show genomic modification for germ line transmission. For high efficient CRISPR, depending on the specific application you can select stable lines containing large indels, which allows genotyping directly by PCR via fragment length polymorphism without further REZ digest.

Acknowledgement

We thank Fargol Mazahari and Frauke van Bebber for critically reading the manuscript. We thank Hannelore Hartmann for contributing to the experience with the CRISPR/Cas9 system and Christian Haass for support.

References

1. Schmid B, Haass C (2013) Genomic editing opens new avenues for zebrafish as a model for neurodegeneration. J Neurochem 127: 461–470

2. Blackburn PR, Campbell JM, Clark KJ et al (2013) The CRISPR system—keeping zebrafish gene targeting fresh. Zebrafish 10: 116–118

3. Hruscha A, Krawitz P, Rechenberg A et al (2013) Efficient CRISPR/Cas9 genome editing with low off-target effects in zebrafish. Development 140:4982–4987

4. Hwang WY, Fu Y, Reyon D et al (2013) Efficient genome editing in zebrafish using a CRISPR-Cas system. Nat Biotechnol 313: 227–229

5. Chang N, Sung C, Gao L et al (2013) Genome editing with RNA-guided Cas9 nuclease in zebrafish embryos. Cell Res 23: 465–472

6. Jiang W, Bikard D, Cox D et al (2013) RNA-guided editing of bacterial genomes using CRISPR-Cas systems. Nat Biotechnol 31: 233–239

7. Hsu PD, Scott DA, Weinstein JA et al (2013) DNA targeting specificity of RNA-guided Cas9 nucleases. Nat Biotechnol 31:827–832

8. Geisler R, Rauch G-J, Geiger-Rudolph S et al (2007) Large-scale mapping of mutations affecting zebrafish development. BMC Genomics 8:11

9. Schmid B, Hruscha A, Hogl S et al (2013) Loss of ALS-associated TDP-43 in zebrafish causes muscle degeneration, vascular dysfunction, and reduced motor neuron axon outgrowth. Proc Natl Acad Sci U S A 110:4986–4991

10. van Bebber F, Hruscha A, Willem M et al (2013) Loss of Bace2 in zebrafish affects melanocyte migration and is distinct from Bace1 knock out phenotypes. J Neurochem 127:471–481

Chapter 25

In Vivo Assessment of Neuronal Cell Death in *Drosophila*

Pierre Dourlen

Abstract

Following neuronal cell death at the cellular level and over several time points is challenging in living animal because of the difficulty of accessing and identifying individual neurons. In the eye of a living *Drosophila*, it is possible to visualize neurons thanks to the cornea neutralization technique. This technique can be coupled to the generation of mosaic clones by the Tomato/GFP-FLP/FRT method to identify a group of photoreceptor neurons at a single-cell resolution. This method has proved to be efficient for the study of photoreceptor development and degeneration. In this chapter, I describe this method and focus on *fatp* mutant photoreceptor neuron degeneration.

Key words Neurodegeneration, *Drosophila*, Photoreceptor, Neuron, Eye, Mosaic, Clone, Cell death

1 Introduction

The *Drosophila* eye is composed of around 800 units called ommatidia. Each ommatidium contains six outer and two inner photoreceptor neurons, equivalent of mammalian rods and cones, surrounded by pigment cells [1, 2]. The repetitive structure of the eye makes it an ideal model for quantification of photoreceptor degeneration. The distal end of photoreceptor neurons is located just under the eye surface, from which it is separated only by the crystalline cone and cornea, two transparent dioptric media. These components can be optically neutralized by covering the corneal surface with a medium such as water whose refractive index approximately matches that of chitin [3, 4]. Therefore, after neutralization of the cornea, one can directly visualize the underlying photoreceptor neurons with a microscope and an immersion objective. After initial establishment, the technique was standardized by attaching the fly on the surface of an agarose gel and immersing the entire fly in water [5]. Further improvement was obtained allowing direct photoreceptor visualization by specifically expressing green fluorescent protein (GFP) in these neurons using a rhodopsin1 driver [5, 6]. The technique was applied on mosaic eyes, composed of clones of

Laura Lossi and Adalberto Merighi (eds.), *Neuronal Cell Death: Methods and Protocols*, Methods in Molecular Biology, vol. 1254, DOI 10.1007/978-1-4939-2152-2_25, © Springer Science+Business Media New York 2015

wild-type and homozygous mutant photoreceptor neurons, to study recessive lethal mutations [5]. By this approach, clones were identified based on the pigmentation of the pigment cells surrounding photoreceptor neurons. A further improvement in clone identification was then obtained by expressing a second fluorescent protein, the red fluorescent protein Tomato, directly in wild-type photoreceptor neurons [7]. This resulted in the unambiguous detection of wild-type and mutant photoreceptor neurons in the so-called Tomato/GFP-FLP/FRT method. Precise detection of the shape of clones allowed us to identify and follow mutant clones in the same eye of living *Drosophila* over weeks [7–9]. It also allowed answering cell autonomy issues and determining if the degeneration was restricted to the mutant cells only or also affected the surrounding wild-type cells.

In this chapter, I present the Tomato/GFP-FLP/FRT method for visualization of photoreceptor neuron degeneration in *Drosophila* retina. I illustrate this method with the example of photoreceptor neuron degeneration in mosaic *fatp^{k10307}* mutant retina.

2 Materials

1. Drosophila strain with the construct of interest recombined on a FRT chromosome. *See* **Note 1**.

2. Tomato/GFP-FLP/FRT lines. *See* **Note 2**.

3. 1.5 % agarose solution: Dissolve 1.5 g agarose in 100 mL distilled water in an Erlenmeyer flask. After boiling, keep the Erlenmeyer in a waterbath at 45 °C for 1 h, allow cooling down and maintain at this temperature. *See* **Note 3**.

4. 35 × 15 mm or 60 × 15 mm Petri dish.

5. Squeeze bottle filled with ice-cold distilled water in an ice bucket.

6. Forceps Dumont #5.

7. Upright fluorescence microscope equipped with a long distance immersion objective (e.g. Zeiss W N-Achroplan 40×/0.75 M27, Carl Zeiss Microscopy, Oberkochen, Germany) and green and red filters, such as GFP and red fluorescent protein (RFP) filters. *See* **Note 4**.

8. 5 × 5 × 10 cm cardboard boxes.

9. Hypodermic needle (18 gauge).

10. Slide.

11. Stereomicroscope.

12. Anesthesia apparatus for *Drosophila*.

13. Incubator.

3   Methods

1. Cross the line carrying the construct of interest with the corresponding Tomato/GFP-FLP/FRT line and sort the progeny with clones. *See* **Note 5**.

2. Anesthetize flies with CO_2 for at least 30 s.

3. Pour the 45 °C agarose solution into a Petri dish and fill half of the Petri dish.

4. Deposit flies on the agarose surface on its side (*see* **Note 6**). Plunge one wing in the agarose with a forceps and pull the wing down to have half of the body embedded in the agarose. Stick the other wing on the surface of the agarose (Fig. 1). *See* **Note 7**.

5. Let the agarose harden. *See* **Note 8**.

6. Once the agarose is solidified, orient the head with forceps in such a way as to have the eye pointing toward the top. *See* **Note 9**.

7. Cover the flies with ice-cold distilled water from the squeeze bottle. *See* **Note 10**.

8. Place the Petri dish in the ice bucket until observation under the microscope. *See* **Note 11**.

9. Lay the Petri dish on a slide attached to the stage of the microscope with the slide holder. *See* **Note 12**.

10. Pull up the stage to plunge the water immersion objective in the water of the Petri dish.
    See **Note 13**.

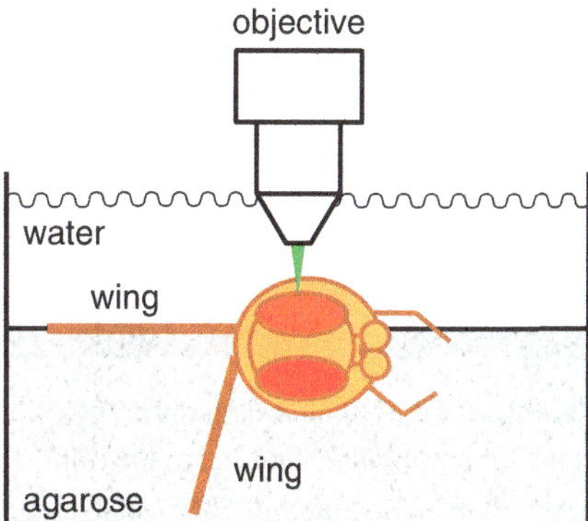

Fig. 1 Scheme of the set-up of the fly under the microscope objective. The *Drosophila* is attached on the surface of a Petri dish filled with agarose. The body is half embedded with one wing plunged into the agarose and the other one on its surface. The head is orientated with one eye toward the top. The fly is covered with water to plunge the water immersion objective

11. Turn on the excitation light with the green filter. Look at the stage and translate the Petri dish to position a fly head under the beam. Move the stage up and down and stop when the beam converges on the surface of the eye (Fig. 1).

12. Look though the eyepiece and see the eye surface. Focus below to "enter" in the eye and see the photoreceptor neurons. All the photoreceptor neurons are green but only the wild-type ones are red. When the channels are merged wild-type photoreceptor neurons are yellow and the photoreceptor neurons carrying the construct are green (Fig. 2).

13. Take images with the green and red filters (*see* **Note 14**). With the red filter, identify a group of photoreceptor with a unique shape. *See* **Note 15**.

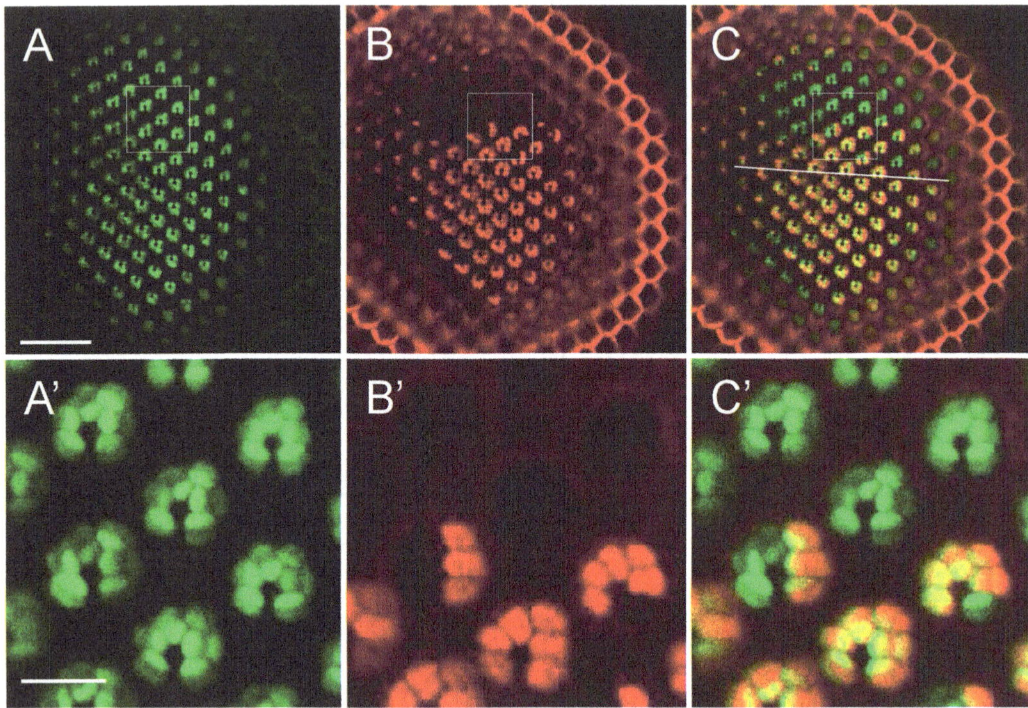

Fig. 2 Visualization of a *fatp* mutant mosaic Drosophila eye with the Tomato/GFP-FLP/FRT method. Retinal *fatp* mutant mosaic photoreceptor neurons observed in the right eye of a 1-day-old fly. (**a**, **a'**) All outer photoreceptor neurons express GFP. In the *left bottom corners*, scale bars measure 50 and 10 μm. (**b**, **b'**) Wild-type photoreceptor neurons express tdTomato. (**c**, **c'**) On the merge image, mutant photoreceptor neurons are labeled in *green* whereas wild-type photoreceptor neurons are labeled in *yellow*. (**a'**, **b'**, **c'**) Close-up views. The six outer photoreceptor neurons of an ommatidium are organized as an asymmetric trapezoid in which the three aligned photoreceptor neurons are on the anterior side (*on the right*) and the two opposite paralleled ones are on the posterior side (*on the left*). All the ommatidia are pointing from the equator (*white horizontal line* in **c**) to the poles, to the ventral pole in the ventral half of the retina (*toward the bottom on the image*) and to the dorsal pole in the dorsal half of the retina (*toward the top on the image*)

14. Remove the Petri dish from the stage of the microscope. Empty the water.

15. Place the Petri dish back under the stereomicroscope to remove the fly from the agarose. Tug gently on the upper wing of the fly with forceps. *See* **Note 16**.

16. Dry the fly on a paper towel and put it back in a food vial at 25 °C in an incubator. *See* **Note 17**.

17. For the next observations, carry out **steps 2–15** again.

18. Attach the fly on the agarose on the same side as the previous observation.

19. Try to orient the head the same way to find again the same group of photoreceptor neurons. When the previous field of photoreceptor neurons is only partially seen, orient the head again to have the field of photoreceptor centered. *See* **Note 18**.

20. Once images of the same eye at different time points have been taken, crop the images on a group of photoreceptor neurons, and construct the kinetics of cell death (Fig. 3).

As you can see from Fig. 3, two time points (day 1 and day 14) are shown for a retina that is mosaic for a *fatp* mutation (*fatp^{k10307}*). At day 14, most of the mutant photoreceptor neurons have disappeared, which shows cell death. The wild-type photoreceptor neurons all remain, which indicates that the neurodegeneration is cell-autonomous.

4 Notes

1. The construct of interest can be a mutation or an overexpression construct (UAS-cDNA). It has to be recombined on an FRT chromosome as clones are generated thanks to FLP/FRT-mediated mitotic recombination [10].

2. Four Tomato/GFP-FLP/FRT lines have been generated to study constructs located on the 2L, 2R, 3L, or 3R chromosomal arms. These lines are available at the Bloomington Drosophila Stock Center (Indiana University) with stock numbers #43345 (2L), #43346 (2R), #43347 (3L), #43348 (3R).

3. If the temperature of the agarose is too low, the agarose hardens too quickly and it is not possible to immerse the fly in it. If the temperature of the agarose is too high, it will damage the cornea.

4. The use of a confocal microscope gives much better images in terms of low autofluorescence background and width of focused photoreceptor neurons.

5. The progeny with clones should have the following genotype (example of a construct of interest on the 2L chromosomal

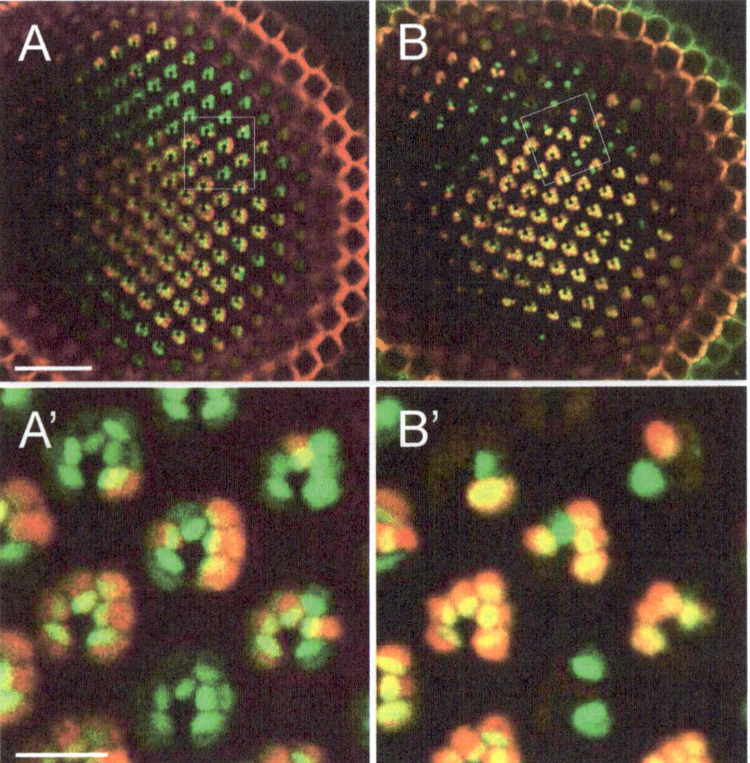

Fig. 3 Visualization of photoreceptor neuron degeneration in *fatp* mutant mosaic *Drosophila* retina with the Tomato/GFP-FLP/FRT method. The same group of mosaic mutant photoreceptor neurons is visualized at day 1 (**a**, **a′**) and day 14 (**b**, **b′**). At day 14, mutant photoreceptor neurons (in *green*, *arrows*) have disappeared indicating neurodegeneration. (**a**, **a′**) In the *left bottom* corners, scale bars measure 50 and 10 μm. (**a′**, **b′**) Close-up views

arm): rh1-Gal4, ey-FLP; construct of interest FRT40A/rh1-tdTomato[ninaC] FRT40A; UAS-GFP[ninaC]. To check the presence of clones, one can visualize the deep pseudopupil under a fluorescence stereomicroscope. With GFP filter, the deep pseudopupil should appear complete whereas with the RFP filter it should be only partially visible. The deep pseudopupil corresponds to the superposition of the images of several ommatidia into a virtual magnified single ommatidium, when focused in the depth of the eye where the center of curvature of the eye is reached [11, 12].

6. When a time-course analysis is planned and flies are supposed to be observed several times, put only one fly per dish to avoid mixing the flies. Deposit the fly always on the same side. Otherwise, it will be impossible to find again the group of photoreceptor neurons visualized in the previous observation. Start with 3 times as many flies as expected at the end. Some

flies die and the same group of photoreceptor neurons is sometimes not found again after successive observations.

7. Pay attention to immerse the *Drosophila* enough in the agarose, otherwise, it will be free to move its head during the visualization.

8. The Petri dish can be put in a chamber filled with CO_2 to keep the fly anesthetized until the agarose has hardened. To do so, use a small cardboard box, put the needle at the end of the gas tube and pierce the cardboard box with the needle.

9. The orientation of the eye is a critical step of the method. The width of the field of focused photoreceptor neurons depends on it. Some parts of the eye are flatter than others. Looking in the middle of the eye (when the pseudopupil is in the middle of the eye) usually gives good results (up to 80 focused ommatidia at once). It is also important to standardize the orientation of the eye to view the same region of the eye and to be able to find the same group of photoreceptors upon successive observations. When orientating the head, check that no prothoracic leg covers the eye.

10. The water neutralizes optically the cornea and the cold temperature of the water maintains the fly anesthetized. Flies can survive up to 2 h in these conditions.

11. If for any reason, the fly cannot be observed, it is possible to empty the Petri dish, keep it in the fridge overnight and observe the fly the day after. However, this will affect the fly survival.

12. This allows to smoothly translate the Petri dish with the X–Y translational control knobs of the stage.

13. Be careful not to plunge the objective too deeply in the Petri dish or it will crush the fly.

14. Focus at different levels to have different fields of the retina. Go as deep as possible to see the widest possible field of photoreceptor neurons. Wide field of photoreceptor neurons increases the chance to find the same photoreceptor neurons in the following observations.

15. This will be useful in the following observations to find the same photoreceptors.

16. This step is critical for the survival of the fly. If the wing of the fly is damaged, you can put your forceps under the body and lift it carefully.

17. First invert the vial upside down to avoid that the fly gets stuck in the food. Return the vial in the right orientation once the fly is awake. To avoid mixing the flies, put one fly per vial.

18. Use the asymmetry of the ommatidia to have positional clues in the retina. The six outer photoreceptor neurons form a trapezoid

in which the three aligned photoreceptor neurons are located on the anterior side, and the opposite two photoreceptor neurons are parallel to the posterior side (Fig. 2). The ommatidium is pointing toward the dorsal pole of the eye in the dorsal half of the eye and toward the ventral pole in the ventral half.

References

1. Ready DF, Hanson TE, Benzer S (1976) Development of the Drosophila retina, a neurocrystalline lattice. Dev Biol 53:217–240

2. Wolff T, Ready DF (1993) Pattern formation in the Drosophila retina. In: Bate M (ed) Development in Drosophila melanogaster. Cold Spring Harbor Laboratory Press, Cold Spring Harbor, pp 1277–1325

3. Franceschini N, Kirschfeld K (1971) In vivo optical study of photoreceptor elements in the compound eye of Drosophila. Kybernetik 8:1–13

4. Franceschini N, Kirschfeld K, Minke B (1981) Fluorescence of photoreceptor cells observed in vivo. Science 213:1264–1267

5. Pichaud F, Desplan C (2001) A new visualization approach for identifying mutations that affect differentiation and organization of the Drosophila ommatidia. Development 128:815–826

6. Mollereau B, Wernet MF, Beaufils P et al (2000) A green fluorescent protein enhancer trap screen in Drosophila photoreceptor cells. Mech Dev 93:151–160

7. Gambis A, Dourlen P, Steller H et al (2011) Two-color in vivo imaging of photoreceptor apoptosis and development in Drosophila. Dev Biol 351:128–134

8. Dourlen P, Bertin B, Chatelain G et al (2012) Drosophila fatty acid transport protein regulates rhodopsin-1 metabolism and is required for photoreceptor neuron survival. PLoS Genet 8:e1002833

9. Dourlen P, Levet C, Mejat A et al (2013) The Tomato/GFP-FLP/FRT method for live imaging of mosaic adult Drosophila photoreceptor cells. J Vis Exp: e50610

10. Golic KG (1991) Site-specific recombination between homologous chromosomes in Drosophila. Science 252:958–961

11. Franceschini N, Kirschfeld K (1971) Pseudopupil phenomena in the compound eye of drosophila. Kybernetik 9:159–182

12. Franceschini N (1972) Pupil and pseudopupil in the compound eye of Drosophila. In: Wehner R (ed) Information processing in the visual system of arthropods. Springer-Verlag, Berlin, pp 75–82

Chapter 26

Drosophila Model for Studying Phagocytosis Following Neuronal Cell Death

Boris Shklyar, Flonia Levy-Adam, and Estee Kurant

Abstract

During central nervous system (CNS) development, a large number of neurons die by apoptosis and are efficiently removed through phagocytosis. Since apoptosis and apoptotic cell clearance are highly conserved in evolution, relatively simple and easily accessible *Drosophila* embryonic CNS provides a good model to study molecular and cellular mechanisms of these processes. Here, we describe how to assess neuronal apoptosis and glial phagocytosis of apoptotic neurons using immunohistochemistry of whole fixed embryos and live imaging of developing embryonic CNS. Combination of these different strategies allows a comprehensive analysis of neuronal cell death in vivo.

Key words *Drosophila*, Apoptosis, Phagocytosis, Embryo, CNS

1   Introduction

During metazoan development, over-produced or damaged cells are eliminated through apoptosis and subsequent phagocytosis of apoptotic cells. An efficient clearance of apoptotic cells is crucial for normal development and occurs in tight correlation to apoptosis such that unengulfed apoptotic cells are barely seen in vivo under normal conditions. The whole process of apoptotic cell clearance may be envisaged to proceed in four steps (Fig. 1): (1) *recruitment* of phagocytes to the apoptotic cell through secreted "find-me" signals, (2) *recognition* of the cell as a target for phagocytosis through exposed "eat-me" signals such as phosphatidylserine (PS) and calreticulin, by specific receptors or secreted bridging molecules on the phagocyte site, (3) *engulfment*, characterized by the formation of the phagocytic cup, followed by (4) *phagosome maturation* and degradation of the apoptotic particle in the phagolysosome [1–3].

Phagocytic cells can be of two types: "professional"—macrophages and immature dendritic cells, and "non-professional"—tissue-resident neighboring cells, such as mammary gland epithelium and Sertoli cells in mammals [4].

Laura Lossi and Adalberto Merighi (eds.), *Neuronal Cell Death: Methods and Protocols*, Methods in Molecular Biology, vol. 1254, DOI 10.1007/978-1-4939-2152-2_26, © Springer Science+Business Media New York 2015

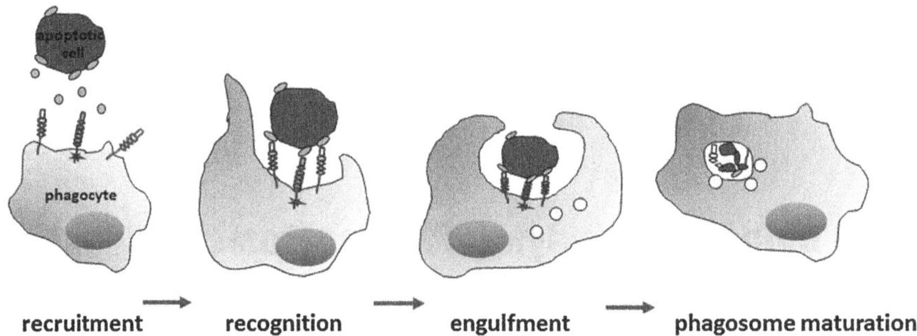

Fig. 1 Schematic presentation of different steps in apoptotic cell clearance

Apoptosis and apoptotic cell clearance are highly conserved in evolution, therefore model organisms such as *Drosophila melanogaster* are used to study the molecular basis of these processes in vivo [5]. During *Drosophila* CNS development, apoptosis of superfluous neurons occurs in three main phases, first in the mid to late embryo, then in pupa, and again in the emerging adult [6–13]. At late embryogenesis most of the cell death takes place in the CNS where more than 30 % of the generated neurons are destined to die [14, 15]. At this stage "professional" macrophages cannot enter the CNS due to a barrier formed by an epithelium of surface glia [16] and the apoptotic neurons are removed by "non-professional" neighboring glia [9]. Compared to vertebrates, the fly nervous system is less complex and much more accessible experimentally making *Drosophila* an excellent model organism to study the molecular and cellular mechanisms of phagocytosis in the nervous system.

In order to monitor the different steps of apoptotic cell clearance and to follow its dynamics in vivo, two populations of cells must be labeled: phagocytic glial cells and apoptotic neurons. In order to specifically label glial cells in fixed embryos as well as for in vivo time lapse recordings, transgenic animals containing a genetically encoded fluorescent reporter (GFP or RFP) driven by glia-specific Gal4 driver are usually used [9, 17].

Drosophila transgenic lines containing *repoGal4::UAScytGFP* (cytoplasmic GFP) or *repoGal4::UASCD8GFP* (membrane GFP) constructs are usually used for labeling embryonic glia. *UASCD8GFP* exhibits stronger signal of GFP, however we prefer *UAScytGFP*, which marks all glial cytoplasm and protrusions and, therefore, enables to detect the inside/outside position of apoptotic particles easier than the membrane patchy *UASCD8GFP*.

Apoptotic neurons are labeled by different markers reflecting specific molecular and morphological alterations typical for distinct steps of apoptosis and clearance. Here we explain the use of three different markers, which together allow a comprehensive analysis of apoptotic cell clearance: (1) fluorescent AnnexinV, which specifically binds to PS, is used as an early marker for apoptosis when it is injected into live embryos [18], (2) a specific antibody against activated caspase 3 (CM1) serves as a later apoptotic marker, and (3) LysoTracker that is used to monitor glial phagocytic activity, which mainly labels lysosomes fused to phagosomes (phagolysosomes).

AnnexinV labeling enables monitoring of engulfment events in real time, which cannot be visualized with CM1 antibody (done in fixed material) or LysoTracker injection that marks glial phagosomes. To label apoptotic cells in live embryos, we inject AnnexinV conjugated to Alexa® Fluor 555. However, CM1 and LysoTracker are useful for quantification of apoptotic particles and evaluation of glial phagosomal activity. To monitor phagocytosis we inject LysoTracker which becomes fluorescent in an acidic environment and is thus a phagolysosome marker. For the immunohistochemistry in fixed embryos we use the CM1 antibody, which detects apoptotic cells inside and outside phagocytes.

Time lapse recordings of apoptotic cell clearance using fluorescent markers add additional information, such as dynamic behavior of phagocytes, which could not be achieved by using conventional immunohistochemistry techniques in fixed embryos.

2   Materials

1. Fly food (for a 3 L preparation): in a pot heat 2 L H_2O + 24 g agar (bacto Agar, Difco™, Becton Dickinson, Franklin Lakes, NY) until it is about to boil. Add 120 mL molasses and cook for 15 min (*see* **Note 1**). In a beaker mix 700 mL H_2O + 300 mL corn meal + 120 mL dry yeast. Add this mixture to the pot and cook for another 25 min, in medium heat. Keep mixing once in a while. Take off from heat and let cool for about 30 min, until the temperature drops to 60 °C (*see* **Note 2**). Then add 11.6 mL propionic acid (Sigma Chemicals, St. Louis, MO) and 31 mL Tegosept (*see* below). Dispense into vials filling to about 1/5. Allow to stand overnight, covered with a breathing cloth. The next day add some dry yeast and close the vials with cotton plugs. Store the food at 4 °C.

2. Tegosept: dissolve 100 g Methyl-4-hydroxybenzoate (Sigma Chemicals) in 1 L absolute ethanol.

3. Collecting agar plates: in a 1 L bottle dissolve 22 g sucrose in 300 mL H_2O by swirling. Then, add 14 g agar, mix intensely and heat in a microwave for 2 min, then mix again and heat for 1 min (*see* **Note 3**). Repeat this once again or until the solution is clear. Let the mixture rest and cool for 10–15 min. When the bottle can be touched with hands add 10 mL absolute ethanol, 5 mL glacial acetic acid and 100 mL grape juice concentrate. Mix intensely, wait a few seconds for bubbles to go up, and pour to the plates using a 25 mL pipet. When the agar is solidified put the plates in a plastic bag and keep them at 4 °C.

4. Yeast paste: mix dry yeast with tap water to get a pasty texture.

5. Collecting cages. *See* **Note 4**.

6. Cell strainers (SPL Life Sciences, Pocheon, Korea).

7. 50 % bleach solution: one part domestic bleach (3 % active chlorine) and one part tap water. *See* **Note 5**.

8. Sprinkler containing tap water.

9. Formaldehyde 37 %, e.g. Sigma Chemicals.

10. n-Heptane, Sigma Chemicals.

11. Methanol absolute.

12. PBT: PBS + 0.1 % Triton® X-100.

13. Normal goat serum.

14. Anti-activated caspase 3 primary antibody (CM1, Becton Dickinson).

15. Cy3-conjugated anti-rabbit secondary antibody (Jackson Laboratories, Bar Harbor, ME).

16. Mounting medium: 80 % glycerol + 20 % PBS, pH 7.4.

17. Heptane glue: unroll a double-sided tape and put as much as you can in a scintillation vial (50 mL volume), fill with heptane, seal the vial with Parafilm, and rock for 24 h. Add heptane if the glue is too thick.

18. Microscope slides (76 × 26 mm) (Marienfeld, Lauda-Königshofen, Germany).

19. Coverslips (24 × 40 mm, #1) (Marienfeld) with glue: using a 200 μL pipette tip disperse a small amount of the heptane glue in one line in the middle of the coverslip. Let it dry for a few seconds. *See* **Note 6**.

20. Dehydration chamber and Silica beads. *See* **Note 7**.

21. Halocarbon oil 700 (Sigma Chemicals).

22. Needles: use thin walled glass filaments (FHC Capillary tubing, Bowdoin, ME) for preparing injection needles using a needle puller. *See* **Note 8**.

23. Microinjection system. *See* **Note 9**.

24. AnnexinV-Alexa® Fluor 555 conjugate (Molecular Probes®, Life Technologies™, Gaithersburg, MD).

25. LysoTracker (Molecular Probes®).

26. Fluorescence dissection microscope.

27. Orbital shaker.

28. Nutating mixer.

29. Micromanipulator.

30. Confocal microscope.

3 Methods

<table>
<tr>
<td valign="top">

3.1 Embryo Preparation

</td>
<td valign="top">

1. Allow the flies to lay eggs on a pre-warmed grape juice agar plate with yeast paste spread in a central area for 2 h at 25 °C, and then keep the plate at 25 °C for 10–12 h (stage 15–16).

2. Collect embryos from agar plates as follows: add tap water to the agar plate (using a sprinkler) and with a clean paint brush help release the embryos from the agar. Pour the water with embryos into a cell strainer which is held over a beaker. Repeat this step if some embryos remained on the plate. *See* **Note 10**.

3. Wash embryos in the cell strainer with tap water using a sprinkler until all the yeast paste is removed.

4. Dry out the excess water by wiping the outside of the strainer using Kimwipes.

5. Dechorionation: place the cell strainer in a clean Petri dish and add enough 50 % bleach solution to cover the embryos. Wait about 2 min. *See* **Note 11**.

6. Remove the cell strainer to a clean Petri dish and rinse extensively with tap water using a sprinkler until embryos loose the bleach odor. *See* **Note 12**.

7. Put the cell strainer into a clean Petri dish and cover the embryos with water to prevent drying up.

8. From here proceed to embryo fixation or to injection procedure following by time lapse experiments.

</td>
</tr>
<tr>
<td valign="top">

3.2 Embryo Fixation

</td>
<td valign="top">

1. Using a paintbrush, transfer the dechorionated embryos to a glass vial containing freshly prepared fixative (4.5 mL PBS, 500 µL 37 % formaldehyde, and 5 mL n-heptane) and incubate for 20 min at room temperature, on an orbital shaker (200 rpm).

2. At the end of fixation, remove the lower phase (formaldehyde) with a Pasteur glass pipette attached to a rubber bulb.

3. Add 5 mL absolute methanol to the vial and shake the mixture vigorously for 1 min, in order to remove the vitelline membrane. Wait a few seconds for embryos to go down to the bottom.

4. Remove the solution by aspiration. Be careful not to get close to the embryos.

5. Wash 3 times with absolute methanol and transfer the embryos to an Eppendorf tube. *See* **Note 13**.

6. Continue with immunostaining or store the embryos in methanol at –20 °C.

</td>
</tr>
</table>

3.2.1 Immunostaining

During incubation and washing steps, tubes are placed on a nutating mixer.

1. Rinse the embryos 4 times and wash once with PBT for 15 min for rehydration.

2. Incubate in 200 μL blocking buffer (PBT + 10 % goat serum) for 15 min.

3. Incubate overnight with the primary antibody in 100 μL PBT + 5 % goat serum, at 4 °C.

4. The next day, wash the embryos 3 times with PBT, 15 min each.

5. Incubate at room temperature for 1 h with fluorescently labeled secondary antibody in 100 μL PBT + 5 % goat serum at a 1:200 dilution.

6. Wash 3 times in PBT, 15 min each.

7. Keep the embryos in mounting medium. *See* **Note 14**.

8. Continue with confocal microscopy.

3.2.2 Imaging of Fixed Embryos

1. Mount the fixed embryos stained with CM1 antibody in a way that their position can be changed under the microscope in order to place the CNS up. To do this put two small (18 × 18 mm, #1.5) coverslips at the edges of a slide by attaching them with a drop of water. In the middle of the slide add 50 μL mounting medium with embryos and then, put a 20 × 40 mm coverslip above the two small slides and the embryos. This prevents pressure on the embryos and allows 3D analysis of glial phagocytosis.

2. Move gently the coverslip to place specific embryos in desired position. Image the embryos on an upright confocal microscope with 20× or 40× objectives.

3.3 Live Imaging

3.3.1 Preparatory Steps for Injection

1. Cut a rectangular piece of grape juice agar and place it on a microscope slide. Transfer the dechorionated embryos from the Petri dish to the agar piece using a clean paint brush.

2. Before starting to arrange the embryos for injection, load 1 μL of the desired reagent into two needles (sometimes a needle can be clogged). It takes about 5 min for the liquid to reach the tip of the needle.

3. Under a fluorescence dissection microscope, select properly staged embryos according to the autofluorescence of the gut. Select embryos expressing the GFP marker in the CNS using a paint brush and place each embryo close to the edge of the agar piece in a row one after another. The embryos should be placed with their ventral side up for imaging phagocytosis by glia in the CNS. Set up about ten embryos in one row.

4. Attach the embryos to a coverslip coated in the middle with a strip of heptane glue. *See* **Note 15**.

5. Place the coverslip with the embryos in the dehydration chamber on the Silica beads for about 5 min (this depends on the humidity of the room and the amount to be injected). *See* **Note 16**.

6. After drying up, cover the embryos with halocarbon oil to avoid further dehydration. *See* **Note 17**.

7. Place the coverslip on a microscope slide with the embryos facing up. Embryos are now ready for injection. Injection location is typically on the lateral side in the middle of the embryo. *See* **Note 18**.

3.3.2 Embryo Injection

1. Attach the needle to a micromanipulator.

2. To break the needle tip, put the edge of the coverslip in the field of view and adjust the needle to the same focal plane. Very carefully move the coverslip until it hits the needle tip and breaks it.

3. Focusing on the embryos, place the needle into the oil and make sure you get a liquid drop from the needle as you activate the pneumatic PicoPump. It means the needle tip has been broken well. *See* **Note 19**.

4. Without touching the needle holder move the embryo into the needle and inject a drop into the embryo. Proceed to the next one until all the embryos are injected.

3.3.3 Time Lapse Recording

1. To image the live embryos, use an inverted confocal microscope with 40× or 100× objectives.

2. Find a nicely positioned embryo with the CNS in the middle, which shows strong GFP expression and a good labeling of apoptotic cells (AnnexinV) or phagolysosomes (LysoTracker).

3. For time lapse recordings of live embryos choose five or six confocal slices (2 μm thick each). Afterwards, make a projection of three slices (6 μm thickness) in order to observe whole cells (*see* **Note 20**). Fluorescent AnnexinV and LysoTracker are stable and do not bleach easily. To avoid GFP bleaching, make recordings in intervals of 60 s.

In Fig. 2, stage 16 embryos stained or injected with different markers for apoptotic cells are depicted. Each picture represents projections from confocal stacks of the CNS. Apoptotic cells (in red) are seen outside and inside glia (in green) as detected with AnnexinV (Fig. 2A–A") or CM1 antibody (Fig. 2B–B"). LysoTracker-labeled phagolysosomes are found exclusively inside glial GFP-positive cells (Fig. 2C–C").

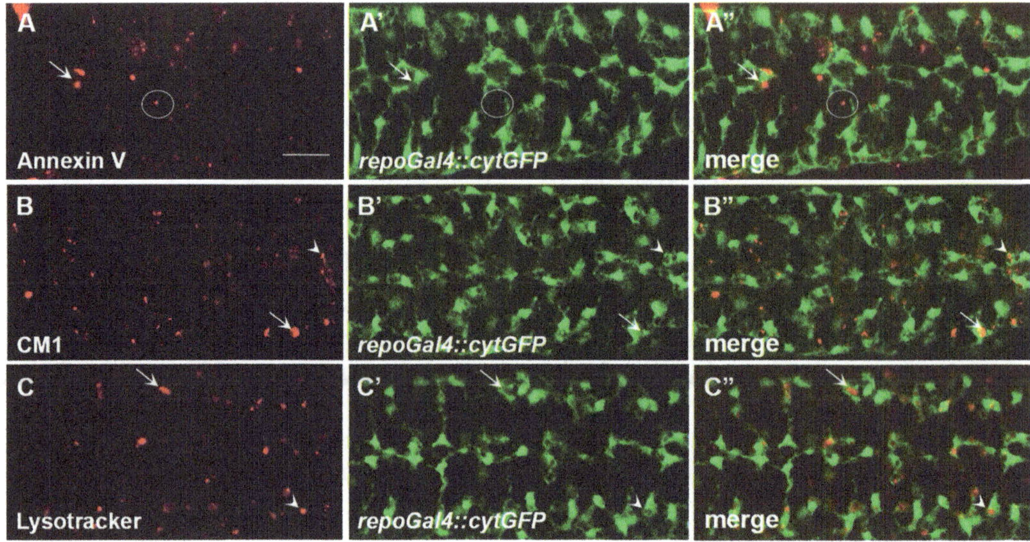

Fig. 2 Phagocytosis of apoptotic neurons in the *Drosophila* embryonic CNS. (**A–C"**) Projections from confocal stacks of the embryonic CNS at stage 16; ventral view. Glia are labeled with *repoG4::UAScytGFP* (*green*). Each frame is a projection of 3 slices of 2 μm each. Bar 20 μm. (**A–A"**) Apoptotic cells are labeled with AnnexinV (*red*). Arrows point at the apoptotic cell inside the corresponding glial cell. The circle highlights AnnexinV positive cell, which is not engulfed yet. (**B–B"**) Apoptotic particles labeled with CM1 (*red*). Arrows and arrowheads mark two representative apoptotic particles inside corresponding glial cells. (**C–C"**) Phagolysosomes are labeled with LysoTracker (*red*). Arrows and arrowheads mark two representative phagolysosomes inside corresponding glial cells

4 Notes

1. Do not cover tightly the pot lid to avoid overflowing of the content.

2. Higher temperature may result in Tegosept and propionic acid degradation.

3. Leave the bottle cap loose to prevent the mixture overflow.

4. For satisfactory amount of laid eggs put around 300 flies in the collecting cage while the ratio males/females is 1:3.

5. Using tap water for bleach dilution is obligatory since distilled water prevents dechorionation of the embryos.

6. The glue line should be transparent as much as possible in order to prevent transition disturbance of the laser light and the emitted signal while imaging by confocal microscope. If the glue is too viscous add more heptane to get appropriate viscosity.

7. Avoid exposure of Silica beads to the air to prevent humidity absorption.

8. We use a model P30 puller (Sutter Instrument Co., Novato, CA) with the following settings Heat 800, Pull 920 to produce

injection needles from 1 mm glass thin borosilicate capillaries with filament (WPI, Sarasota, FL).

9. We use a pneumatic PicoPump Model PV 820 (WPI) with the following settings: Hold pressure 3, Eject pressure 20.

10. Disperse the yeast paste since it also contains a considerable amount of laid eggs.

11. Swirl the cell strainer occasionally after adding bleach, 2–3 times during the 2 min of the procedure.

12. Perform five rinses by a strong stream of tap water.

13. Transfer the embryos using edge cut 1 mL tip to avoid their damage.

14. Wait for the embryos to sink completely before proceeding with microscopy.

15. This step and **step 6** are critical for the success of in vivo imaging. Under a fluorescent dissection microscope slowly and carefully put the coverslip with the glue on top of the embryos. As you touched the embryos push the coverslip gently for good adherence.

16. This step may be skipped in case it took too much time to arrange the embryos.

17. To prevent changing of embryo position put a small drop of the oil directly on each embryo while holding the coverslip horizontally. Wait till the drops are scattered.

18. Put a drop of water on the microscope slide to prevent moving of the coverslip.

19. If the needle tip was broken correctly, the size of the drop should be 2–3 % of embryo volume.

20. In some cases embryos may start rolling during the time of video recording. Therefore, we take a look once in a while at the recording in order to stop and start over with another good positioned embryo.

Acknowledgments

The work was supported by the Israel Science Foundation (grant no. 427/11).

References

1. Ravichandran KS (2010) Find-me and eat-me signals in apoptotic cell clearance: progress and conundrums. J Exp Med 207:1807–1817

2. Ravichandran KS (2011) Beginnings of a good apoptotic meal: the find-me and eat-me signaling pathways. Immunity 35:445–455

3. Ravichandran KS, Lorenz U (2007) Engulfment of apoptotic cells: signals for a good meal. Nat Rev Immunol 7:964–974

4. Henson PM, Hume DA (2006) Apoptotic cell removal in development and tissue homeostasis. Trends Immunol 27:244–250

5. Kinchen JM (2010) A model to die for: signaling to apoptotic cell removal in worm, fly and mouse. Apoptosis 15:998–1006

6. Freeman MR (2006) Sculpting the nervous system: glial control of neuronal development. Curr Opin Neurobiol 16:119–125

7. White K, Steller H (1995) The control of apoptosis in Drosophila. Trends Cell Biol 5:74–78

8. Kurant E (2011) Keeping the CNS clear: glial phagocytic functions in Drosophila. Glia 59:1304–1311

9. Kurant E, Axelrod S, Leaman D et al (2008) Six-microns-under acts upstream of Draper in the glial phagocytosis of apoptotic neurons. Cell 133:498–509

10. Choi YJ, Lee G, Park JH (2006) Programmed cell death mechanisms of identifiable peptidergic neurons in Drosophila melanogaster. Development 133:2223–2232

11. Peterson C, Carney GE, Taylor BJ (2002) Reaper is required for neuroblast apoptosis during Drosophila development. Development 129:1467–1476

12. Winbush A, Weeks JC (2011) Steroid-triggered, cell-autonomous death of a Drosophila motoneuron during metamorphosis. Neural Dev 6:15

13. Steller H (2008) Regulation of apoptosis in Drosophila. Cell Death Differ 15:1132–1138

14. Abrams JM, White K, Fessler LI et al (1993) Programmed cell death during Drosophila embryogenesis. Development 117:29–43

15. Rogulja-Ortmann A, Luer K, Seibert J et al (2007) Programmed cell death in the embryonic central nervous system of Drosophila melanogaster. Development 134:105–116

16. Schwabe T, Bainton RJ, Fetter RD et al (2005) GPCR signaling is required for blood-brain barrier formation in drosophila. Cell 123:133–144

17. Shklyar B, Levy-Adam F, Mishnaevski K et al (2013) Caspase activity is required for engulfment of apoptotic cells. Mol Cell Biol 33:3191–3201

18. van den Eijnde SM, Boshart L, Baehrecke EH (1998) Cell surface exposure of phosphatidylserine during apoptosis is phylogenetically conserved. Apoptosis 3:9–16

A

Laura Lossi and Adalberto Merighi (eds.), *Neuronal Cell Death: Methods and Protocols*, Methods in Molecular Biology,
vol. 1254, DOI 10.1007/978-1-4939-2152-2, © Springer Science+Business Media New York 2015

F

U

CPSIA information can be obtained at www.ICGtesting.com
Printed in the USA
LVOW02*1814260215

428498LV00002B/49/P